T0137486

Lecture Notes in Applied and Computational Mechanics

Volume 89

Series Editors

This series aims to report new developments in applied and computational mechanics—quickly, informally and at a high level. This includes the fields of fluid, solid and structural mechanics, dynamics and control, and related disciplines. The applied methods can be of analytical, numerical and computational nature. The series scope includes monographs, professional books, selected contributions from specialized conferences or workshops, edited volumes, as well as outstanding advanced textbooks.

Indexed by EI-Compendex, SCOPUS, Zentralblatt Math, Ulrich's, Current Mathematical Publications, Mathematical Reviews and MetaPress.

More information about this series at http://www.springer.com/series/4623

Hector Jensen · Costas Papadimitriou

Sub-structure Coupling for Dynamic Analysis

Application to Complex Simulation-Based Problems Involving Uncertainty

Hector Jensen
Federico Santa María Technical University
Valparaiso, Chile

Costas Papadimitriou
University of Thessaly
Volos, Greece

ISSN 1613-7736 ISSN 1860-0816 (electronic)
Lecture Notes in Applied and Computational Mechanics
ISBN 978-3-030-12821-0 ISBN 978-3-030-12819-7 (eBook)
https://doi.org/10.1007/978-3-030-12819-7

Library of Congress Control Number: 2019930965

This Springer imprint is published by the registered company Springer Nature Switzerland AG
The registered company address is: Gewerbestrasse 11, 6330 Cham, Switzerland

Preface

The solution to many complex simulation-based problems involving finite element models requires a large number of dynamic re-analyses. This class of problems includes Bayesian uncertainty quantification, structural dynamic simulation, model updating, reliability analysis, sensitivity analysis, uncertainty management in structural dynamics, reliability-based design optimization, global optimization, and so on. These problems have a wide range of important applications in several engineering fields. The corresponding computational demand depends on the number of finite element analyses and on the time taken to perform an individual analysis. For problems involving finite element models with a medium/large number of degrees of freedom, the computational effort may be excessive. To cope with this difficulty, model reduction techniques are used to generate reduced-order models in order to carry out the different analyses in a significantly reduced space of generalized coordinates. In particular, a method based on a type of model reduction techniques, known as substructure coupling for dynamic analysis, is used in the context of this monograph. The method involves dividing the structure into a number of substructures, obtaining reduced-order models of the substructures and then assembling a reduced-order model of the entire structure. While the use of reduced-order models alleviates part of the computational effort, their repetitive generation during simulation processes can be computationally expensive, due to the substantial computational overhead that arises at the substructure level. In this regard, an efficient finite element model parametrization scheme is implemented. When the division of the structural model is guided by such a parametrization scheme, a drastic reduction in computational effort is achieved without compromising the accuracy of the results. The capabilities of the developed procedures are demonstrated in a number of simulation-based problems involving uncertainty. These include reliability analysis, reliability sensitivity analysis, reliability-based design optimization, and Bayesian model updating.

Valparaiso, Chile
Volos, Greece

Hector Jensen
Costas Papadimitriou

Acknowledgements

The assistance of our students from the Department of Civil Engineering at the Federico Santa Maria Technical University and from the Department of Mechanical Engineering at the University of Thessaly is gratefully acknowledged.

Contents

Part II Application to Reliability Problems

Part I
Reduced-Order Models

Chapter 1
Model Reduction Techniques for Structural Dynamic Analyses

Abstract This chapter presents a model reduction technique based on substructure coupling for dynamic analysis. The dynamic behavior of the substructures is described by a set of dominant fixed-interface normal modes along with a set of interface constraint modes that account for the coupling at each interface where the substructures are connected. Based on these modes, the corresponding reduced-order matrices are derived. The internal dynamic behavior of the substructures is then enhanced by consideration of the contribution of residual fixed-interface normal modes. Next, the interface degrees of freedom are reduced by consideration of a small number of characteristic constraint modes. Pseudo-codes are provided in order to illustrate how the reduced-order matrices are constructed, by including dominant and residual fixed-interface normal modes as well as interface reduction. Finally, the dynamic response of reduced-order models is discussed.

1.1 Structural Model

Attention is focused on a general class of structural dynamical systems, with localized nonlinearities characterized by multi-degrees of freedom models that satisfy the equation of motion

$$\mathbf{M}\ddot{\mathbf{u}}(t) + \mathbf{C}\dot{\mathbf{u}}(t) + \mathbf{K}\mathbf{u}(t) = \mathbf{f}_{NL}(\mathbf{u}(t), \dot{\mathbf{u}}(t), \mathbf{y}(t)) + \mathbf{f}(t) \tag{1.1}$$

where $\mathbf{u}(t)$ denotes the displacement vector of dimension n, $\dot{\mathbf{u}}(t)$ the velocity vector, $\ddot{\mathbf{u}}(t)$ the acceleration vector, $\mathbf{f}_{NL}(\mathbf{u}(t), \dot{\mathbf{u}}(t), \mathbf{y}(t))$ the vector of nonlinear restoring forces, $\mathbf{y}(t)$ the vector of a set of variables that describes the state of the nonlinear components, and $\mathbf{f}(t)$ the external force vector. The matrices $\mathbf{M}$, $\mathbf{C}$, and $\mathbf{K}$, which are assumed to be symmetric, describe the mass, damping, and stiffness, respectively. The evolution of the set of variables $\mathbf{y}(t)$ is described through an appropriate nonlinear model that depends on the nature of the nonlinearity. The equation of motion for the

H. Jensen and C. Papadimitriou, *Sub-structure Coupling for Dynamic Analysis*, Lecture Notes in Applied and Computational Mechanics 89,
https://doi.org/10.1007/978-3-030-12819-7_1

displacement vector $\mathbf{u}(t)$ and the equation for the evolution of the set of variables $\mathbf{y}(t)$ constitute a system of coupled nonlinear equations. This characterization of the dynamical system allows the description of different types of models commonly used for nonlinearities, such as hysteresis, degradation, plasticity, and other types of nonlinearities [2, 3, 27].

In the dynamic characterization of this class of structural dynamical systems, it is often inefficient to carry out a finite element analysis of the entire model. In fact, in many dynamic analysis problems, the lower frequencies and the corresponding modes tend to dominate the dynamic behavior of the structure. It is also common for component structures to be analyzed independently in complex systems, which makes it more convenient to perform a dynamic analysis at the substructure level. In this framework, model reduction techniques have been developed as practical and efficient tools to model and analyze the dynamics of complex structural systems [13]. The objective of model reduction techniques is to obtain reduced-order models that can be solved significantly faster than the original high-fidelity model, incorporating the important dynamics of the analyzed system so that the results from the reduced-order models are sufficiently accurate. A class of model reduction techniques known as component mode synthesis (CMS), or substructure coupling for dynamic analysis [5, 14, 15, 24, 26], is briefly reviewed in this chapter.

1.2 Substructure Modes

Substructure coupling involves dividing the structure into a number of linear and nonlinear substructures (or components), obtaining reduced-order models of the linear substructures, and then assembling a reduced-order model of the entire structure. Specifically, after dividing the structure into substructures, the model reduction technique involves two basic steps: defining sets of substructure modes, and coupling the substructure mode models to form a reduced-order system model. Substructure modes include normal, constraint, rigid-body, and attachment modes [12, 15].

Depending on the substructure modes under consideration, substructuring can be grouped into fixed-interface, free-interface, and loaded-interface methods [7, 12, 13, 16, 18, 25, 28]. Among these, the Craig-Bampton method [12, 15], a fixed-interface technique, is widely used for its simplicity and computational stability. In this approach, which is used in the present work, the substructure modes correspond to fixed-interface normal modes and interface constraint modes. In this manner, the dynamic behavior of the linear components of the structural system is described by a set of normal modes of individual substructures along with a set of constraint modes that account for the coupling at each interface where the substructures are connected.

1.2.1 Fixed-Interface Normal Modes

To introduce the substructure modes, the following partitioned form of the mass matrix $\mathbf{M}^s \in R^{n^s \times n^s}$ and the stiffness matrix $\mathbf{K}^s \in R^{n^s \times n^s}$ of the substructure s, $s = 1, \ldots, N_s$, are considered

$$\mathbf{M}^s = \begin{bmatrix} \mathbf{M}^s_{ii} & \mathbf{M}^s_{ib} \\ \mathbf{M}^s_{bi} & \mathbf{M}^s_{bb} \end{bmatrix} \tag{1.2}$$

$$\mathbf{K}^s = \begin{bmatrix} \mathbf{K}^s_{ii} & \mathbf{K}^s_{ib} \\ \mathbf{K}^s_{bi} & \mathbf{K}^s_{bb} \end{bmatrix} \tag{1.3}$$

where n^s is the number of degrees of freedom of substructure s, N_s is the total number of linear substructures, and the indices i and b are sets containing the internal and boundary degrees of freedom, respectively, of substructure s. The internal degrees of freedom, which are not shared with any adjacent substructures, are kept in the vector $\mathbf{u}^s_i(t) \in R^{n^s_i}$, while all boundary degrees of freedom are kept in the vector $\mathbf{u}^s_b(t) \in R^{n^s_b}$. The boundary degrees of freedom include only those that are in common with the interface degrees of freedom of adjacent substructures. Note that $n^s = n^s_i + n^s_b$. The fixed-interface normal modes are obtained by restraining all boundary degrees of freedom and solving the eigenvalue problem [15]

$$\mathbf{K}^s_{ii}\boldsymbol{\Phi}^s_{ii} - \mathbf{M}^s_{ii}\boldsymbol{\Phi}^s_{ii}\boldsymbol{\Lambda}^s_{ii} = \mathbf{0} \quad , \quad s = 1, \ldots, N_s \tag{1.4}$$

where the matrix $\boldsymbol{\Phi}^s_{ii}$ contains the complete set of n^s_i fixed-interface normal modes and $\boldsymbol{\Lambda}^s_{ii}$ is the corresponding diagonal matrix containing the eigenvalues. The fixed-interface normal modes are normalized with respect to the mass matrix $\mathbf{M}^s_{ii}$, that is,

$$\boldsymbol{\Phi}^{s\,T}_{ii}\mathbf{M}^s_{ii}\boldsymbol{\Phi}^s_{ii} = \mathbf{I}^s_{ii} \tag{1.5}$$

and

$$\boldsymbol{\Phi}^{s\,T}_{ii}\mathbf{K}^s_{ii}\boldsymbol{\Phi}^s_{ii} = \boldsymbol{\Lambda}^s_{ii} \tag{1.6}$$

where $\mathbf{I}^s_{ii} \in R^{n^s_i \times n^s_i}$ is the identity matrix.

1.2.2 Interface Constraint Modes

The interface constraint modes are defined as the static deformation of the substructure when a unit displacement is applied at one coordinate of vector $\mathbf{u}^s_b(t)$ and zero displacement at the remaining interface degrees of freedom, while the internal degrees of freedom are force free [15]. Then, the interface constraint modes matrix is

$$\boldsymbol{\Psi}^s = \begin{bmatrix} \boldsymbol{\Psi}^s_{ib} \\ \mathbf{I}^s_{bb} \end{bmatrix} \in R^{n^s \times n^s_b} \quad (1.7)$$

where $\boldsymbol{\Psi}^s_{ib} \in R^{n^s_i \times n^s_b}$ is the interior partition of the interface constraint modes matrix and $\mathbf{I}^s_{bb} \in R^{n^s_b \times n^s_b}$ is the identity matrix, which satisfy [15]

$$\begin{bmatrix} \mathbf{K}^s_{ii} & \mathbf{K}^s_{ib} \\ \mathbf{K}^s_{bi} & \mathbf{K}^s_{bb} \end{bmatrix} \begin{bmatrix} \boldsymbol{\Psi}^s_{ib} \\ \mathbf{I}^s_{bb} \end{bmatrix} = \begin{bmatrix} \mathbf{0}^s_{ib} \\ \mathbf{R}^s_{bb} \end{bmatrix} \quad (1.8)$$

where $\mathbf{0}^s_{ib} \in R^{n^s_i \times n^s_b}$ is the null matrix, and $\mathbf{R}^s_{bb} \in R^{n^s_b \times n^s_b}$ is the corresponding matrix of interface forces. Solving the first block of Eq. (1.8), the interior partition of the interface constraint modes matrix takes the form

$$\boldsymbol{\Psi}^s_{ib} = -\mathbf{K}^{s\,-1}_{ii} \mathbf{K}^s_{ib} \quad (1.9)$$

and therefore, the interface constraint modes matrix is given by

$$\boldsymbol{\Psi}^s = \begin{bmatrix} \boldsymbol{\Psi}^s_{ib} \\ \mathbf{I}^s_{bb} \end{bmatrix} = \begin{bmatrix} -\mathbf{K}^{s\,-1}_{ii} \mathbf{K}^s_{ib} \\ \mathbf{I}^s_{bb} \end{bmatrix} \quad (1.10)$$

For illustration purposes, Figs. 1.1 and 1.2 show the fixed-interface normal modes and the interface constraint modes, respectively, for the substructure in Fig. 1.3, which is composed of three beam elements. The variables $u_i(t)$, $i = 1, \ldots, 4$, correspond to the internal degrees of freedom, while $u_i(t)$, $i = 5, \ldots, 8$, are the interface degrees of freedom.

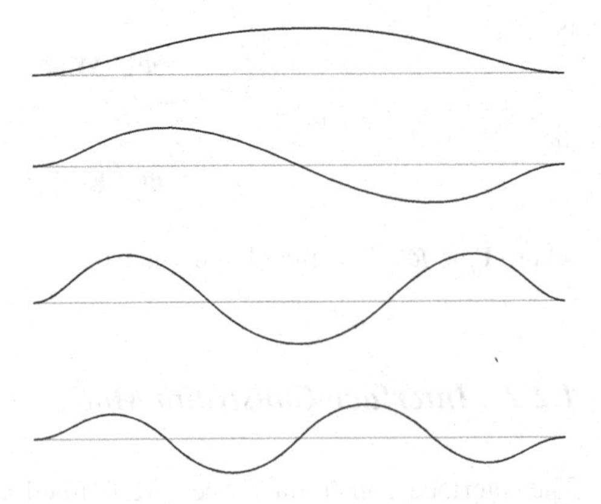

Fig. 1.1 Substructure fixed-interface normal modes

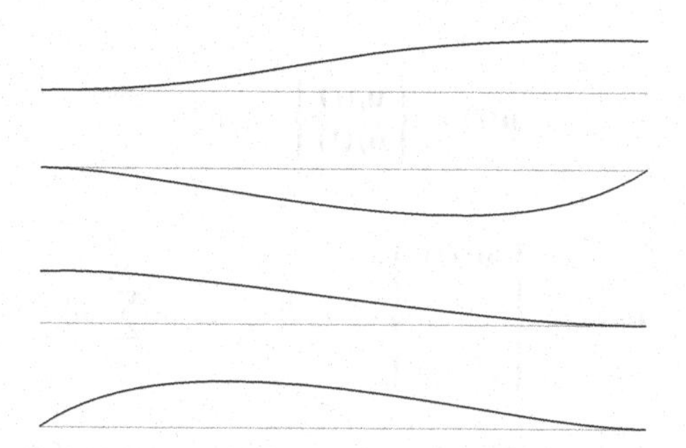

Fig. 1.2 Substructure interface constraint modes

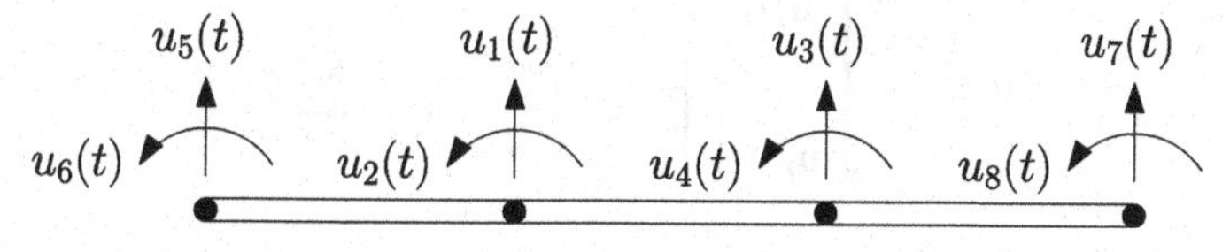

Fig. 1.3 Eight degrees of freedom beam substructure

1.3 Reduced-Order Model: Standard Formulation

As mentioned above, the objective of model reduction techniques is to characterize the dynamic behavior of the system by a reduced number of generalized coordinates. In the standard formulation of component mode synthesis, the dynamic of the system is described by a number of generalized coordinates, which includes a fraction of the fixed-interface modal coordinates of each substructure and the physical interface coordinates. The derivation of the corresponding reduced-order model is presented in this section.

1.3.1 Transformation Matrix

The fixed-interface normal modes and the interior partition of the interface constraint modes are used to define a transformation matrix $\mathbf{T}_D$ (Craig-Bampton transformation matrix), which relates the vector of physical coordinates of all substructures $\bar{\mathbf{u}}(t)$ to the vector of generalized coordinates $\mathbf{q}(t)$ as

$$\bar{\mathbf{u}}(t) = \mathbf{T}_D \mathbf{q}(t) \tag{1.11}$$

where

$$\bar{\mathbf{u}}(t) = \begin{Bmatrix} \mathbf{u}_i(t) \\ \mathbf{u}_I(t) \end{Bmatrix} \in R^n \tag{1.12}$$

in which

$$\mathbf{u}_i(t) = \begin{Bmatrix} \mathbf{u}_i^1(t) \\ \vdots \\ \mathbf{u}_i^{N_s}(t) \end{Bmatrix} \in R^{n_i}, \quad n_i = \sum_{s=1}^{N_s} n_i^s \tag{1.13}$$

is the vector of physical coordinates at the internal degrees of freedom of all substructures,

$$\mathbf{u}_I(t) = \begin{Bmatrix} \mathbf{u}_I^1(t) \\ \vdots \\ \mathbf{u}_I^{N_I}(t) \end{Bmatrix} \in R^{n_I}, \quad n_I = \sum_{l=1}^{N_I} n_I^l \tag{1.14}$$

is the vector of physical coordinates at the N_I independent interfaces, where n_I^l is the number of degrees of freedom at the interface l,

$$\mathbf{T}_D = \begin{bmatrix} [\boldsymbol{\Phi}_{id}^1, \ldots, \boldsymbol{\Phi}_{id}^{N_s}] & [\boldsymbol{\Psi}_{ib}^1, \ldots, \boldsymbol{\Psi}_{ib}^{N_s}]\tilde{\mathbf{T}} \\ \mathbf{0} & \mathbf{I} \end{bmatrix} \in R^{n \times n_D} \tag{1.15}$$

is the Craig-Bampton transformation matrix, where $\boldsymbol{\Phi}_{id}^s \in R^{n_i^s \times n_{id}^s}$, $n_{id}^s \ll n_i^s$, are the kept fixed-interface normal modes of each substructure, $n_D = n_{id} + n_I$, in which $n_{id} = \sum_{s=1}^{N_s} n_{id}^s$, $\mathbf{0} \in R^{n_I \times n_{id}}$ is the null matrix, $\mathbf{I} \in R^{n_I \times n_I}$ is the identity matrix, $[\cdot, \ldots, \cdot]$ is a block diagonal matrix that has the matrices inside the square brackets as diagonal blocks, $\tilde{\mathbf{T}} \in R^{n_b \times n_I}$ is a transformation matrix consisting of zeros and ones that maps the vector $\mathbf{u}_I(t)$ of independent interface coordinates to the vector of boundary coordinates of all substructures $\mathbf{u}_b(t)$, that is

$$\mathbf{u}_b(t) = \tilde{\mathbf{T}}\mathbf{u}_I(t) \tag{1.16}$$

where

$$\mathbf{u}_b(t) = \begin{Bmatrix} \mathbf{u}_b^1(t) \\ \vdots \\ \mathbf{u}_b^{N_s}(t) \end{Bmatrix} \in R^{n_b}, \quad n_b = \sum_{s=1}^{N_s} n_b^s \tag{1.17}$$

and where the vector of generalized coordinates is defined as

$$\mathbf{q}(t) = \begin{Bmatrix} \boldsymbol{\eta}(t) \\ \mathbf{u}_I(t) \end{Bmatrix} \in R^{n_D} \tag{1.18}$$

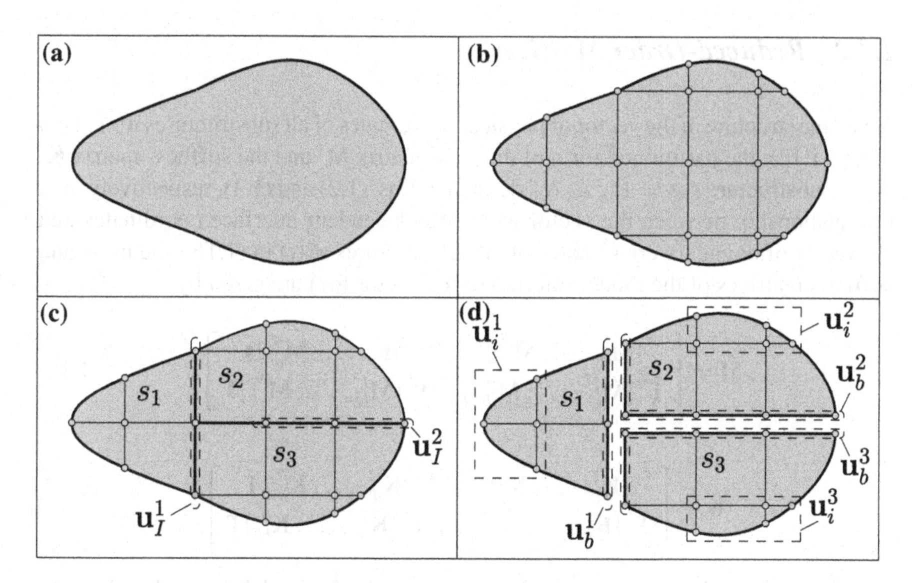

Fig. 1.4 Schematic illustration of notation used for physical coordinates

$\eta(t)$ being the vector of kept fixed-interface modal coordinates of all substructures. The kept fixed-interface normal modes of each substructure $\boldsymbol{\Phi}_{id}^s$ are referred to as dominant fixed-interface normal modes. General guidelines on how to choose the number n_{id}^s for each substructure are left for subsequent chapters with applications.

Figure 1.4 illustrates the different previously-defined vectors that group the physical coordinates of a particular problem. In Fig. 1.4a, an arbitrary structural system is depicted, which is analyzed using the discretized model shown in Fig. 1.4b. The nodes are designated with dots, and the different elements are delimited with thin lines. To analyze the discretized model, three substructures are considered, which are termed s_1, s_2, and s_3 in Fig. 1.4c. In addition, Fig. 1.4c also indicates the different nodes that are associated with the physical coordinates of the independent interfaces of the model $\mathbf{u}_I^1$ and $\mathbf{u}_I^2$, respectively. The nodes associated with each independent interface can be selected arbitrarily, as long as a node is associated with a single independent interface. Finally, Fig. 1.4d indicates the nodes associated with the boundary and the internal degrees of freedom of each substructure. The nodes associated with vectors $\mathbf{u}_i^s$, $s = 1, 2, 3$, do not contain repeated entries between themselves; in contrast, the nodes associated with vectors $\mathbf{u}_b^s$, $s = 1, 2, 3$, do contain repeated entries between themselves. The boundary coordinates of a substructure s, $\mathbf{u}_b^s(t)$, may contain one or more independent interface coordinates $\mathbf{u}_I^l(t)$ $(l = 1, \ldots, N_I)$. The particular structure of the transformation matrix $\tilde{\mathbf{T}}$ depends on the definition of the independent interface coordinates $\mathbf{u}_I^l(t)$, $l = 1, \ldots, N_I$.

1.3.2 Reduced-Order Matrices

Given the structure of the vector of physical coordinates of all substructures $\bar{\mathbf{u}}(t)$ given in Eq. (1.12); the partitioned form of the mass matrix $\mathbf{M}^s$ and the stiffness matrix $\mathbf{K}^s$ of the substructure s, $s = 1, \ldots, N_s$, defined in Eqs. (1.2) and (1.3), respectively; and the relationship between the vector $\mathbf{u}_I(t)$ of independent interface coordinates and the vector of boundary coordinates of all substructures $\mathbf{u}_b(t)$ in (1.16); the mass and stiffness matrices of the model referred to the vector $\bar{\mathbf{u}}(t)$ are given by

$$\hat{\mathbf{M}} = \begin{bmatrix} [\mathbf{M}_{ii}^1, \ldots, \mathbf{M}_{ii}^{N_s}] & [\mathbf{M}_{ib}^1, \ldots, \mathbf{M}_{ib}^{N_s}]\tilde{\mathbf{T}} \\ \tilde{\mathbf{T}}^T [\mathbf{M}_{ib}^{1^T}, \ldots, \mathbf{M}_{ib}^{N_s^T}] & \tilde{\mathbf{T}}^T [\mathbf{M}_{bb}^1, \ldots, \mathbf{M}_{bb}^{N_s}]\tilde{\mathbf{T}} \end{bmatrix}, \tag{1.19}$$

and

$$\hat{\mathbf{K}} = \begin{bmatrix} [\mathbf{K}_{ii}^1, \ldots, \mathbf{K}_{ii}^{N_s}] & [\mathbf{K}_{ib}^1, \ldots, \mathbf{K}_{ib}^{N_s}]\tilde{\mathbf{T}} \\ \tilde{\mathbf{T}}^T [\mathbf{K}_{ib}^{1^T}, \ldots, \mathbf{K}_{ib}^{N_s^T}] & \tilde{\mathbf{T}}^T [\mathbf{K}_{bb}^1, \ldots, \mathbf{K}_{bb}^{N_s}]\tilde{\mathbf{T}} \end{bmatrix} \tag{1.20}$$

The corresponding mass and stiffness matrices of the model referred to the generalized coordinates $\mathbf{q}(t)$ take the form

$$\hat{\mathbf{M}}_D = \mathbf{T}_D^T \, \hat{\mathbf{M}} \, \mathbf{T}_D \tag{1.21}$$

and

$$\hat{\mathbf{K}}_D = \mathbf{T}_D^T \, \hat{\mathbf{K}} \, \mathbf{T}_D \tag{1.22}$$

where the dimension of these matrices is equal to $n_D \times n_D$. Carrying out the previous products yields

$$\hat{\mathbf{M}}_D = \begin{bmatrix} \mathbf{I} & [\hat{\mathbf{M}}_{ib}^1, \ldots, \hat{\mathbf{M}}_{ib}^{N_s}]\tilde{\mathbf{T}} \\ \tilde{\mathbf{T}}^T [\hat{\mathbf{M}}_{ib}^{1^T}, \ldots, \hat{\mathbf{M}}_{ib}^{N_s^T}] & \tilde{\mathbf{T}}^T [\hat{\mathbf{M}}_{bb}^1, \ldots, \hat{\mathbf{M}}_{bb}^{N_s}]\tilde{\mathbf{T}} \end{bmatrix} \tag{1.23}$$

and

$$\hat{\mathbf{K}}_D = \begin{bmatrix} [\boldsymbol{\Lambda}_{id}^1, \ldots, \boldsymbol{\Lambda}_{id}^{N_s}] & \mathbf{0} \\ \mathbf{0} & \tilde{\mathbf{T}}^T [\hat{\mathbf{K}}_{bb}^1, \ldots, \hat{\mathbf{K}}_{bb}^{N_s}]\tilde{\mathbf{T}} \end{bmatrix} \tag{1.24}$$

with

$$\hat{\mathbf{M}}_{ib}^s = \boldsymbol{\Phi}_{id}^{s^T} \mathbf{M}_{ii}^s \boldsymbol{\Psi}_{ib}^s + \boldsymbol{\Phi}_{id}^{s^T} \mathbf{M}_{ib}^s, \tag{1.25}$$

$$\hat{\mathbf{K}}_{bb}^s = \mathbf{K}_{ib}^{s^T} \boldsymbol{\Psi}_{ib}^s + \mathbf{K}_{bb}^s = -\mathbf{K}_{ib}^{s^T} \mathbf{K}_{ii}^{s\,-1} \mathbf{K}_{ib}^s + \mathbf{K}_{bb}^s \tag{1.26}$$

and

$$\hat{\mathbf{M}}^s_{bb} = (\boldsymbol{\Psi}^{s^T}_{ib}\mathbf{M}^s_{ii} + \mathbf{M}^{s^T}_{ib})\boldsymbol{\Psi}^s_{ib} + \boldsymbol{\Psi}^{s^T}_{ib}\mathbf{M}^s_{ib} + \mathbf{M}^s_{bb}\ ,\ s = 1, \ldots, N_s \tag{1.27}$$

The identity matrix $\mathbf{I}$ is of dimension equal to $n_{id} \times n_{id}$, and the diagonal matrices $\boldsymbol{\Lambda}^s_{id}, s = 1, \ldots, N_s$, contain the eigenvalues of the dominant fixed-interface normal modes for each substructure. It can be seen that the reduced-order matrices $\hat{\mathbf{M}}_D$ and $\hat{\mathbf{K}}_D$ are symmetric. The dimension of the reduced-order matrices can be substantially smaller than the dimension of the unreduced matrices, that is, $n_D \ll n$. In a similar manner, the damping matrix $\hat{\mathbf{C}}_D$ can be defined directly in terms of the damping matrix $\bar{\mathbf{C}}$ of the unreduced model referred to the vector of physical coordinates of all substructures $\bar{\mathbf{u}}(t)$. In general, however, damping is treated as modal damping, imposed on the modes of the reduced-order system model [15].

1.4 Reduced-Order Model: Improved Formulation

According to the transformation matrix $\mathbf{T}_D$, introduced in Eq. (1.15), the vector of physical coordinates at the internal degrees of freedom of all substructures $\mathbf{u}_i(t)$ is approximated as

$$\mathbf{u}_i(t) = [\boldsymbol{\Phi}^1_{id}, \ldots, \boldsymbol{\Phi}^{N_s}_{id}]\eta(t) + [\boldsymbol{\Psi}^1_{ib}, \ldots, \boldsymbol{\Psi}^{N_s}_{ib}]\tilde{\mathbf{T}}\mathbf{u}_I(t) \tag{1.28}$$

where $\boldsymbol{\Phi}^s_{id}$ contains the n^s_{id} dominant fixed-interface normal modes of substructure s. Thus, the $n^s_i - n^s_{id}$ residual normal modes are not considered in the approximation of $\mathbf{u}_i(t)$. To improve the accuracy of this approximation, the static contribution of the residual normal modes of each substructure can be considered explicitly in the analysis [8, 20, 21], as shown below.

1.4.1 Static Correction

To consider the static contribution of the residual normal modes, the undamped free vibration of the linear components of Eq. (1.1), referred to the set of generalized coordinates $\mathbf{q}(t)$, is first considered. The equation of motion reads

$$\hat{\mathbf{M}}_D \begin{Bmatrix} \ddot{\eta}(t) \\ \ddot{\mathbf{u}}_I(t) \end{Bmatrix} + \hat{\mathbf{K}}_D \begin{Bmatrix} \eta(t) \\ \mathbf{u}_I(t) \end{Bmatrix} = \mathbf{0} \tag{1.29}$$

From the first block of this equation and the characterization of the matrices $\hat{\mathbf{M}}_D$ and $\hat{\mathbf{K}}_D$, given in (1.23) and (1.24), respectively, the vector of fixed-interface modal coordinates of all substructures $\boldsymbol{\eta}(t)$ satisfies the equation

$$\ddot{\boldsymbol{\eta}}(t) + [\boldsymbol{\Lambda}_{id}^1, \ldots, \boldsymbol{\Lambda}_{id}^{N_s}]\, \boldsymbol{\eta}(t) = -[\hat{\mathbf{M}}_{ib}^1, \ldots, \hat{\mathbf{M}}_{ib}^{N_s}]\tilde{\mathbf{T}}\, \ddot{\mathbf{u}}_I(t) \tag{1.30}$$

In view of the definition of $\hat{\mathbf{M}}_{ib}^s$, $s = 1, \ldots, N_s$, given in Eq. (1.25) and the definition of the interior partition of the interface constraint modes $\boldsymbol{\Psi}_{ib}^s$, $s = 1, \ldots, N_s$, given in Eq. (1.9), (1.30) can be rewritten as

$$\ddot{\boldsymbol{\eta}}(t) + [\boldsymbol{\Lambda}_{id}^1, \ldots, \boldsymbol{\Lambda}_{id}^{N_s}]\, \boldsymbol{\eta}(t) = -[\boldsymbol{\Phi}_{id}^{1^T}, \ldots, \boldsymbol{\Phi}_{id}^{N_s^T}]\tilde{\mathbf{M}}_{ib}\tilde{\mathbf{T}}\, \ddot{\mathbf{u}}_I(t) \tag{1.31}$$

where

$$\tilde{\mathbf{M}}_{ib} = [\mathbf{M}_{ib}^1 - \mathbf{M}_{ii}^1\mathbf{K}_{ii}^{1^{-1}}\mathbf{K}_{ib}^1, \ldots, \mathbf{M}_{ib}^{N_s} - \mathbf{M}_{ii}^{N_s}\mathbf{K}_{ii}^{N_s^{-1}}\mathbf{K}_{ib}^{N_s}] \tag{1.32}$$

Using the relationship between the vector $\mathbf{u}_I(t)$ and the vector of boundary coordinates of all substructures $\mathbf{u}_b(t)$ given in Eq. (1.16), the equation of motion (1.31) at the substructure level becomes

$$\ddot{\boldsymbol{\eta}}^s(t) + \boldsymbol{\Lambda}_{id}^s\, \boldsymbol{\eta}^s(t) = -\boldsymbol{\Phi}_{id}^{s^T}\tilde{\mathbf{M}}_{ib}^s\ddot{\mathbf{u}}_b^s(t)\ , \ s = 1, \ldots, N_s \tag{1.33}$$

where

$$\tilde{\mathbf{M}}_{ib}^s = \mathbf{M}_{ib}^s - \mathbf{M}_{ii}^s\mathbf{K}_{ii}^{s^{-1}}\mathbf{K}_{ib}^s \tag{1.34}$$

The static contribution of the residual fixed-interface normal modes to the response of the physical coordinates at the internal degrees of freedom $\mathbf{u}_i^s(t)$, due to the interface load $\tilde{\mathbf{M}}_{ib}^s\ddot{\mathbf{u}}_b^s(t)$, is approximated by using static correction [6, 19, 20, 29]. Due to the structure of the equation of motion (1.33), the static correction $\mathbf{u}_{i,\text{static}}^s(t)$ takes the form

$$\mathbf{u}_{i,\text{static}}^s(t) = -\bar{\mathbf{F}}^s\tilde{\mathbf{M}}_{ib}^s\ddot{\mathbf{u}}_b^s(t) \tag{1.35}$$

where $\bar{\mathbf{F}}^s$ is the residual flexibility matrix corresponding to the fixed-interface normal modes problem of substructure s, given by

$$\bar{\mathbf{F}}^s = \mathbf{K}_{ii}^{s^{-1}} - \boldsymbol{\Phi}_{id}^s\boldsymbol{\Lambda}_{id}^{s^{-1}}\boldsymbol{\Phi}_{id}^{s^T} \tag{1.36}$$

The previous approximation is reasonable, since high-frequency modes (residual fixed-interface normal modes) react essentially in a static manner when excited by low frequencies. Then, following Eqs. (1.16) and (1.28), the vector of physical coordinates at the internal degrees of freedom of substructure s can be expressed as

$$\begin{aligned}\mathbf{u}_i^s(t) &= \boldsymbol{\Phi}_{id}^s\, \eta^s(t) + \boldsymbol{\Psi}_{ib}^s \mathbf{u}_b^s(t) + \mathbf{u}_{i,\text{static}}^s(t) \\ &= \boldsymbol{\Phi}_{id}^s\, \eta^s(t) + \boldsymbol{\Psi}_{ib}^s \mathbf{u}_b^s(t) - \bar{\mathbf{F}}^s \tilde{\mathbf{M}}_{ib}^s \ddot{\mathbf{u}}_b^s(t)\ ,\ s = 1, \ldots, N_s\end{aligned} \tag{1.37}$$

If the relationship between the vectors $\mathbf{u}_I(t)$ and $\mathbf{u}_b(t)$ is once again considered, the vector of physical coordinates at the internal degrees of freedom of all substructures $\mathbf{u}_i(t)$ can be written in the form

$$\mathbf{u}_i(t) = [\boldsymbol{\Phi}_{id}^1, \ldots, \boldsymbol{\Phi}_{id}^{N_s}]\eta(t) + [\boldsymbol{\Psi}_{ib}^1, \ldots, \boldsymbol{\Psi}_{ib}^{N_s}]\tilde{\mathbf{T}}\mathbf{u}_I(t) - \bar{\mathbf{F}}\tilde{\mathbf{M}}_{ib}\tilde{\mathbf{T}}\, \ddot{\mathbf{u}}_I(t) \tag{1.38}$$

where $\bar{\mathbf{F}}$ is a block diagonal matrix containing the residual flexibility matrix of all substructures, that is,

$$\bar{\mathbf{F}} = [\bar{\mathbf{F}}^1, \ldots, \bar{\mathbf{F}}^{N_s}] \tag{1.39}$$

1.4.2 *Improved Transformation Matrix*

If the approximation of $\mathbf{u}_i(t)$ given in Eq. (1.38) and the definition of the vector of physical coordinates of all substructures $\bar{\mathbf{u}}(t)$ in Eq. (1.12) are used, it follows that

$$\bar{\mathbf{u}}(t) = \begin{bmatrix} [\boldsymbol{\Phi}_{id}^1, \ldots, \boldsymbol{\Phi}_{id}^{N_s}] & [\boldsymbol{\Psi}_{ib}^1, \ldots, \boldsymbol{\Psi}_{ib}^{N_s}]\tilde{\mathbf{T}} \\ \mathbf{0} & \mathbf{I} \end{bmatrix} \begin{Bmatrix} \eta(t) \\ \mathbf{u}_I(t) \end{Bmatrix} + \begin{bmatrix} \mathbf{0} & -\bar{\mathbf{F}}\tilde{\mathbf{M}}_{ib}\tilde{\mathbf{T}} \\ \mathbf{0} & \mathbf{0} \end{bmatrix} \begin{Bmatrix} \ddot{\eta}(t) \\ \ddot{\mathbf{u}}_I(t) \end{Bmatrix} \tag{1.40}$$

The relationship between the vector of generalized coordinates $\mathbf{q}(t)$ and its second derivative is obtained directly from Eq. (1.29), that is,

$$\begin{Bmatrix} \ddot{\eta}(t) \\ \ddot{\mathbf{u}}_I(t) \end{Bmatrix} = -\hat{\mathbf{M}}_D^{-1}\ \hat{\mathbf{K}}_D \begin{Bmatrix} \eta(t) \\ \mathbf{u}_I(t) \end{Bmatrix} \tag{1.41}$$

and therefore, Eq. (1.40) can be rewritten in the form

$$\bar{\mathbf{u}}(t) = \{\mathbf{T}_D + \mathbf{T}_R\} \begin{Bmatrix} \eta(t) \\ \mathbf{u}_I(t) \end{Bmatrix} \tag{1.42}$$

where $\mathbf{T}_D$ is the transformation matrix defined in Eq. (1.15) that includes the contribution of dominant fixed-interface normal modes, and

$$\mathbf{T}_R = \begin{bmatrix} \mathbf{0} & \bar{\mathbf{F}}\tilde{\mathbf{M}}_{ib}\tilde{\mathbf{T}} \\ \mathbf{0} & \mathbf{0} \end{bmatrix} \hat{\mathbf{M}}_D^{-1} \hat{\mathbf{K}}_D \tag{1.43}$$

is the transformation matrix that accounts for the contribution of the residual fixed-interface normal modes. Given the structure of the reduced-order mass matrix $\hat{\mathbf{M}}_D$, it follows that its inverse takes the form [17, 22]

$$\hat{\mathbf{M}}_D^{-1} = \begin{bmatrix} \mathbf{I} + \mathbf{M}_{iI}(\mathbf{M}_I - \mathbf{M}_{iI}^T\mathbf{M}_{iI})^{-1}\mathbf{M}_{iI}^T & -\mathbf{M}_{iI}(\mathbf{M}_I - \mathbf{M}_{iI}^T\mathbf{M}_{iI})^{-1} \\ -(\mathbf{M}_I - \mathbf{M}_{iI}^T\mathbf{M}_{iI})^{-1}\mathbf{M}_{iI}^T & (\mathbf{M}_I - \mathbf{M}_{iI}^T\mathbf{M}_{iI})^{-1} \end{bmatrix} \tag{1.44}$$

where

$$\mathbf{M}_I = \tilde{\mathbf{T}}^T [\hat{\mathbf{M}}_{bb}^1, \ldots, \hat{\mathbf{M}}_{bb}^{N_s}]\tilde{\mathbf{T}} \tag{1.45}$$

and

$$\mathbf{M}_{iI} = [\hat{\mathbf{M}}_{ib}^1, \ldots, \hat{\mathbf{M}}_{ib}^{N_s}]\tilde{\mathbf{T}} \tag{1.46}$$

If the definitions of $\hat{\mathbf{M}}_D^{-1}$ and $\hat{\mathbf{K}}_D$ are considered, and if the product between the 2×2 partition matrix in Eq. (1.43) and the matrix $\hat{\mathbf{M}}_D^{-1}\hat{\mathbf{K}}_D$ is performed, the transformation matrix $\mathbf{T}_R$ is expressed as

$$\mathbf{T}_R = \begin{bmatrix} -\bar{\mathbf{F}}\tilde{\mathbf{M}}_{ib}\tilde{\mathbf{T}}(\mathbf{M}_I - \mathbf{M}_{iI}^T\mathbf{M}_{iI})^{-1}\mathbf{M}_{iI}^T\boldsymbol{\Lambda} & \bar{\mathbf{F}}\tilde{\mathbf{M}}_{ib}\tilde{\mathbf{T}}(\mathbf{M}_I - \mathbf{M}_{iI}^T\mathbf{M}_{iI})^{-1}\mathbf{K}_I \\ \mathbf{0} & \mathbf{0} \end{bmatrix} \tag{1.47}$$

where

$$\boldsymbol{\Lambda} = [\boldsymbol{\Lambda}_{id}^1, \ldots, \boldsymbol{\Lambda}_{id}^{N_s}] \tag{1.48}$$

and

$$\mathbf{K}_I = \tilde{\mathbf{T}}^T [\hat{\mathbf{K}}_{bb}^1, \ldots, \hat{\mathbf{K}}_{bb}^{N_s}]\tilde{\mathbf{T}} \tag{1.49}$$

The matrix $\mathbf{T}_D + \mathbf{T}_R$ represents an improved transformation matrix that explicitly incorporates the contribution of the substructures' residual modes into the analysis. All the basic matrices involved in the definition of $\mathbf{T}_R$ are already available from the formulation based on dominant fixed-interface normal modes.

1.4.3 Enhanced Reduced-Order Matrices

Based on the improved transformation matrix $\mathbf{T}_D + \mathbf{T}_R$, the enhanced reduced-order mass matrix $\hat{\mathbf{M}}_R \in R^{n_D \times n_D}$ and the reduced-order stiffness matrix $\hat{\mathbf{K}}_R \in R^{n_D \times n_D}$ are defined as

$$\begin{aligned}\hat{\mathbf{M}}_R &= (\mathbf{T}_D + \mathbf{T}_R)^T \hat{\mathbf{M}} (\mathbf{T}_D + \mathbf{T}_R) \\ &= \mathbf{T}_D^T \hat{\mathbf{M}} \mathbf{T}_D + \mathbf{T}_R^T \hat{\mathbf{M}} \mathbf{T}_D + \mathbf{T}_D^T \hat{\mathbf{M}} \mathbf{T}_R + \mathbf{T}_R^T \hat{\mathbf{M}} \mathbf{T}_R \\ &= \hat{\mathbf{M}}_D + \mathbf{T}_R^T \hat{\mathbf{M}} \mathbf{T}_D + \mathbf{T}_D^T \hat{\mathbf{M}} \mathbf{T}_R + \mathbf{T}_R^T \hat{\mathbf{M}} \mathbf{T}_R\end{aligned} \tag{1.50}$$

and

$$\begin{aligned}\hat{\mathbf{K}}_R &= (\mathbf{T}_D + \mathbf{T}_R)^T \hat{\mathbf{K}} (\mathbf{T}_D + \mathbf{T}_R) \\ &= \mathbf{T}_D^T \hat{\mathbf{K}} \mathbf{T}_D + \mathbf{T}_R^T \hat{\mathbf{K}} \mathbf{T}_D + \mathbf{T}_D^T \hat{\mathbf{K}} \mathbf{T}_R + \mathbf{T}_R^T \hat{\mathbf{K}} \mathbf{T}_R \\ &= \hat{\mathbf{K}}_D + \mathbf{T}_R^T \hat{\mathbf{K}} \mathbf{T}_D + \mathbf{T}_D^T \hat{\mathbf{K}} \mathbf{T}_R + \mathbf{T}_R^T \hat{\mathbf{K}} \mathbf{T}_R\end{aligned} \tag{1.51}$$

It is expected that the reduced-order matrices $\hat{\mathbf{M}}_R$ and $\hat{\mathbf{K}}_R$ are more precisely constructed than the reduced-order matrices obtained from the formulation based on dominant fixed-interface normal modes because of the explicit contribution of the residual modes of the substructures in the transformation matrix $\mathbf{T}_R$. Thus, the effect of considering residual normal modes is to enhance the dynamic behavior of the internal degrees of freedom at the substructure level and, consequently, at the global level.

1.4.4 Remarks on the Use of Residual Modes

As previously mentioned, if the residual fixed-interface normal modes are considered explicitly in the analysis, the approximation of the response at the substructure level is expected to improve. In this context, it is noted that reduced-order models based only on dominant fixed-interface normal modes can also be used if the level of accuracy is appropriate. Note, however, that the number of fixed-interface normal modes included in the analysis should be increased if the same level of accuracy is required as the one obtained using residual normal modes [20]. The use or non-use of the contribution of the residual normal modes in the analysis is problem-dependent.

1.5 Numerical Implementation: Pseudo-Code No. 1

Based on the formulation presented, the following pseudo-code illustrates how the reduced mass and stiffness matrices are evaluated.

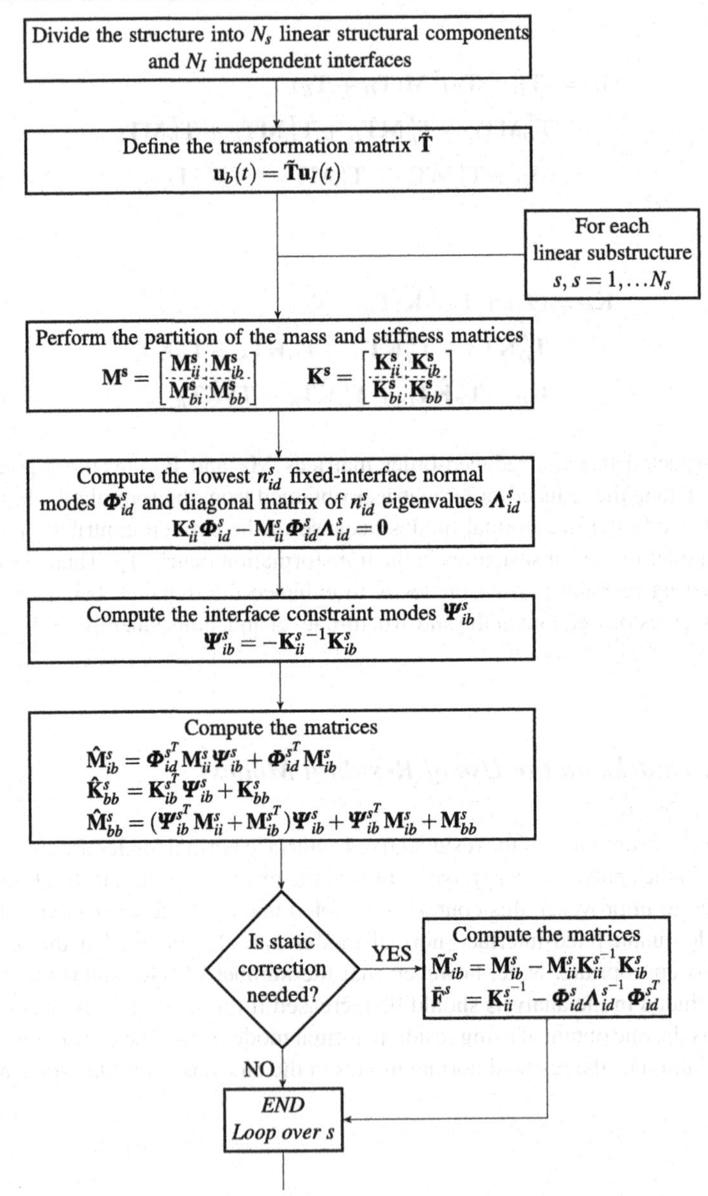

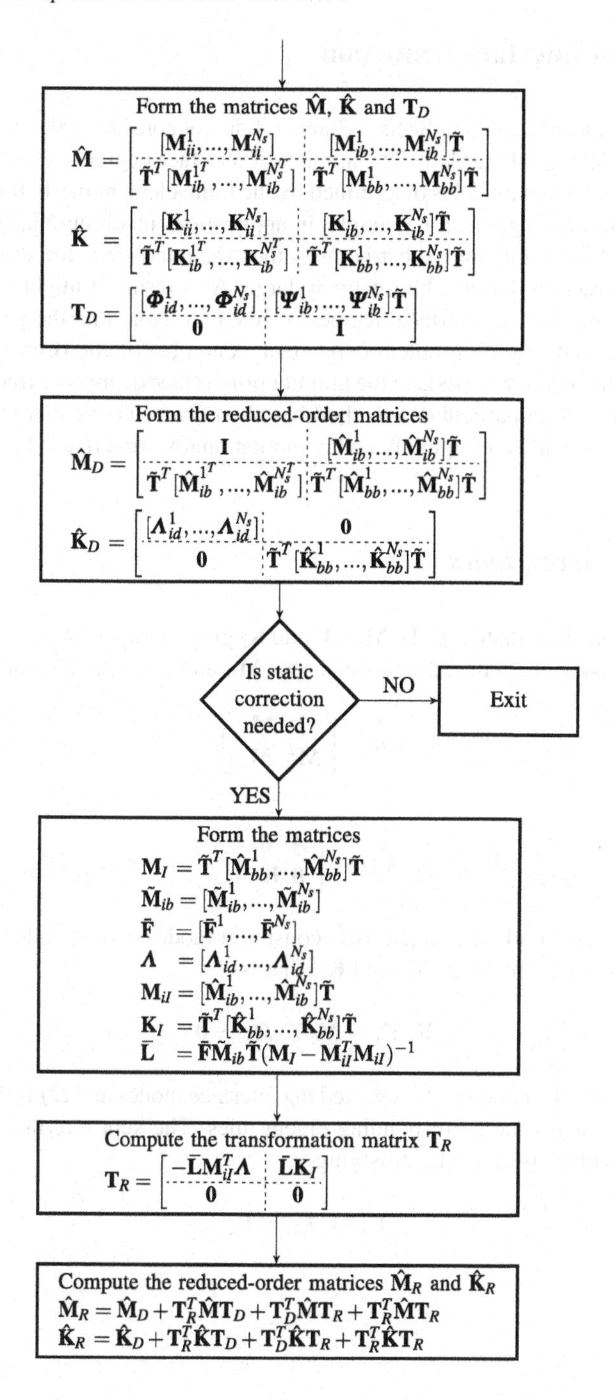
Form the matrices $\hat{\mathbf{M}}$, $\hat{\mathbf{K}}$ and $\mathbf{T}_D$
Form the reduced-order matrices
Is static correction needed?
NO
Exit
YES
Form the matrices
Compute the transformation matrix $\mathbf{T}_R$
Compute the reduced-order matrices $\hat{\mathbf{M}}_R$ and $\hat{\mathbf{K}}_R$
$\hat{\mathbf{M}}_R = \hat{\mathbf{M}}_D + \mathbf{T}_R^T\hat{\mathbf{M}}\mathbf{T}_D + \mathbf{T}_D^T\hat{\mathbf{M}}\mathbf{T}_R + \mathbf{T}_R^T\hat{\mathbf{M}}\mathbf{T}_R$
$\hat{\mathbf{K}}_R = \hat{\mathbf{K}}_D + \mathbf{T}_R^T\hat{\mathbf{K}}\mathbf{T}_D + \mathbf{T}_D^T\hat{\mathbf{K}}\mathbf{T}_R + \mathbf{T}_R^T\hat{\mathbf{K}}\mathbf{T}_R$

1.6 Global Interface Reduction

The formulations presented in Sects. 1.3 and 1.4 do not consider order reduction for the interface degrees of freedom. The number of interface degrees of freedom in the original finite element model is determined by the finite element mesh. If the mesh is fine in the interface regions, for example, in applications involving line and surface coupling, or if there are many substructures, then the interface matrices involved in the reduced-order model may be relatively large. As a result, it might be desirable to reduce the number of interface degrees of freedom. Note that the pertinence of using interface reduction is problem-dependent. A number of approaches have been proposed in the literature to reduce the number of interface degrees of freedom [1, 4, 9, 11, 30, 31, 33]. In particular, a method based on characteristic constraint modes has proven to be quite conceptually and computationally attractive [11].

1.6.1 Interface Modes

If the definition of the matrices $\mathbf{M}_I$, $\mathbf{M}_{iI}$, $\boldsymbol{\Lambda}$, and $\mathbf{K}_I$ given in Eqs. (1.45), (1.46), (1.48), and (1.49) is used, the reduced-order matrices $\hat{\mathbf{M}}_D$ and $\hat{\mathbf{K}}_D$ in (1.23) and (1.24) can be rewritten as

$$\hat{\mathbf{M}}_D = \begin{bmatrix} \mathbf{I} & \mathbf{M}_{iI} \\ \mathbf{M}_{iI}^T & \mathbf{M}_I \end{bmatrix} \tag{1.52}$$

and

$$\hat{\mathbf{K}}_D = \begin{bmatrix} \boldsymbol{\Lambda} & \mathbf{0} \\ \mathbf{0} & \mathbf{K}_I \end{bmatrix} \tag{1.53}$$

The approach based on characteristic constraint modes corresponds to an eigenvalue problem of the matrices $\mathbf{M}_I$ and $\mathbf{K}_I$, that is,

$$\mathbf{K}_I \boldsymbol{\Upsilon}_I - \mathbf{M}_I \boldsymbol{\Upsilon}_I \boldsymbol{\Omega}_I = \mathbf{0} \tag{1.54}$$

where the matrix $\boldsymbol{\Upsilon}_I$ contains the selected n_{IR} interface modes and $\boldsymbol{\Omega}_I$ is the diagonal matrix that contains the corresponding eigenvalues. The kept interface modes are normalized with respect to $\mathbf{M}_I$, satisfying

$$\boldsymbol{\Upsilon}_I^T \mathbf{M}_I \boldsymbol{\Upsilon}_I = \mathbf{I}_I \tag{1.55}$$

and

$$\boldsymbol{\Upsilon}_I^T \mathbf{K}_I \boldsymbol{\Upsilon}_I = \boldsymbol{\Omega}_I \tag{1.56}$$

where $\mathbf{I}_I$ denotes the identity matrix. The number of selected interface modes can be relatively small compared with the number of interface degrees of freedom, that is, $n_{IR} \ll n_I$.

1.6.2 Reduced-Order Matrices Based on Dominant Fixed-Interface Modes

The kept set of interface modes is used to represent the vector of physical coordinates at the N_I independent interfaces $\mathbf{u}_I(t)$, expressed as

$$\mathbf{u}_I(t) = \boldsymbol{\Upsilon}_I \gamma(t) \tag{1.57}$$

where $\gamma(t)$ are the interface modal coordinates. With this representation, the vector of physical coordinates of all substructures $\bar{\mathbf{u}}(t)$ can be expressed as

$$\bar{\mathbf{u}}(t) = \mathbf{T}_D \begin{bmatrix} \mathbf{I} & \mathbf{0} \\ \mathbf{0} & \boldsymbol{\Upsilon}_I \end{bmatrix} \mathbf{q}_I(t) = \mathbf{T}_{DI} \mathbf{q}_I(t) \tag{1.58}$$

where $\mathbf{q}_I(t)$ is the vector of generalized coordinates

$$\mathbf{q}_I(t) = \begin{Bmatrix} \eta(t) \\ \gamma(t) \end{Bmatrix} \tag{1.59}$$

and

$$\mathbf{T}_{DI} = \begin{bmatrix} [\boldsymbol{\Phi}_{id}^1, \ldots, \boldsymbol{\Phi}_{id}^{N_s}] & [\boldsymbol{\Psi}_{ib}^1, \ldots, \boldsymbol{\Psi}_{ib}^{N_s}] \tilde{\mathbf{T}} \boldsymbol{\Upsilon}_I \\ \mathbf{0} & \boldsymbol{\Upsilon}_I \end{bmatrix} \tag{1.60}$$

is the transformation matrix that considers the effect of the dominant fixed-interface normal modes and interface reduction. The corresponding mass and stiffness matrices of the model, referred to the fixed-interface modal coordinates of all substructures $\eta(t)$ and the interface modal coordinates $\gamma(t)$, are defined as

$$\hat{\mathbf{M}}_{DI} = \mathbf{T}_{DI}^T \, \hat{\mathbf{M}} \, \mathbf{T}_{DI} \tag{1.61}$$

and

$$\hat{\mathbf{K}}_{DI} = \mathbf{T}_{DI}^T \, \hat{\mathbf{K}} \, \mathbf{T}_{DI} \tag{1.62}$$

which yield

$$\hat{\mathbf{M}}_{DI} = \begin{bmatrix} \mathbf{I} & [\hat{\mathbf{M}}_{ib}^1, \ldots, \hat{\mathbf{M}}_{ib}^{N_s}]\tilde{\mathbf{T}}\boldsymbol{\Upsilon}_I \\ \boldsymbol{\Upsilon}_I^T \tilde{\mathbf{T}}^T [\hat{\mathbf{M}}_{ib}^{1^T}, \ldots, \hat{\mathbf{M}}_{ib}^{N_s^T}] & \mathbf{I}_I \end{bmatrix} \tag{1.63}$$

and

$$\hat{\mathbf{K}}_{DI} = \begin{bmatrix} [\boldsymbol{\Lambda}_{id}^1, \ldots, \boldsymbol{\Lambda}_{id}^{N_s}] & \mathbf{0} \\ \mathbf{0} & \boldsymbol{\Omega}_I \end{bmatrix} \tag{1.64}$$

where the dimension of these matrices is equal to $n_{DI} \times n_{DI}$, $n_{DI} = \sum_{s=1}^{N_s} n_{id}^s + n_{IR}$. It can be seen that the reduced stiffness matrix is diagonal, while the reduced mass matrix contains off-diagonal terms. The matrix

$$\boldsymbol{\Upsilon}_{II} = [\boldsymbol{\Psi}_{ib}^1, \ldots, \boldsymbol{\Psi}_{ib}^{N_s}]\tilde{\mathbf{T}}\boldsymbol{\Upsilon}_I \quad \in R^{n_i \times n_{IR}} \tag{1.65}$$

involved in defining the transformation matrix $\mathbf{T}_{DI}$ represents an eigenvector-based linear combination of the interior partition of the interface constraint modes $\boldsymbol{\Psi}_{ib}^s$, $s = 1, \ldots, N_s$, referred to the physical coordinates of the internal degrees of freedom of all substructures $\mathbf{u}_i(t)$. The columns of the augmented matrix $\boldsymbol{\Upsilon}_{CC} \in R^{n \times n_{IR}}$, defined in terms of $\boldsymbol{\Upsilon}_{II}$ and $\boldsymbol{\Upsilon}_I$ as

$$\boldsymbol{\Upsilon}_{CC} = \begin{bmatrix} \boldsymbol{\Upsilon}_{II} \\ \boldsymbol{\Upsilon}_I \end{bmatrix} \tag{1.66}$$

correspond to the characteristic constraint modes [10, 11]. These modes provide a significant physical insight into the motion of the interface and the transmission of vibration between the substructures, since they contain in an approximate sense the principal modes of deformation for the interface [11, 31]. The dominant fixed-interface normal modes and the characteristic constraint modes may define a reduced-order model with a greatly-reduced number of degrees of freedom, that is, $n_{DI} \ll n$. Therefore, the effect of interface reduction is the reduction of the dimension of the model even further.

1.6.3 Reduced-Order Matrices Based on Residual Fixed-Interface Modes

If the relationship between the vector of physical coordinates at the independent interfaces $\mathbf{u}_I(t)$ and the interface modal coordinates $\gamma(t)$ given in Eq. (1.57) is considered, Eq. (1.40) can be rewritten as

$$\bar{\mathbf{u}}(t) = \mathbf{T}_D \begin{bmatrix} \mathbf{I} & \mathbf{0} \\ \mathbf{0} & \boldsymbol{\Upsilon}_I \end{bmatrix} \begin{Bmatrix} \eta(t) \\ \gamma(t) \end{Bmatrix} + \begin{bmatrix} \mathbf{0} & -\bar{\mathbf{F}}\tilde{\mathbf{M}}_{ib}\tilde{\mathbf{T}} \\ \mathbf{0} & \mathbf{0} \end{bmatrix} \begin{bmatrix} \mathbf{I} & \mathbf{0} \\ \mathbf{0} & \boldsymbol{\Upsilon}_I \end{bmatrix} \begin{Bmatrix} \ddot{\eta}(t) \\ \ddot{\gamma}(t) \end{Bmatrix} \tag{1.67}$$

On the other hand, the relationship between the vector $\mathbf{q}_I(t)$, given in Eq. (1.59), and its second derivative can be obtained directly from the equation corresponding to the undamped free vibration of the linear components of Eq. (1.1) referred to the set of generalized coordinates $\mathbf{q}_I(t)$. Similar to Eq. (1.41), the second derivative of $\mathbf{q}_I(t)$ is given by

$$\begin{Bmatrix} \ddot{\eta}(t) \\ \ddot{\gamma}(t) \end{Bmatrix} = -\hat{\mathbf{M}}_{DI}^{-1} \ \hat{\mathbf{K}}_{DI} \begin{Bmatrix} \eta(t) \\ \gamma(t) \end{Bmatrix} \tag{1.68}$$

Based on the previous relationships and applying the derivation presented in Sect. 1.4.2, it follows that the transformation matrix $\mathbf{T}_R$ in Eq. (1.43), that includes the contribution of the residual fixed-interface normal modes, becomes

$$\mathbf{T}_{RI} = \begin{bmatrix} \mathbf{0} & \bar{\mathbf{F}}\tilde{\mathbf{M}}_{ib}\tilde{\mathbf{T}} \\ \mathbf{0} & \mathbf{0} \end{bmatrix} \begin{bmatrix} \mathbf{I} & \mathbf{0} \\ \mathbf{0} & \boldsymbol{\Upsilon}_I \end{bmatrix} \hat{\mathbf{M}}_{DI}^{-1} \ \hat{\mathbf{K}}_{DI} = \begin{bmatrix} \mathbf{0} & \bar{\mathbf{F}}\tilde{\mathbf{M}}_{ib}\tilde{\mathbf{T}}\boldsymbol{\Upsilon}_I \\ \mathbf{0} & \mathbf{0} \end{bmatrix} \hat{\mathbf{M}}_{DI}^{-1} \ \hat{\mathbf{K}}_{DI} \tag{1.69}$$

where the inverse of the reduced-order mass matrix $\hat{\mathbf{M}}_{DI}$ is given by [17, 22]

$$\hat{\mathbf{M}}_{DI}^{-1} = \begin{bmatrix} \mathbf{I} + \mathbf{M}_{iIR}(\mathbf{I}_I - \mathbf{M}_{iIR}^T\mathbf{M}_{iIR})^{-1}\mathbf{M}_{iIR}^T & -\mathbf{M}_{iIR}(\mathbf{I}_I - \mathbf{M}_{iIR}^T\mathbf{M}_{iIR})^{-1} \\ -(\mathbf{I}_I - \mathbf{M}_{iIR}^T\mathbf{M}_{iIR})^{-1}\mathbf{M}_{iIR}^T & (\mathbf{I}_I - \mathbf{M}_{iIR}^T\mathbf{M}_{iIR})^{-1} \end{bmatrix} \tag{1.70}$$

where

$$\mathbf{M}_{iIR} = [\hat{\mathbf{M}}_{ib}^{1}, \ldots, \hat{\mathbf{M}}_{ib}^{N_s}]\tilde{\mathbf{T}}\boldsymbol{\Upsilon}_I \tag{1.71}$$

Taking into account the definition of $\hat{\mathbf{M}}_{DI}^{-1}$ and $\hat{\mathbf{K}}_{DI}$ and carrying out the corresponding products between the matrices in (1.69), the transformation matrix $\mathbf{T}_{RI}$ is written as

$$\mathbf{T}_{RI} = \begin{bmatrix} -\bar{\mathbf{F}}\tilde{\mathbf{M}}_{ib}\tilde{\mathbf{T}}\boldsymbol{\Upsilon}_I(\mathbf{I}_I - \mathbf{M}_{iIR}^T\mathbf{M}_{iIR})^{-1}\mathbf{M}_{iIR}^T\boldsymbol{\Lambda} & \bar{\mathbf{F}}\tilde{\mathbf{M}}_{ib}\tilde{\mathbf{T}}\boldsymbol{\Upsilon}_I(\mathbf{I}_I - \mathbf{M}_{iIR}^T\mathbf{M}_{iIR})^{-1}\boldsymbol{\Omega}_I \\ \mathbf{0} & \mathbf{0} \end{bmatrix} \tag{1.72}$$

Then, the relationship between the vector of physical coordinates of all substructures and the vector of generalized coordinates given in Eq. (1.58) is redefined as

$$\bar{\mathbf{u}}(t) = (\mathbf{T}_{DI} + \mathbf{T}_{RI})\,\mathbf{q}_I(t) \tag{1.73}$$

Consequently, the enhanced reduced-order mass matrix $\hat{\mathbf{M}}_{RI} \in R^{n_{DI} \times n_{DI}}$ and reduced-order stiffness matrix $\hat{\mathbf{K}}_{RI} \in R^{n_{DI} \times n_{DI}}$ that consider interface reduction and the contribution of residual fixed-interface normal modes, take the form

$$\begin{aligned}\hat{\mathbf{M}}_{RI} &= (\mathbf{T}_{DI} + \mathbf{T}_{RI})^T \hat{\mathbf{M}} (\mathbf{T}_{DI} + \mathbf{T}_{RI}) \\ &= \mathbf{T}_{DI}^T \hat{\mathbf{M}} \mathbf{T}_{DI} + \mathbf{T}_{RI}^T \hat{\mathbf{M}} \mathbf{T}_{DI} + \mathbf{T}_{DI}^T \hat{\mathbf{M}} \mathbf{T}_{RI} + \mathbf{T}_{RI}^T \hat{\mathbf{M}} \mathbf{T}_{RI} \\ &= \hat{\mathbf{M}}_{DI} + \mathbf{T}_{RI}^T \hat{\mathbf{M}} \mathbf{T}_{DI} + \mathbf{T}_{DI}^T \hat{\mathbf{M}} \mathbf{T}_{RI} + \mathbf{T}_{RI}^T \hat{\mathbf{M}} \mathbf{T}_{RI}\end{aligned} \tag{1.74}$$

and

$$\begin{aligned}\hat{\mathbf{K}}_{RI} &= (\mathbf{T}_{DI} + \mathbf{T}_{RI})^T \hat{\mathbf{K}} (\mathbf{T}_{DI} + \mathbf{T}_{RI}) \\ &= \mathbf{T}_{DI}^T \hat{\mathbf{K}} \mathbf{T}_{DI} + \mathbf{T}_{RI}^T \hat{\mathbf{K}} \mathbf{T}_{DI} + \mathbf{T}_{DI}^T \hat{\mathbf{K}} \mathbf{T}_{RI} + \mathbf{T}_{RI}^T \hat{\mathbf{K}} \mathbf{T}_{RI} \\ &= \hat{\mathbf{K}}_{DI} + \mathbf{T}_{RI}^T \hat{\mathbf{K}} \mathbf{T}_{DI} + \mathbf{T}_{DI}^T \hat{\mathbf{K}} \mathbf{T}_{RI} + \mathbf{T}_{RI}^T \hat{\mathbf{K}} \mathbf{T}_{RI}\end{aligned} \tag{1.75}$$

The term $\mathbf{I}_I - \mathbf{M}_{iIR}^T \mathbf{M}_{iIR}$, that appears in the transformation matrix $\mathbf{T}_{RI}$ and has to be inverted, is a matrix of dimension $n_{IR} \times n_{IR}$, being n_{IR} the selected number of interface modal coordinates as previously pointed out. Thus, if the number of selected interface modes is small the computational cost of inverting this matrix is not high.

1.7 Numerical Implementation: Pseudo-Code No. 2

Based on the formulation presented, the following pseudo-code illustrates how the reduced mass and stiffness matrices are evaluated by including global interface reduction.

Define the matrices

$\mathbf{M}_I = \tilde{\mathbf{T}}^T [\hat{\mathbf{M}}_{bb}^1, ..., \hat{\mathbf{M}}_{bb}^{N_s}] \tilde{\mathbf{T}} \qquad \mathbf{K}_I = \tilde{\mathbf{T}}^T [\hat{\mathbf{K}}_{bb}^1, ..., \hat{\mathbf{K}}_{bb}^{N_s}] \tilde{\mathbf{T}}$

see Pseudo-Code No.1 for the definition of $\tilde{\mathbf{T}}$, $\hat{\mathbf{M}}_{bb}^s$ and $\hat{\mathbf{K}}_{bb}^s$, $s = 1, ..., N_s$

↓

Compute the lowest n_{IR} interface modes $\boldsymbol{\Upsilon}_I$ and diagonal matrix of n_{IR} eigenvalues $\boldsymbol{\Omega}_I$

$\mathbf{K}_I \boldsymbol{\Upsilon}_I - \mathbf{M}_I \boldsymbol{\Upsilon}_I \boldsymbol{\Omega}_I = \mathbf{0}$

↓

Form the transformation matrix

$$\mathbf{T}_{DI} = \left[\begin{array}{c|c} [\boldsymbol{\Phi}_{id}^1, ..., \boldsymbol{\Phi}_{id}^{N_s}] & [\boldsymbol{\Psi}_{ib}^1, ..., \boldsymbol{\Psi}_{ib}^{N_s}] \tilde{\mathbf{T}} \boldsymbol{\Upsilon}_I \\ \hline \mathbf{0} & \boldsymbol{\Upsilon}_I \end{array} \right]$$

see Pseudo-Code No.1 for the calculation of $\boldsymbol{\Phi}_{id}^s$ and $\boldsymbol{\Psi}_{ib}^s$, $s = 1, ..., N_s$.

↓

Form the reduced-order mass and stiffness matrices

$$\hat{\mathbf{M}}_{DI} = \left[\begin{array}{c|c} \mathbf{I} & [\hat{\mathbf{M}}_{ib}^1, ..., \hat{\mathbf{M}}_{ib}^{N_s}] \tilde{\mathbf{T}} \boldsymbol{\Upsilon}_I \\ \hline \boldsymbol{\Upsilon}_I^T \tilde{\mathbf{T}}^T [\hat{\mathbf{M}}_{ib}^{1^T}, ..., \hat{\mathbf{M}}_{ib}^{N_s^T}] & \mathbf{I}_I \end{array} \right]$$

$$\hat{\mathbf{K}}_{DI} = \left[\begin{array}{c|c} [\boldsymbol{\Lambda}_{id}^1, ..., \boldsymbol{\Lambda}_{id}^{N_s}] & \mathbf{0} \\ \hline \mathbf{0} & \boldsymbol{\Omega}_I \end{array} \right]$$

see Pseudo-Code No.1 for the construction of $\hat{\mathbf{M}}_{ib}^s$ and $\boldsymbol{\Lambda}_{id}^s$ $s = 1, \ldots, N_s$.

↓

Is static correction needed? — NO → Exit

↓ YES

Compute the matrices

$\mathbf{M}_{iIR} = \mathbf{M}_{iI} \boldsymbol{\Upsilon}_I$

$\bar{\mathbf{L}} = \bar{\mathbf{F}} \tilde{\mathbf{M}}_{ib} \tilde{\mathbf{T}} \boldsymbol{\Upsilon}_I (\mathbf{I}_I - \mathbf{M}_{iIR}^T \mathbf{M}_{iIR})^{-1}$

see Pseudo-Code No.1 for the construction of $\mathbf{M}_{iI}$, $\bar{\mathbf{F}}$ and $\tilde{\mathbf{M}}_{ib}$

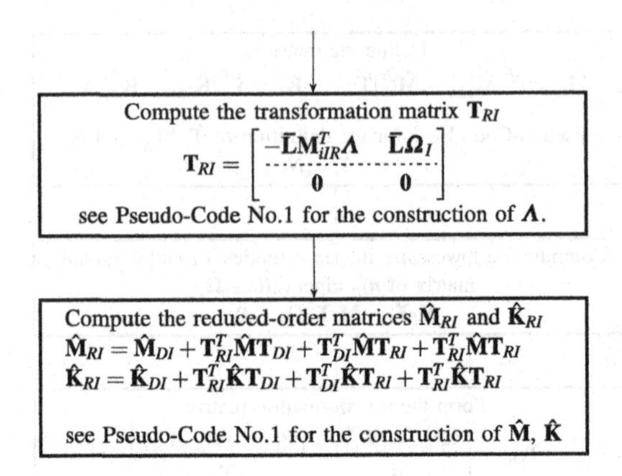

Compute the transformation matrix $\mathbf{T}_{RI}$
$\mathbf{T}_{RI} = \begin{bmatrix} -\bar{\mathbf{L}}\mathbf{M}_{iIR}^T \boldsymbol{\Lambda} & \bar{\mathbf{L}}\boldsymbol{\Omega}_I \\ \mathbf{0} & \mathbf{0} \end{bmatrix}$
see Pseudo-Code No.1 for the construction of $\boldsymbol{\Lambda}$.
Compute the reduced-order matrices $\hat{\mathbf{M}}_{RI}$ and $\hat{\mathbf{K}}_{RI}$
$\hat{\mathbf{M}}_{RI} = \hat{\mathbf{M}}_{DI} + \mathbf{T}_{RI}^T \hat{\mathbf{M}} \mathbf{T}_{DI} + \mathbf{T}_{DI}^T \hat{\mathbf{M}} \mathbf{T}_{RI} + \mathbf{T}_{RI}^T \hat{\mathbf{M}} \mathbf{T}_{RI}$
$\hat{\mathbf{K}}_{RI} = \hat{\mathbf{K}}_{DI} + \mathbf{T}_{RI}^T \hat{\mathbf{K}} \mathbf{T}_{DI} + \mathbf{T}_{DI}^T \hat{\mathbf{K}} \mathbf{T}_{RI} + \mathbf{T}_{RI}^T \hat{\mathbf{K}} \mathbf{T}_{RI}$
see Pseudo-Code No.1 for the construction of $\hat{\mathbf{M}}$, $\hat{\mathbf{K}}$

1.8 Local Interface Reduction

The previous treatment of interface reduction is based on global interface modes. Alternatively, the interface modes can be considered at the local level. To this end, let $\mathbf{K}_{Ill} \in R^{n_I^l \times n_I^l}$ and $\mathbf{M}_{Ill} \in R^{n_I^l \times n_I^l}$ be the partitions of the interface matrices $\mathbf{K}_I$ and $\mathbf{M}_I$, respectively, associated with the physical coordinates at interface l, $l = 1, \ldots, N_l$, i.e. $\mathbf{u}_I^l(t)$. The interface modes corresponding to interface l satisfy the eigenvalue problem

$$\mathbf{K}_{Ill}\boldsymbol{\Upsilon}_{Ill} - \mathbf{M}_{Ill}\boldsymbol{\Upsilon}_{Ill}\boldsymbol{\Omega}_{Ill} = \mathbf{0} \tag{1.76}$$

where $\boldsymbol{\Upsilon}_{Ill}$ contains the selected n_{IR}^l local interface modes and $\boldsymbol{\Omega}_{Ill}$ is the diagonal matrix that contains the corresponding eigenvalues. The local interface modes are normalized with respect to $\mathbf{M}_{Ill}$ in the form

$$\boldsymbol{\Upsilon}_{Ill}^T\mathbf{M}_{Ill}\boldsymbol{\Upsilon}_{Ill} = \mathbf{I}_{Ill} \tag{1.77}$$

and

$$\boldsymbol{\Upsilon}_{Ill}^T\mathbf{K}_{Ill}\boldsymbol{\Upsilon}_{Ill} = \boldsymbol{\Omega}_{Ill} \tag{1.78}$$

where $\mathbf{I}_{Ill}$ denotes the identity matrix. The previous modes are used to represent the vector of physical coordinates $\mathbf{u}_I^l(t)$ at interface l in terms of the local interface modal coordinates $\gamma_l(t)$ in the form

$$\mathbf{u}_I^l(t) = \boldsymbol{\Upsilon}_{Ill}\gamma_l(t) \tag{1.79}$$

Similarly to Eq. (1.58) the vector of physical coordinates of all substructures $\bar{\mathbf{u}}(t)$ can be written as

$$\bar{\mathbf{u}}(t) = \mathbf{T}_D \begin{bmatrix} \mathbf{I} & \mathbf{0} \\ \mathbf{0} & [\boldsymbol{\Upsilon}_{I11}, \ldots, \boldsymbol{\Upsilon}_{IN_lN_l}] \end{bmatrix} \mathbf{q}_{IL}(t) = \mathbf{T}_{DIL}\mathbf{q}_{IL}(t) \tag{1.80}$$

where $\mathbf{q}_{IL}(t)$ is the vector of generalized coordinates

$$\mathbf{q}_{IL}(t) = \left\{ \begin{array}{c} \eta(t) \\ \gamma_L(t) \end{array} \right\}, \tag{1.81}$$

$\gamma_L(t) =< \gamma_1(t)^T, \ldots, \gamma_{N_l}(t) >^T$ is the vector of local interface modal coordinates of all independent interfaces, and

$$\mathbf{T}_{DIL} = \begin{bmatrix} [\boldsymbol{\Phi}_{id}^1, \ldots, \boldsymbol{\Phi}_{id}^{N_s}] & [\boldsymbol{\Psi}_{ib}^1, \ldots, \boldsymbol{\Psi}_{ib}^{N_s}]\tilde{\mathbf{T}}[\boldsymbol{\Upsilon}_{I11}, \ldots, \boldsymbol{\Upsilon}_{IN_lN_l}] \\ \mathbf{0} & [\boldsymbol{\Upsilon}_{I11}, \ldots, \boldsymbol{\Upsilon}_{IN_lN_l}] \end{bmatrix} \tag{1.82}$$

is the transformation matrix that accounts for the effect of the dominant fixed-interface normal modes and the local interface normal modes. The related reduced-order mass and stiffness matrices are defined as

$$\hat{\mathbf{M}}_{DIL} = \mathbf{T}_{DIL}^T \, \hat{\mathbf{M}} \, \mathbf{T}_{DIL} \tag{1.83}$$

and

$$\hat{\mathbf{K}}_{DIL} = \mathbf{T}_{DIL}^T \, \hat{\mathbf{K}} \, \mathbf{T}_{DIL} \tag{1.84}$$

which give

$$\hat{\mathbf{M}}_{DIL} = \begin{bmatrix} \mathbf{I} & [\hat{\mathbf{M}}_{ib}^1, \ldots, \hat{\mathbf{M}}_{ib}^{N_s}]\tilde{\mathbf{T}}[\boldsymbol{\Upsilon}_{I11}, \ldots, \boldsymbol{\Upsilon}_{IN_lN_l}] \\ [\boldsymbol{\Upsilon}_{I11}^T, \ldots, \boldsymbol{\Upsilon}_{IN_lN_l}^T]\tilde{\mathbf{T}}^T[\hat{\mathbf{M}}_{ib}^{1^T}, \ldots, \hat{\mathbf{M}}_{ib}^{N_s^T}] & [\boldsymbol{\Upsilon}_{I11}^T, \ldots, \boldsymbol{\Upsilon}_{IN_lN_l}^T]\mathbf{M}_I[\boldsymbol{\Upsilon}_{I11}, \ldots, \boldsymbol{\Upsilon}_{IN_lN_l}] \end{bmatrix} \tag{1.85}$$

and

$$\hat{\mathbf{K}}_{DIL} = \begin{bmatrix} [\boldsymbol{\Lambda}_{id}^1, \ldots, \boldsymbol{\Lambda}_{id}^{N_s}] & \mathbf{0} \\ \mathbf{0} & [\boldsymbol{\Upsilon}_{I11}^T, \ldots, \boldsymbol{\Upsilon}_{IN_lN_l}^T]\mathbf{K}_I[\boldsymbol{\Upsilon}_{I11}, \ldots, \boldsymbol{\Upsilon}_{IN_lN_l}] \end{bmatrix} \tag{1.86}$$

where these matrices are of dimension equal to $n_{DIL} \times n_{DIL}$, $n_{DIL} = \sum_{s=1}^{N_s} n_{id}^s + \sum_{l=1}^{N_l} n_{IR}^l$.

On the other hand, the transformation matrix that includes the contribution of the residual fixed-interface normal modes changes to

$$\mathbf{T}_{RIL} = \begin{bmatrix} \mathbf{T}_{RIL1} & \mathbf{T}_{RIL2} \\ \mathbf{0} & \mathbf{0} \end{bmatrix} \tag{1.87}$$

where

$$\begin{aligned} \mathbf{T}_{RIL1} = & -\bar{\mathbf{F}}\bar{\mathbf{M}}_{ib}\tilde{\mathbf{T}}[\boldsymbol{\Upsilon}_{I11}, \ldots, \boldsymbol{\Upsilon}_{IN_lN_l}] \\ & \times \left([\boldsymbol{\Upsilon}_{I11}^T, \ldots, \boldsymbol{\Upsilon}_{IN_lN_l}^T]\mathbf{M}_I[\boldsymbol{\Upsilon}_{I11}, \ldots, \boldsymbol{\Upsilon}_{IN_lN_l}] - \mathbf{M}_{iIRL}^T\mathbf{M}_{iIRL} \right)^{-1} \mathbf{M}_{iIRL}^T \boldsymbol{\Lambda} \end{aligned} \tag{1.88}$$

and

$$\mathbf{T}_{RIL2} = \bar{\mathbf{F}}\bar{\mathbf{M}}_{ib}\tilde{\mathbf{T}}[\boldsymbol{\Upsilon}_{I11}, \ldots, \boldsymbol{\Upsilon}_{IN_lN_l}] \times \left([\boldsymbol{\Upsilon}_{I11}^T, \ldots, \boldsymbol{\Upsilon}_{IN_lN_l}^T]\mathbf{M}_I[\boldsymbol{\Upsilon}_{I11}, \ldots, \boldsymbol{\Upsilon}_{IN_lN_l}] - \mathbf{M}_{iIRL}^T\mathbf{M}_{iIRL}\right)^{-1} \times [\boldsymbol{\Upsilon}_{I11}^T, \ldots, \boldsymbol{\Upsilon}_{IN_lN_l}^T]\mathbf{K}_I[\boldsymbol{\Upsilon}_{I11}, \ldots, \boldsymbol{\Upsilon}_{IN_lN_l}] \tag{1.89}$$

with

$$\mathbf{M}_{iIRL} = [\hat{\mathbf{M}}_{ib}^1, \ldots, \hat{\mathbf{M}}_{ib}^{N_s}]\tilde{\mathbf{T}}[\boldsymbol{\Upsilon}_{I11}, \ldots, \boldsymbol{\Upsilon}_{IN_lN_l}] \tag{1.90}$$

Due to the normalization of the local interface modes, the diagonal blocks of the matrices $[\boldsymbol{\Upsilon}_{I11}^T, \ldots, \boldsymbol{\Upsilon}_{IN_lN_l}^T]\mathbf{M}_I[\boldsymbol{\Upsilon}_{I11}, \ldots, \boldsymbol{\Upsilon}_{IN_lN_l}]$ and $[\boldsymbol{\Upsilon}_{I11}^T, \ldots, \boldsymbol{\Upsilon}_{IN_lN_l}^T]$ $\mathbf{K}_I[\boldsymbol{\Upsilon}_{I11}, \ldots, \boldsymbol{\Upsilon}_{IN_lN_l}]$ are given by I_{Ill} and $\boldsymbol{\Omega}_{Ill}, l = 1, \ldots, N_l$, respectively. The matrix

$$[\boldsymbol{\Upsilon}_{I11}^T, \ldots, \boldsymbol{\Upsilon}_{IN_lN_l}^T]\mathbf{M}_I[\boldsymbol{\Upsilon}_{I11}, \ldots, \boldsymbol{\Upsilon}_{IN_lN_l}] - \mathbf{M}_{iIRL}^T\mathbf{M}_{iIRL} \tag{1.91}$$

which has to be inverted in the transformation matrix $\mathbf{T}_{RIL}$, has a dimension equal to $n_{IRL} \times n_{IRL}$, where $n_{IRL} = \sum_{l=1}^{N_l} n_{IR}^l$. If the contribution of the residual fixed-interface normal modes is considered, Eq. (1.80) changes to

$$\bar{\mathbf{u}}(t) = (\mathbf{T}_{DIL} + \mathbf{T}_{RIL})\,\mathbf{q}_{IL}(t) \tag{1.92}$$

from which the associated enhanced reduced-order mass and stiffness matrices are given by

$$\begin{aligned}\hat{\mathbf{M}}_{RIL} &= (\mathbf{T}_{DIL} + \mathbf{T}_{RIL})^T\hat{\mathbf{M}}(\mathbf{T}_{DIL} + \mathbf{T}_{RIL}) \\ &= \mathbf{T}_{DIL}^T\hat{\mathbf{M}}\mathbf{T}_{DIL} + \mathbf{T}_{RIL}^T\hat{\mathbf{M}}\mathbf{T}_{DIL} + \mathbf{T}_{DIL}^T\hat{\mathbf{M}}\mathbf{T}_{RIL} + \mathbf{T}_{RIL}^T\hat{\mathbf{M}}\mathbf{T}_{RIL} \\ &= \hat{\mathbf{M}}_{DIL} + \mathbf{T}_{RIL}^T\hat{\mathbf{M}}\mathbf{T}_{DIL} + \mathbf{T}_{DIL}^T\hat{\mathbf{M}}\mathbf{T}_{RIL} + \mathbf{T}_{RIL}^T\hat{\mathbf{M}}\mathbf{T}_{RIL}\end{aligned} \tag{1.93}$$

and

$$\begin{aligned}\hat{\mathbf{K}}_{RIL} &= (\mathbf{T}_{DIL} + \mathbf{T}_{RIL})^T\hat{\mathbf{K}}(\mathbf{T}_{DIL} + \mathbf{T}_{RIL}) \\ &= \mathbf{T}_{DIL}^T\hat{\mathbf{K}}\mathbf{T}_{DIL} + \mathbf{T}_{RIL}^T\hat{\mathbf{K}}\mathbf{T}_{DIL} + \mathbf{T}_{DIL}^T\hat{\mathbf{K}}\mathbf{T}_{RIL} + \mathbf{T}_{RIL}^T\hat{\mathbf{K}}\mathbf{T}_{RIL} \\ &= \hat{\mathbf{K}}_{DIL} + \mathbf{T}_{RIL}^T\hat{\mathbf{K}}\mathbf{T}_{DIL} + \mathbf{T}_{DIL}^T\hat{\mathbf{K}}\mathbf{T}_{RIL} + \mathbf{T}_{RIL}^T\hat{\mathbf{K}}\mathbf{T}_{RIL}\end{aligned} \tag{1.94}$$

The choice of using global or local interface modes depends on the problem at hand. In other words, one particular approach is not necessarily better than the other.

1.9 Numerical Implementation: Pseudo-Code No. 3

Based on the formulation presented, the following pseudo-code illustrates how the reduced mass and stiffness matrices are evaluated by including local interface reduction.

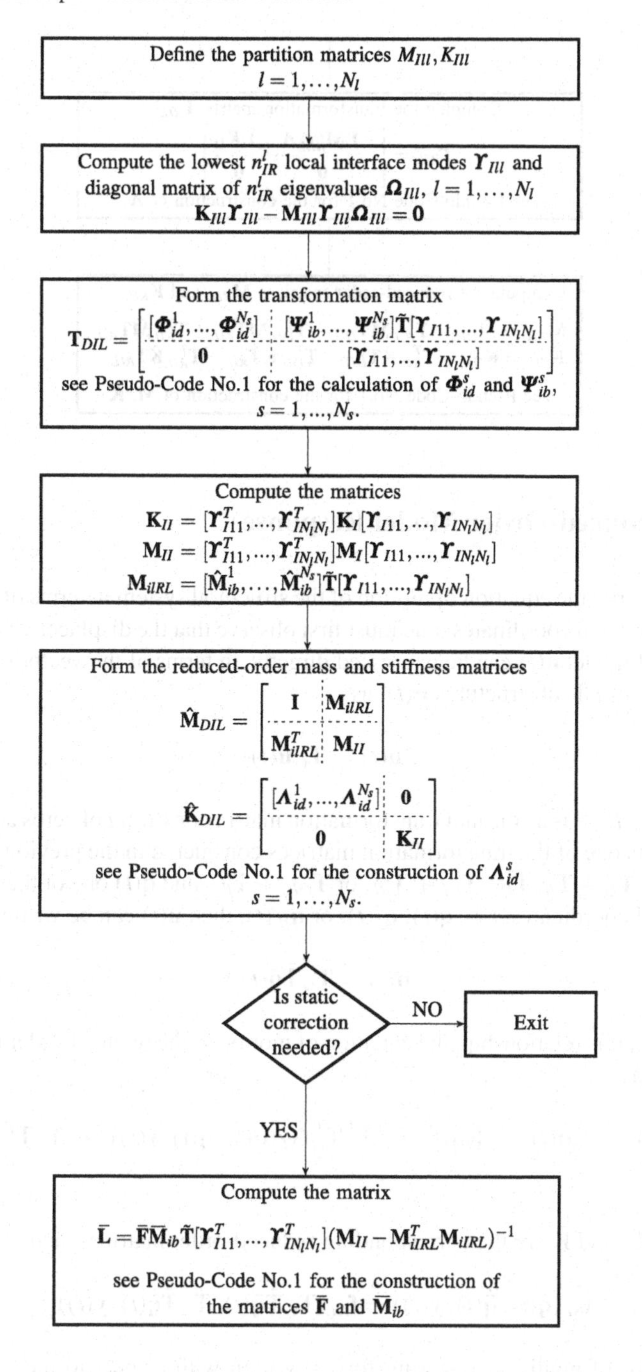

Define the partition matrices $M_{Il l}, K_{Il l}$
$l = 1, \ldots, N_l$
Compute the lowest n_{IR}^l local interface modes $\boldsymbol{\Upsilon}_{Il l}$ and diagonal matrix of n_{IR}^l eigenvalues $\boldsymbol{\Omega}_{Il l}$, $l = 1, \ldots, N_l$
$\mathbf{K}_{Il l}\boldsymbol{\Upsilon}_{Il l} - \mathbf{M}_{Il l}\boldsymbol{\Upsilon}_{Il l}\boldsymbol{\Omega}_{Il l} = \mathbf{0}$
Form the transformation matrix
$\mathbf{T}_{DIL} = \begin{bmatrix} [\boldsymbol{\Phi}_{id}^1, \ldots, \boldsymbol{\Phi}_{id}^{N_s}] & [\boldsymbol{\Psi}_{ib}^1, \ldots, \boldsymbol{\Psi}_{ib}^{N_s}]\tilde{\mathbf{T}}[\boldsymbol{\Upsilon}_{I11}, \ldots, \boldsymbol{\Upsilon}_{IN_lN_l}] \\ \mathbf{0} & [\boldsymbol{\Upsilon}_{I11}, \ldots, \boldsymbol{\Upsilon}_{IN_lN_l}] \end{bmatrix}$
see Pseudo-Code No.1 for the calculation of $\boldsymbol{\Phi}_{id}^s$ and $\boldsymbol{\Psi}_{ib}^s$, $s = 1, \ldots, N_s$.
Compute the matrices
$\mathbf{K}_{II} = [\boldsymbol{\Upsilon}_{I11}^T, \ldots, \boldsymbol{\Upsilon}_{IN_lN_l}^T]\mathbf{K}_I[\boldsymbol{\Upsilon}_{I11}, \ldots, \boldsymbol{\Upsilon}_{IN_lN_l}]$
$\mathbf{M}_{II} = [\boldsymbol{\Upsilon}_{I11}^T, \ldots, \boldsymbol{\Upsilon}_{IN_lN_l}^T]\mathbf{M}_I[\boldsymbol{\Upsilon}_{I11}, \ldots, \boldsymbol{\Upsilon}_{IN_lN_l}]$
$\mathbf{M}_{iIRL} = [\hat{\mathbf{M}}_{ib}^1, \ldots, \hat{\mathbf{M}}_{ib}^{N_s}]\tilde{\mathbf{T}}[\boldsymbol{\Upsilon}_{I11}, \ldots, \boldsymbol{\Upsilon}_{IN_lN_l}]$
Form the reduced-order mass and stiffness matrices
$\hat{\mathbf{M}}_{DIL} = \begin{bmatrix} \mathbf{I} & \mathbf{M}_{iIRL} \\ \mathbf{M}_{iIRL}^T & \mathbf{M}_{II} \end{bmatrix}$
$\hat{\mathbf{K}}_{DIL} = \begin{bmatrix} [\boldsymbol{\Lambda}_{id}^1, \ldots, \boldsymbol{\Lambda}_{id}^{N_s}] & \mathbf{0} \\ \mathbf{0} & \mathbf{K}_{II} \end{bmatrix}$
see Pseudo-Code No.1 for the construction of $\boldsymbol{\Lambda}_{id}^s$
$s = 1, \ldots, N_s$.
Is static correction needed?
NO
Exit
YES
Compute the matrix
$\bar{\mathbf{L}} = \bar{\mathbf{F}}\bar{\mathbf{M}}_{ib}\tilde{\mathbf{T}}[\boldsymbol{\Upsilon}_{I11}^T, \ldots, \boldsymbol{\Upsilon}_{IN_lN_l}^T](\mathbf{M}_{II} - \mathbf{M}_{iIRL}^T\mathbf{M}_{iIRL})^{-1}$
see Pseudo-Code No.1 for the construction of the matrices $\bar{\mathbf{F}}$ and $\bar{\mathbf{M}}_{ib}$

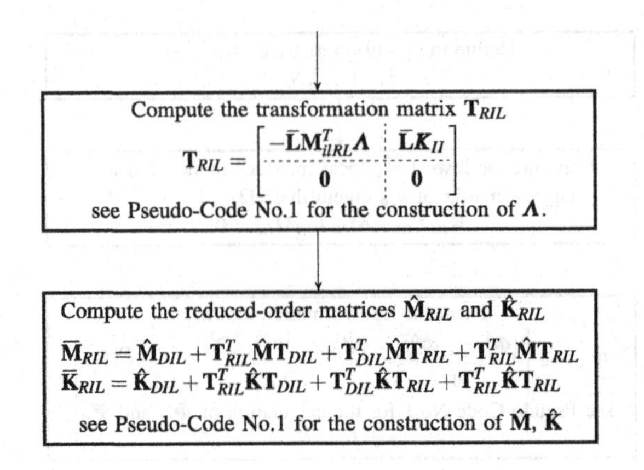

Compute the transformation matrix $\mathbf{T}_{RIL}$
$\mathbf{T}_{RIL} = \begin{bmatrix} -\bar{\mathbf{L}}\mathbf{M}_{iIRL}^T\mathbf{\Lambda} & \bar{\mathbf{L}}\mathbf{K}_{II} \\ \mathbf{0} & \mathbf{0} \end{bmatrix}$
see Pseudo-Code No.1 for the construction of $\mathbf{\Lambda}$.
Compute the reduced-order matrices $\hat{\mathbf{M}}_{RIL}$ and $\hat{\mathbf{K}}_{RIL}$
$\bar{\mathbf{M}}_{RIL} = \hat{\mathbf{M}}_{DIL} + \mathbf{T}_{RIL}^T\hat{\mathbf{M}}\mathbf{T}_{DIL} + \mathbf{T}_{DIL}^T\hat{\mathbf{M}}\mathbf{T}_{RIL} + \mathbf{T}_{RIL}^T\hat{\mathbf{M}}\mathbf{T}_{RIL}$
$\bar{\mathbf{K}}_{RIL} = \hat{\mathbf{K}}_{DIL} + \mathbf{T}_{RIL}^T\hat{\mathbf{K}}\mathbf{T}_{DIL} + \mathbf{T}_{DIL}^T\hat{\mathbf{K}}\mathbf{T}_{RIL} + \mathbf{T}_{RIL}^T\hat{\mathbf{K}}\mathbf{T}_{RIL}$
see Pseudo-Code No.1 for the construction of $\hat{\mathbf{M}}$, $\hat{\mathbf{K}}$

1.10 Reduced-Order Model Response

In order to write the equation of motion of the structural system in terms of a reduced set of generalized coordinates, one must first observe that the displacement vector of the original model $\mathbf{u}(t)$ can be expressed directly, in terms of the vector of physical coordinates of all substructures $\bar{\mathbf{u}}(t)$, as

$$\mathbf{u}(t) = \mathbf{T}_G\bar{\mathbf{u}}(t) \tag{1.95}$$

where $\mathbf{T}_G \in R^{n\times n}$ is a constant transformation matrix consisting of zeros and ones. If $\bar{\mathbf{T}}$ represents one of the transformation matrices considered in the previous sections, that is, $\mathbf{T}_D$, $\mathbf{T}_D + \mathbf{T}_R$, $\mathbf{T}_{DI}$, $\mathbf{T}_{DI} + \mathbf{T}_{RI}$, or $\mathbf{T}_{DIL} + \mathbf{T}_{RIL}$ and $\bar{\mathbf{q}}(t)$ one of the vectors of generalized coordinates, i.e., $\mathbf{q}(t)$, $\mathbf{q}_I(t)$, or $\mathbf{q}_{IL}(t)$, then $\mathbf{u}(t)$ can be written in terms of $\bar{\mathbf{q}}(t)$ as

$$\mathbf{u}(t) = \mathbf{T}_G\bar{\mathbf{T}}\bar{\mathbf{q}}(t) \tag{1.96}$$

Based on this relationship, the equation of motion of the reduced-order model can be written as

$$\bar{\mathbf{M}}\ddot{\bar{\mathbf{q}}}(t) + \bar{\mathbf{C}}\dot{\bar{\mathbf{q}}}(t) + \bar{\mathbf{K}}\bar{\mathbf{q}}(t) = \bar{\mathbf{T}}^T\mathbf{T}_G^T\bar{\mathbf{f}}_{NL}(\bar{\mathbf{q}}(t), \dot{\bar{\mathbf{q}}}(t), \mathbf{y}(t)) + \bar{\mathbf{T}}^T\mathbf{T}_G^T\mathbf{f}(t) \tag{1.97}$$

where $\bar{\mathbf{M}}$, $\bar{\mathbf{C}}$, and $\bar{\mathbf{K}}$ are the corresponding reduced-order matrices, and

$$\bar{\mathbf{f}}_{NL}(\bar{\mathbf{q}}(t), \dot{\bar{\mathbf{q}}}(t), \mathbf{y}(t)) = \mathbf{f}_{NL}(\mathbf{T}_G\bar{\mathbf{T}}\bar{\mathbf{q}}(t), \mathbf{T}_G\bar{\mathbf{T}}\dot{\bar{\mathbf{q}}}(t), \mathbf{y}(t)) \tag{1.98}$$

is the vector of nonlinear restoring forces written with respect to the generalized coordinates $\bar{\mathbf{q}}(t)$, its derivative, and the set of variables $\mathbf{y}(t)$. Specifically, if $\bar{\mathbf{T}}$ is the transformation matrix $\mathbf{T}_D + \mathbf{T}_R$, then the matrices $\bar{\mathbf{M}}$ and $\bar{\mathbf{K}}$ are given by $\hat{\mathbf{M}}_R$

and $\hat{\mathbf{K}}_R$ in Eqs. (1.50) and (1.51), respectively. The matrix $\bar{\mathbf{C}}$ is the damping matrix referred to the generalized coordinates $\bar{\mathbf{q}}(t)$, which can be defined in terms of the damping matrices at the substructure level. The damping matrix $\hat{\mathbf{C}}$ related to the vector of physical coordinates $\bar{\mathbf{u}}(t)$ has a similar structure to $\hat{\mathbf{M}}$ and $\hat{\mathbf{K}}$. The equation (1.97), together with the equation for the evolution of the set of variables $\mathbf{y}(t)$, can be integrated efficiently by an appropriate step-by-step integration scheme [6] or by modal analysis. If modal analysis is used, the response is represented by a linear combination of the mode shapes as

$$\mathbf{u}(t) = \mathbf{T}_G \bar{\mathbf{T}} \sum_{r=1}^{m} \boldsymbol{\varphi}_r \upsilon_r(t) \tag{1.99}$$

where m is the number of modes considered in the reduced-order model, $\upsilon_r(t)$, $r = 1, \ldots, m$, are the modal response functions, and $\boldsymbol{\varphi}_r$, $r = 1, \ldots, m$, are the mode shapes of the reduced-order model, which are obtained by solving the eigenvalue problem

$$(\bar{\mathbf{K}} - \omega_r^2 \bar{\mathbf{M}})\boldsymbol{\varphi}_r = \mathbf{0} \ , \quad r = 1, \ldots, m \tag{1.100}$$

where ω_r, $r = 1, \ldots, m$, are the modal frequencies of the reduced-order model. If a classically-damped model is assumed for the linear components of the reduced-order model, the modal response functions satisfy the nonlinear differential equations

$$\ddot{\upsilon}_r(t) + 2\xi_r \omega_r \dot{\upsilon}_r(t) + \omega_r^2 \upsilon_r(t) = \boldsymbol{\varphi}_r^T \bar{\mathbf{T}}^T \mathbf{T}_G^T \tilde{\mathbf{f}}_{NL}(\boldsymbol{\upsilon}(t), \dot{\boldsymbol{\upsilon}}(t), \mathbf{y}(t)) \ + \ \boldsymbol{\varphi}_r^T \bar{\mathbf{T}}^T \mathbf{T}_G^T \mathbf{f}(t)$$
$$r = 1, \ldots, m \tag{1.101}$$

where ξ_r, $r = 1, \ldots, m$, are the assigned damping ratios, and

$$\tilde{\mathbf{f}}_{NL}(\boldsymbol{\upsilon}(t), \dot{\boldsymbol{\upsilon}}(t), \mathbf{y}(t)) = \mathbf{f}_{NL}\left(\mathbf{T}_G \bar{\mathbf{T}} \sum_{r=1}^{m} \boldsymbol{\varphi}_r \upsilon_r(t), \mathbf{T}_G \bar{\mathbf{T}} \sum_{r=1}^{m} \boldsymbol{\varphi}_r \dot{\upsilon}_r(t), \mathbf{y}(t)\right) \tag{1.102}$$

is the vector of nonlinear restoring forces written with respect to the modal response vector $\boldsymbol{\upsilon}(t) =< \upsilon_1^T(t), \ldots, \upsilon_m^T(t) >^T$, its derivative, and the set of variables $\mathbf{y}(t)$. In the previous equations, it was assumed that the mode shapes of the reduced-order model have been normalized with respect to $\bar{\mathbf{M}}$. The solution for the modal responses, together with the equation for the evolution of the set of variables $\mathbf{y}(t)$, can be obtained by any suitable integration scheme [6]. The previous representation of the solution can be extended to the case of non-classically damped systems either by recasting the equation of motion of the reduced-order system into a first-order state-space form or by solving a quadratic eigenvalue problem [23, 32].

References

1. Y. Aoyama, G. Yagawa, Component mode synthesis for large scale structural eigenanalysis. Comput. Struct. **79**(6), 605–615 (2001)
2. T.T. Baber, M.N. Noori, Modeling general hysteresis behavior and random vibration applications. J. Vib. Acoust. Stress Reliab. Des. ASCE **108**, 411–420 (1986)
3. T.T. Baber, Y. Wen, Random vibration hysteretic, degrading systems. J. Eng. Mech. Div. **107**(6), 1069–1087 (1981)
4. E. Balmes, Use of generalized interface degrees of freedom in component mode synthesis, in *Proceedings of the 14th International Modal Analysis Conference*. Bethel, CT., vol. 1 (1996), pp. 204–210
5. R.M. Bampton, A modal combination program for dynamic analysis of structures. Technical memorandum 33–920, Jet Propulsion Laboratory, Pasadena, CA, 1 July 1967
6. K.J. Bathe, *Finite Element Procedures* (Prentice Hall, New Jersey, 2006)
7. W.A. Benfield, R.F. Hruda, Vibration analysis of structures by component mode substitution. AIAA J. **9**(7), 1255–1261 (1971)
8. S.H. Boo, J.G. Kim, P.S. Lee, Error estimation of the automated multi-level substructuring method. Int. J. Numer. Methods Eng. **106**, 927–950 (2016)
9. F. Bourquin, F. d'Hennezel, Numerical study of an intrinsic component mode synthesis method. Comput. Methods Appl. Mech. Eng. **97**, 49–76 (1992)
10. F. Bourquin, F. d'Hennezel, Intrinsic component mode synthesis and plate vibration. Comput. Struct. **44**(1–2), 315–324 (1992)
11. M.P. Castanier, Y.C. Tan, C. Pierre, Characteristic constraint modes for component mode synthesis. AIAA J. **39**(6), 1182–1186 (2001)
12. R.R. Craig Jr., M.C.C. Bampton, Coupling of substructures for dynamic analysis. AIAA J. **6**(7), 1313–1319 (1968)
13. R.R. Craig Jr., *Structural Dynamics, An Introduction to Computer Methods* (Wiley, New York, 1981)
14. R.R. Craig Jr., Coupling of structures for dynamic analysis: an overview, in *AIAA Dynamic Specialist Conference*, Paper AIAA-2000-1573, Atlanta, GA, 5–6 April 2000
15. R.R. Craig Jr., A.J. Kurdila, *Fundamentals of Structural Dynamics*, 2nd edn. (Wiley, Hoboken, New Jersey, 2006)
16. R.L. Goldman, Vibration analysis by dynamic partitioning. AIAA J. **7**, 1152–1154 (1969)
17. R.A. Horn, C.R. Johnson, *Matrix Analysis* (University Press, Cambridge, 1985)
18. W.C. Hurty, Dynamic analysis of structural systems using component modes. AIAA J. **3**(4), 678–685 (1965)
19. H.A. Jensen, A. Marillanca, O. Peñaloza, A computational procedure for response statistics-based optimization of stochastic non-linear FE-models. Comput. Methods Appl. Mech. Eng. **198**, 125–137 (2008)
20. H.A. Jensen, A. Muñoz, C. Papadimitriou, C. Vergara, An enhanced substructure coupling technique for dynamic re-analyses: application to simulation-based problems. Comput. Methods Appl. Mech. Eng. **307**, 215–234 (2016)
21. J.G. Kim, P.S. Lee, An enhanced Craig-Bampton method. Int. J. Numer. Methods Eng. **103**, 79–93 (2015)
22. T.T. Lu, S.H. Shiou, Inverses of 2×2 block matrices. Int. J. Comput. Math. Appl. **43**, 119–129 (2002)
23. F. Ma, M. Morzfeld, A general methodology for decoupling damped linear systems. Procedia Eng. **314**, 2498–2502 (2011)
24. N.M.M. Maia, J.M.M. Silva (eds.), *Theoretical and Experimental Modal Analysis* (Research Studies Press, Baldock, Hertfordshire, England, 1997)
25. R.H. MacNeal, A hybrid method of component mode synthesis. Comput. Struct. **1**(4), 581–601 (1971)
26. L. Meirovitch, *Computational Methods in Structural Dynamics* (Sijthoff and Noordhoff, Rockville, MD., 1980)

27. Y.J. Park, Y.K. Wen, A.H. Ang, Random vibration of hysteretic systems under bi-directional ground motions. Earthq. Eng. Struct. Dyn. **14**(4), 543–557 (1986)
28. D.J. Rixen, A dual Craig Bampton method for dynamic substructuring. J. Comput. Appl. Math. **168**(1–2), 383–391 (2002)
29. C.A. Schenk, H.J. Pradlwarter, G.I. Schuëller, Non-stationary response of large, non-linear finite element systems under stochastic loading. Comput. Struct. **83**, 1086–1102 (2005)
30. Y.C Tan, M.P. Castanier, C. Pierre, Characteristic mode based component mode synthesis for power flow analysis in complex structures, in *Proceedings of the 41st AIAA/ASME/ASCE/AHS/ASC structures, Structural Dynamics and Materials Conference and Exhibit*, Reston VA, 1908–1917 (2000)
31. Y.C Tan, P. Castanier, C. Pierre, Efficient reduced order modeling of low to mid frequency vibration and power flow in complex structures, in *Proceedings of the 19th International Modal Analysis Conference*, Bethel, CT, vol. 2 (2001), pp. 1070–1076
32. F. Tisseur, K. Meerbergen, The quadratic eigenvalue problem. SIAM Rev. **34**(2), 235–286 (2001)
33. D.M. Tran, Component mode synthesis using interface modes. Application to structures with cyclic symmetry. Comput. Struct. **79**, 209–222 (2001)

Chapter 2
Parametrization of Reduced-Order Models Based on Normal Modes

Abstract This chapter deals with the parametrization of reduced-order models based on dominant and residual fixed-interface normal modes, in terms of model parameters. The division of the original structure is guided by a parametrization scheme, which assumes that the substructure matrices for each of the introduced linear substructures depend on only one of the model parameters. Based on this assumption, a global parametrization of the reduced-order matrices is provided. Invariant issues are discussed that are related to the matrices that account for the contribution of residual normal modes. A pseudo-code is then provided in order to illustrate how the parametrization of the reduced-order matrices is constructed.

2.1 Motivation

The solution of complex simulation-based problems involving uncertainty such as, Bayesian finite element model updating, reliability and sensitivity analysis of dynamic systems, reliability-based design optimization, uncertainty quantification, propagation analysis, etc., is computationally very demanding, due to the large number of dynamic analyses required during the corresponding simulation processes [1, 3–5, 7, 8, 10, 12, 15–17, 21, 25–31]. In fact, the solution of these types of problems requires evaluating the system response at a large number of samples in the uncertain parameter space (of the order of hundreds or thousands) [2, 9, 13, 14, 23]. Consequently, the computational cost may become excessive when the computational time for performing a dynamic analysis is substantial. Certainly, part of the computational effort is alleviated by dividing the original model into substructures and reducing the number of physical coordinates to a much-smaller number of generalized coordinates. However, the construction of reduced-order models at each sample implies re-computing the fixed-interface normal modes and the interface constraint modes for each substructure as well as for the interface modes. This procedure can be very expensive computationally, due to the substantial computational overhead that arises at the substructure level [11, 22]. To cope with this difficulty, an efficient finite element model parametrization scheme can be considered. When

H. Jensen and C. Papadimitriou, *Sub-structure Coupling for Dynamic Analysis*,
Lecture Notes in Applied and Computational Mechanics 89,
https://doi.org/10.1007/978-3-030-12819-7_2

dividing the structure into substructures is guided by such a parametrization scheme, dramatic computational savings can be achieved.

In the framework of this chapter, it is assumed that the finite element model is parametrized by a set of parameters $\boldsymbol{\theta} \in \Omega_{\theta} \subset R^{n_\theta}$, which are considered uncertain. These parameters are modeled using a probability density function $q(\boldsymbol{\theta})$ that indicates the relative plausibility of the possible values of the parameters $\boldsymbol{\theta} \in \Omega_{\theta}$. The specific characterization of the probability density function is problem-dependent and is further discussed in subsequent chapters with applications.

It should be noted that the set of parameters $\boldsymbol{\theta}$ could also be associated with design variables, not only with uncertain parameters. Thus, in deterministic and reliability-based design optimization one can take advantage of the formulation presented in this chapter and Chap. 3 and drastically reduce the computational effort that comes from the large number of re-analyses required to reach the final design.

2.2 Parametrization Scheme

2.2.1 *Substructure Matrices*

The division of the original structure is guided by a parametrization scheme that assumes that the substructure matrices for each of the introduced linear substructures depend on only one of the model parameters. In this setting, the stiffness and the mass matrices of a substructure s, $s = 1, \ldots, N_s$, depend on only one of the model parameters. The case in which the stiffness and mass matrices of a substructure s do not depend on the model parameters is also included in this parametrization scheme. Specifically, let S_0 be the set of substructures that do not depend on the vector of model parameters $\boldsymbol{\theta}$. In this case, the substructure mass and stiffness matrices are written as

$$\mathbf{M}^s = \bar{\mathbf{M}}^s \tag{2.1}$$

and

$$\mathbf{K}^s = \bar{\mathbf{K}}^s \tag{2.2}$$

It is noted that the substructure fixed-interface normal modes and interface constraint modes are independent of the system parameters' value for these substructures. Thus, only a single analysis is required to estimate the normal and interface modes for the substructure $s \in S_0$. This implies that the solution of the eigenvalue problem to compute the fixed-interface normal modes $\boldsymbol{\Phi}^s_{ii}$, and the solution of the linear system to compute the interface constraint modes $\boldsymbol{\Psi}^s_{ib}$ for a component $s \in S_0$, are carried out once.

On the other hand, let S_j be the set of substructures that depend on the model parameter θ_j. For substructure $s \in S_j$, it is assumed that the mass and stiffness matrices take the general form

$$\mathbf{M}^s = \bar{\mathbf{M}}^s \, g^j(\theta_j) \tag{2.3}$$

and

$$\mathbf{K}^s = \bar{\mathbf{K}}^s \, h^j(\theta_j) \tag{2.4}$$

where the matrices $\bar{\mathbf{M}}^s$ and $\bar{\mathbf{K}}^s$ are independent of θ_j, and $g^j(\theta_j)$ and $h^j(\theta_j)$ are linear or nonlinear functions of the model parameter θ_j. These matrices are obtained from the reference model by setting $g^j(\theta_j) = 1$ and $h^j(\theta_j) = 1$. It is clear that the partitions of the mass matrix $\mathbf{M}^s$ and of the stiffness matrix $\mathbf{K}^s$ in Eqs. (1.2) and (1.3) admit the same parametrization, that is,

$$\begin{aligned}
\mathbf{M}^s_{ii} &= \bar{\mathbf{M}}^s_{ii} \, g^j(\theta_j) \\
\mathbf{M}^s_{ib} &= \bar{\mathbf{M}}^s_{ib} \, g^j(\theta_j) \\
\mathbf{M}^s_{bi} &= \bar{\mathbf{M}}^s_{bi} \, g^j(\theta_j) \\
\mathbf{M}^s_{bb} &= \bar{\mathbf{M}}^s_{bb} \, g^j(\theta_j)
\end{aligned} \tag{2.5}$$

and

$$\begin{aligned}
\mathbf{K}^s_{ii} &= \bar{\mathbf{K}}^s_{ii} \, h^j(\theta_j) \\
\mathbf{K}^s_{ib} &= \bar{\mathbf{K}}^s_{ib} \, h^j(\theta_j) \\
\mathbf{K}^s_{bi} &= \bar{\mathbf{K}}^s_{bi} \, h^j(\theta_j) \\
\mathbf{K}^s_{bb} &= \bar{\mathbf{K}}^s_{bb} \, h^j(\theta_j)
\end{aligned} \tag{2.6}$$

where the matrices $\bar{\mathbf{M}}^s_{ii}$, $\bar{\mathbf{M}}^s_{ib}$, $\bar{\mathbf{M}}^s_{bi}$, $\bar{\mathbf{M}}^s_{bb}$, $\bar{\mathbf{K}}^s_{ii}$, $\bar{\mathbf{K}}^s_{ib}$, $\bar{\mathbf{K}}^s_{bi}$, and $\bar{\mathbf{K}}^s_{bb}$ are independent of θ_j. The previous expressions correspond to parametrization of the mass and stiffness matrices in terms of model parameters. This type of parametrization is often encountered in structural systems modeled by standard finite elements. For example, parameters such as the modulus of elasticity, dimension of cross sections, thicknesses, etc. may correspond to coefficients associated with finite elements, such as trusses, bending beams, certain class of plates and shells, solids, etc. [6, 15, 17, 19, 20, 23, 24].

Some guidelines are provided in Chap. 3 on how to proceed in cases when the mass or stiffness matrices of a substructure depend on more than one parameter.

2.2.2 Normal Modes and Interface Constraint Modes

With the use of the parametrization of $\mathbf{M}^s_{ii}$ in Eq. (2.5), and of the normalization of the fixed-interface normal modes $\boldsymbol{\Phi}^s_{ii}$ with respect to the mass matrix $\mathbf{M}^s_{ii}$, given in Eq. (1.5), it follows that the normal modes $\boldsymbol{\Phi}^s_{ii}$, $s \in S_j$, can be expressed as [15, 23]

$$\boldsymbol{\Phi}^s_{ii} = \bar{\boldsymbol{\Phi}}^s_{ii} \frac{1}{\sqrt{g^j(\theta_j)}} \tag{2.7}$$

where the matrix $\bar{\boldsymbol{\Phi}}^s_{ii}$ is independent of the model parameter θ_j. Next, if the previous parametrization of the matrices $\mathbf{M}^s_{ii}$, $\mathbf{K}^s_{ii}$ and $\boldsymbol{\Phi}^s_{ii}$ in the eigenvalue problem (1.4) is considered, the matrix $\boldsymbol{\Lambda}^s_{ii}$, $s \in S_j$, containing the corresponding eigenvalues allows the parametrization [15, 23]

$$\boldsymbol{\Lambda}^s_{ii} = \bar{\boldsymbol{\Lambda}}^s_{ii} \frac{h^j(\theta_j)}{g^j(\theta_j)} \tag{2.8}$$

where the matrix $\bar{\boldsymbol{\Lambda}}^s_{ii}$ is independent of θ_j. From the above parametrization of $\boldsymbol{\Phi}^s_{ii}$ and $\boldsymbol{\Lambda}^s_{ii}$, it follows that the matrices $\bar{\boldsymbol{\Lambda}}^s_{ii}$ and $\bar{\boldsymbol{\Phi}}^s_{ii}$ are the solution of the eigenvalue problem

$$\bar{\mathbf{K}}^s_{ii} \bar{\boldsymbol{\Phi}}^s_{ii} - \bar{\mathbf{M}}^s_{ii} \bar{\boldsymbol{\Phi}}^s_{ii} \bar{\boldsymbol{\Lambda}}^s_{ii} = \mathbf{0} \tag{2.9}$$

where the mode shapes $\bar{\boldsymbol{\Phi}}^s_{ii}$ satisfy the orthogonal conditions

$$\bar{\boldsymbol{\Phi}}^{s^T}_{ii} \bar{\mathbf{M}}^s_{ii} \bar{\boldsymbol{\Phi}}^s_{ii} = \mathbf{I}^s_{ii} \tag{2.10}$$

and

$$\bar{\boldsymbol{\Phi}}^{s^T}_{ii} \bar{\mathbf{K}}^s_{ii} \bar{\boldsymbol{\Phi}}^s_{ii} = \bar{\boldsymbol{\Lambda}}^s_{ii} \tag{2.11}$$

Furthermore, the interface constraint modes $\boldsymbol{\Psi}^s_{ib}$, $s \in S_j$, given in (1.9), are also independent of θ_j, since

$$\boldsymbol{\Psi}^s_{ib} = -\mathbf{K}^{s\,-1}_{ii} \mathbf{K}^s_{ib} = -\bar{\mathbf{K}}^{s\,-1}_{ii}\, h^{j^{-1}}(\theta_j) \bar{\mathbf{K}}^s_{ib}\, h^j(\theta_j) = -\bar{\mathbf{K}}^{s\,-1}_{ii} \bar{\mathbf{K}}^s_{ib} = \bar{\boldsymbol{\Psi}}^s_{ib} \tag{2.12}$$

It is clear from the previous equations that a single eigen-analysis (given in (2.9)) for each substructure is required to provide the exact estimate of the fixed-interface normal modes (in (2.7) and (2.8)) for any value of the model parameter $\boldsymbol{\theta}$. In other words, the computationally-intensive re-analyses that estimate the normal and constraint modes of each substructure for different values of $\boldsymbol{\theta}$, required during the corresponding simulation processes, can be completely avoided.

2.3 Parametrization of Reduced-Order Matrices

In what follows, the parametrization of the different matrices involved in the characterization of reduced-order models based on dominant and residual fixed-interface normal modes is derived.

2.3.1 Unreduced Matrices

If the parametrization of the partitions of the mass matrix $\mathbf{M}^s$ and stiffness matrix $\mathbf{K}^s$ are used, the mass matrix $\hat{\mathbf{M}}$ and the stiffness matrix $\hat{\mathbf{K}}$ of the unreduced model referred to the vector of physical coordinates of all substructures $\bar{\mathbf{u}}(t)$, given in Eqs. (1.19) and (1.20), respectively, take the form

$$\hat{\mathbf{M}}(\boldsymbol{\theta}) = \begin{bmatrix} [\bar{\mathbf{M}}_{ii}^1\delta_{10},\ldots,\bar{\mathbf{M}}_{ii}^{N_s}\delta_{N_s0}] & [\bar{\mathbf{M}}_{ib}^1\delta_{10},\ldots,\bar{\mathbf{M}}_{ib}^{N_s}\delta_{N_s0}]\tilde{\mathbf{T}} \\ \tilde{\mathbf{T}}^T[\bar{\mathbf{M}}_{ib}^{1^T}\delta_{10},\ldots,\bar{\mathbf{M}}_{ib}^{N_s^T}\delta_{N_s0}] & \tilde{\mathbf{T}}^T[\bar{\mathbf{M}}_{bb}^1\delta_{10},\ldots,\bar{\mathbf{M}}_{bb}^{N_s}\delta_{N_s0}]\tilde{\mathbf{T}} \end{bmatrix}$$
$$+\sum_{j=1}^{n_\theta}\begin{bmatrix} [\bar{\mathbf{M}}_{ii}^1\delta_{1j},\ldots,\bar{\mathbf{M}}_{ii}^{N_s}\delta_{N_sj}] & [\bar{\mathbf{M}}_{ib}^1\delta_{1j},\ldots,\bar{\mathbf{M}}_{ib}^{N_s}\delta_{N_sj}]\tilde{\mathbf{T}} \\ \tilde{\mathbf{T}}^T[\bar{\mathbf{M}}_{ib}^{1^T}\delta_{1j},\ldots,\bar{\mathbf{M}}_{ib}^{N_s^T}\delta_{N_sj}] & \tilde{\mathbf{T}}^T[\bar{\mathbf{M}}_{bb}^1\delta_{10},\ldots,\bar{\mathbf{M}}_{bb}^{N_s}\delta_{N_sj}]\tilde{\mathbf{T}} \end{bmatrix} g^j(\theta_j) \tag{2.13}$$

and

$$\hat{\mathbf{K}}(\boldsymbol{\theta}) = \begin{bmatrix} [\bar{\mathbf{K}}_{ii}^1\delta_{10},\ldots,\bar{\mathbf{K}}_{ii}^{N_s}\delta_{N_s0}] & [\bar{\mathbf{K}}_{ib}^1\delta_{10},\ldots,\bar{\mathbf{K}}_{ib}^{N_s}\delta_{N_s0}]\tilde{\mathbf{T}} \\ \tilde{\mathbf{T}}^T[\bar{\mathbf{K}}_{ib}^{1^T}\delta_{10},\ldots,\bar{\mathbf{K}}_{ib}^{N_s^T}\delta_{N_s0}] & \tilde{\mathbf{T}}^T[\bar{\mathbf{K}}_{bb}^1\delta_{1j},\ldots,\bar{\mathbf{K}}_{bb}^{N_s}\delta_{N_s0}]\tilde{\mathbf{T}} \end{bmatrix}$$
$$+\sum_{j=1}^{n_\theta}\begin{bmatrix} [\bar{\mathbf{K}}_{ii}^1\delta_{1j},\ldots,\bar{\mathbf{K}}_{ii}^{N_s}\delta_{N_sj}] & [\bar{\mathbf{K}}_{ib}^1\delta_{1j},\ldots,\bar{\mathbf{K}}_{ib}^{N_s}\delta_{N_sj}]\tilde{\mathbf{T}} \\ \tilde{\mathbf{T}}^T[\bar{\mathbf{K}}_{ib}^{1^T}\delta_{1j},\ldots,\bar{\mathbf{K}}_{ib}^{N_s^T}\delta_{N_sj}] & \tilde{\mathbf{T}}^T[\bar{\mathbf{K}}_{bb}^1\delta_{1j},\ldots,\bar{\mathbf{K}}_{bb}^{N_s}\delta_{N_sj}]\tilde{\mathbf{T}} \end{bmatrix} h^j(\theta_j) \tag{2.14}$$

where

$$\delta_{s0} = \begin{cases} 1 \text{ if } s \in S_0 \\ 0 \text{ otherwise} \end{cases}, \quad s = 1,\ldots,N_s \tag{2.15}$$

and

$$\delta_{sj} = \begin{cases} 1 \text{ if } s \in S_j \\ 0 \text{ otherwise} \end{cases}, \quad s = 1,\ldots,N_s \tag{2.16}$$

2.3.2 Transformation Matrix $\mathbf{T}_D$

Following the parametrization of the fixed-interface normal modes $\boldsymbol{\Phi}_{ii}^s$, given in Eq. (2.7), and the independence of the interface constraint modes $\bar{\boldsymbol{\Psi}}_{ib}^s$ on the model parameters, the transformation matrix $\mathbf{T}_D$ defined in Eq. (1.15) can be expressed as

$$
\begin{aligned}
\mathbf{T}_D(\boldsymbol{\theta}) = & \begin{bmatrix} [\bar{\boldsymbol{\Phi}}^1_{id}\delta_{10}, \ldots, \bar{\boldsymbol{\Phi}}^{N_s}_{id}\delta_{N_s0}] & [\bar{\boldsymbol{\Psi}}^1_{ib}\delta_{10}, \ldots, \bar{\boldsymbol{\Psi}}^{N_s}_{ib}\delta_{N_s0}]\tilde{\mathbf{T}} \\ \mathbf{0} & \mathbf{I} \end{bmatrix} \\
& + \sum_{j=1}^{n_\theta} \left\{ \begin{bmatrix} [\bar{\boldsymbol{\Phi}}^1_{id}\delta_{1j}, \ldots, \bar{\boldsymbol{\Phi}}^{N_s}_{id}\delta_{N_sj}] & \mathbf{0} \\ \mathbf{0} & \mathbf{0} \end{bmatrix} \frac{1}{\sqrt{g^j(\theta_j)}} \right. \\
& \left. + \begin{bmatrix} \mathbf{0} & [\bar{\boldsymbol{\Psi}}^1_{ib}\delta_{1j}, \ldots, \bar{\boldsymbol{\Psi}}^{N_s}_{ib}\delta_{N_sj}]\tilde{\mathbf{T}} \\ \mathbf{0} & \mathbf{0} \end{bmatrix} \right\}
\end{aligned} \tag{2.17}
$$

where all terms have been previously defined.

2.3.3 Reduced-Order Matrices $\hat{M}_D$ and $\hat{K}_D$

The mass matrix $\hat{\mathbf{M}}_D$ and the stiffness matrix $\hat{\mathbf{K}}_D$ of the reduced-order model based on dominant fixed-interface normal modes and interface coordinates are defined in Eqs. (1.23) and (1.24), respectively.

If the parametrization of the partitions of the mass matrix $\mathbf{M}^s$ and the stiffness matrix $\mathbf{K}^s$, the parametrization of the dominant fixed-interface normal modes $\boldsymbol{\Phi}^s_{id}$, and the independence of the interface constraint modes $\bar{\boldsymbol{\Psi}}^s_{ib}$ on the model parameters are considered, then the matrices $\hat{\mathbf{M}}^s_{ib}$, $\hat{\mathbf{M}}^s_{bb}$, and $\hat{\mathbf{K}}^s_{bb}$, which are involved in the definition of $\hat{\mathbf{M}}_D$ and $\hat{\mathbf{K}}_D$ and are given in Eqs. (1.25), (1.26), and (1.27), respectively, can be expressed as

$$
\begin{aligned}
\hat{\mathbf{M}}^s_{ib} &= \hat{\bar{\mathbf{M}}}^s_{ib}\sqrt{g^j(\theta_j)} \\
\hat{\mathbf{K}}^s_{bb} &= \hat{\bar{\mathbf{K}}}^s_{bb}\, h^j(\theta_j) \\
\hat{\mathbf{M}}^s_{bb} &= \hat{\bar{\mathbf{M}}}^s_{bb}\, g^j(\theta_j) \ , \quad s \in S_j
\end{aligned} \tag{2.18}
$$

where

$$
\begin{aligned}
\hat{\bar{\mathbf{M}}}^s_{ib} &= \bar{\boldsymbol{\Phi}}^{s^T}_{id}\bar{\mathbf{M}}^s_{ii}\bar{\boldsymbol{\Psi}}^s_{ib} + \bar{\boldsymbol{\Phi}}^{s^T}_{id}\bar{\mathbf{M}}^s_{ib} \\
\hat{\bar{\mathbf{K}}}^s_{bb} &= \bar{\mathbf{K}}^{s^T}_{ib}\bar{\boldsymbol{\Psi}}^s_{ib} + \bar{\mathbf{K}}^s_{bb} \\
\hat{\bar{\mathbf{M}}}^s_{bb} &= (\bar{\boldsymbol{\Psi}}^{s^T}_{ib}\bar{\mathbf{M}}^s_{ii} + \bar{\mathbf{M}}^{s^T}_{ib})\bar{\boldsymbol{\Psi}}^s_{ib} + \bar{\boldsymbol{\Psi}}^{s^T}_{ib}\bar{\mathbf{M}}^s_{ib} + \bar{\mathbf{M}}^s_{bb}
\end{aligned} \tag{2.19}
$$

If the above expansions and the parametrization of the matrices of eigenvalues $\boldsymbol{\Lambda}^s_{id}$, $s = 1, \ldots, N_s$, in (2.8) are considered, the matrices $\hat{\mathbf{M}}_D$ and $\hat{\mathbf{K}}_D$ can be written explicitly in terms of the model parameters $\boldsymbol{\theta}$. In fact, the mass and stiffness matrices of the reduced-order model can be expressed as [14, 15, 17, 23]

$$\hat{\mathbf{M}}_D(\boldsymbol{\theta}) = \begin{bmatrix} [\mathbf{I}\delta_{10}, \ldots, \mathbf{I}\delta_{N_s0}] & [\hat{\bar{\mathbf{M}}}_{ib}^1\delta_{10}, \ldots, \hat{\bar{\mathbf{M}}}_{ib}^{N_s}\delta_{N_s0}]\tilde{\mathbf{T}} \\ \tilde{\mathbf{T}}^T[\hat{\bar{\mathbf{M}}}_{ib}^{1^T}\delta_{10}, \ldots, \hat{\bar{\mathbf{M}}}_{ib}^{N_s^T}\delta_{N_s0}] & \tilde{\mathbf{T}}^T[\hat{\bar{\mathbf{M}}}_{bb}^1\delta_{10}, \ldots, \hat{\bar{\mathbf{M}}}_{bb}^{N_s}\delta_{N_s0}]\tilde{\mathbf{T}} \end{bmatrix}$$
$$+\sum_{j=1}^{n_\theta}\left\{\begin{bmatrix} [\mathbf{I}\delta_{1j}, \ldots, \mathbf{I}\delta_{N_sj}] & \mathbf{0} \\ \mathbf{0} & \mathbf{0} \end{bmatrix}\right.$$
$$+\begin{bmatrix} \mathbf{0} & [\hat{\bar{\mathbf{M}}}_{ib}^1\delta_{1j}, \ldots, \hat{\bar{\mathbf{M}}}_{ib}^{N_s}\delta_{N_sj}]\tilde{\mathbf{T}} \\ \tilde{\mathbf{T}}^T[\hat{\bar{\mathbf{M}}}_{ib}^{1^T}\delta_{1j}, \ldots, \hat{\bar{\mathbf{M}}}_{ib}^{N_s^T}\delta_{N_sj}] & \mathbf{0} \end{bmatrix}\sqrt{g^j(\theta_j)}$$
$$\left.+\begin{bmatrix} \mathbf{0} & \mathbf{0} \\ \mathbf{0} & \tilde{\mathbf{T}}^T[\hat{\bar{\mathbf{M}}}_{bb}^1\delta_{1j}, \ldots, \hat{\bar{\mathbf{M}}}_{bb}^{N_s}\delta_{N_sj}]\tilde{\mathbf{T}} \end{bmatrix} g^j(\theta_j)\right\} \tag{2.20}$$

and

$$\hat{\mathbf{K}}_D(\boldsymbol{\theta}) = \begin{bmatrix} [\bar{\boldsymbol{\Lambda}}_{id}^1\delta_{10}, \ldots, \bar{\boldsymbol{\Lambda}}_{id}^{N_s}\delta_{N_s0}] & \mathbf{0} \\ \mathbf{0} & \tilde{\mathbf{T}}^T[\hat{\bar{\mathbf{K}}}_{bb}^1\delta_{10}, \ldots, \hat{\bar{\mathbf{K}}}_{bb}^{N_s}\delta_{N_s0}]\tilde{\mathbf{T}} \end{bmatrix}$$
$$+\sum_{j=1}^{n_\theta}\left\{\begin{bmatrix} [\bar{\boldsymbol{\Lambda}}_{id}^1\delta_{1j}, \ldots, \bar{\boldsymbol{\Lambda}}_{id}^{N_s}\delta_{N_sj}] & \mathbf{0} \\ \mathbf{0} & \mathbf{0} \end{bmatrix}\frac{h^j(\theta_j)}{g^j(\theta_j)}\right.$$
$$\left.+\begin{bmatrix} \mathbf{0} & \mathbf{0} \\ \mathbf{0} & \tilde{\mathbf{T}}^T[\hat{\bar{\mathbf{K}}}_{bb}^1\delta_{1j}, \ldots, \hat{\bar{\mathbf{K}}}_{bb}^{N_s}\delta_{N_sj}]\tilde{\mathbf{T}} \end{bmatrix} h^j(\theta_j)\right\} \tag{2.21}$$

where all terms have been previously defined. The matrix ($\mathbf{I}$) that appears in Eq. (2.20) represents a generic identity matrix of dimension equal to $n_{id}^s \times n_{id}^s$, $s = 1, \ldots, N_s$. It should be noted that all the matrices involved in defining the reduced-order matrices $\hat{\mathbf{M}}_D$ and $\hat{\mathbf{K}}_D$ are computed and assembled one time from the reference model, and it is therefore unnecessary to re-compute these matrices for different values of $\boldsymbol{\theta}$ encountered during the corresponding simulation processes. Thus, the expansions (2.17), (2.20), and (2.21) provide very efficient expressions for computing the transformation matrix and the reduced mass and stiffness matrices, required in model reduction for different values of the model parameters $\boldsymbol{\theta}$. Such computation is carried out without a need to re-assemble these matrices from the fixed-interface normal modes and the interface constraint modes. Moreover, the above formulation guarantees that the reduced-order model is based on the exact substructure modes for all values of the model parameters.

2.3.4 *Transformation Matrix $\mathbf{T}_R$*

The transformation matrix $\mathbf{T}_R$ that accounts for the contribution of the residual fixed-interface normal modes is given in Eq. (1.47). This matrix is defined in terms of the transformation matrix $\tilde{\mathbf{T}}$ and the matrices $\bar{\mathbf{F}}$, $\tilde{\mathbf{M}}_{ib}$, $\mathbf{M}_I$, $\mathbf{M}_{iI}$, $\boldsymbol{\Lambda}$, and $\mathbf{K}_I$. If the definitions of the previous matrices given in Eqs. (1.39), (1.32), (1.45), (1.46), (1.48), and (1.49) are considered, along with the parametrization of the matrices $\mathbf{M}^s$, $\mathbf{K}^s$, $\boldsymbol{\Phi}^s_{id}$, $\boldsymbol{\Lambda}^s_{id}$, $\hat{\mathbf{M}}^s_{ib}$, $\hat{\mathbf{K}}^s_{bb}$, and $\hat{\mathbf{M}}_{bb}$ (given in Eqs. (2.3), (2.4), (2.7), (2.8), and (2.18), respectively), it follows that each of the matrices involved in the definition of $\mathbf{T}_R$ can be written explicitly in term of the system parameters $\boldsymbol{\theta}$. In fact, they can be written as [15]

$$\bar{\mathbf{F}}(\boldsymbol{\theta}) = [(\bar{\mathbf{K}}_{ii}^{1^{-1}} - \bar{\boldsymbol{\Phi}}_{id}^{1}\bar{\boldsymbol{\Lambda}}_{id}^{1^{-1}}\bar{\boldsymbol{\Phi}}_{id}^{1^T})\delta_{10}, \ldots, (\bar{\mathbf{K}}_{ii}^{N_s^{-1}} - \bar{\boldsymbol{\Phi}}_{id}^{N_s}\bar{\boldsymbol{\Lambda}}_{id}^{N_s^{-1}}\bar{\boldsymbol{\Phi}}_{id}^{N_s^T})\delta_{N_s0}] + \sum_{j=1}^{n_\theta}[(\bar{\mathbf{K}}_{ii}^{1^{-1}} - \bar{\boldsymbol{\Phi}}_{id}^{1}\bar{\boldsymbol{\Lambda}}_{id}^{1^{-1}}\bar{\boldsymbol{\Phi}}_{id}^{1^T})\delta_{1j}, \ldots, (\bar{\mathbf{K}}_{ii}^{N_s^{-1}} - \bar{\boldsymbol{\Phi}}_{id}^{N_s}\bar{\boldsymbol{\Lambda}}_{id}^{N_s^{-1}}\bar{\boldsymbol{\Phi}}_{id}^{N_s^T})\delta_{N_sj}]\frac{1}{h^j(\theta_j)} \tag{2.22}$$

$$\tilde{\mathbf{M}}_{ib}(\boldsymbol{\theta}) = [(\bar{\mathbf{M}}_{ib}^{1} - \bar{\mathbf{M}}_{ii}^{1}\bar{\mathbf{K}}_{ii}^{1^{-1}}\bar{\mathbf{K}}_{ib}^{1})\delta_{10}, \ldots, (\bar{\mathbf{M}}_{ib}^{N_s} - \bar{\mathbf{M}}_{ii}^{N_s}\bar{\mathbf{K}}_{ii}^{N_s^{-1}}\bar{\mathbf{K}}_{ib}^{N_s})\delta_{N_s0}] + \sum_{j=1}^{n_\theta}[(\bar{\mathbf{M}}_{ib}^{1} - \bar{\mathbf{M}}_{ii}^{1}\bar{\mathbf{K}}_{ii}^{1^{-1}}\bar{\mathbf{K}}_{ib}^{1})\delta_{1j}, \ldots, (\bar{\mathbf{M}}_{ib}^{N_s} - \bar{\mathbf{M}}_{ii}^{N_s}\bar{\mathbf{K}}_{ii}^{N_s^{-1}}\bar{\mathbf{K}}_{ib}^{N_s})\delta_{N_sj}]g^j(\theta_j) \tag{2.23}$$

$$\mathbf{M}_I(\boldsymbol{\theta}) = \tilde{\mathbf{T}}^T[\hat{\bar{\mathbf{M}}}_{bb}^{1}\delta_{10}, \ldots, \hat{\bar{\mathbf{M}}}_{bb}^{N_s}\delta_{N_s0}]\tilde{\mathbf{T}} + \sum_{j=1}^{n_\theta}\tilde{\mathbf{T}}^T[\hat{\bar{\mathbf{M}}}_{bb}^{1}\delta_{1j}, \ldots, \hat{\bar{\mathbf{M}}}_{bb}^{N_s}\delta_{N_sj}]\tilde{\mathbf{T}}g^j(\theta_j) \tag{2.24}$$

$$\mathbf{M}_{iI}(\boldsymbol{\theta}) = [\hat{\bar{\mathbf{M}}}_{ib}^{1}\delta_{10}, \ldots, \hat{\bar{\mathbf{M}}}_{ib}^{N_s}\delta_{N_s0}]\tilde{\mathbf{T}} + \sum_{j=1}^{n_\theta}[\hat{\bar{\mathbf{M}}}_{ib}^{1}\delta_{1j}, \ldots, \hat{\bar{\mathbf{M}}}_{ib}^{N_s}\delta_{N_sj}]\tilde{\mathbf{T}}\sqrt{g^j(\theta_j)} \tag{2.25}$$

$$\boldsymbol{\Lambda}(\boldsymbol{\theta}) = [\bar{\boldsymbol{\Lambda}}_{id}^{1}\delta_{10}, \ldots, \bar{\boldsymbol{\Lambda}}_{id}^{N_s}\delta_{N_s0}] + \sum_{j=1}^{n_\theta}[\bar{\boldsymbol{\Lambda}}_{id}^{1}\delta_{1j}, \ldots, \bar{\boldsymbol{\Lambda}}_{id}^{N_s}\delta_{N_sj}]\frac{h^j(\theta_j)}{g^j(\theta_j)} \tag{2.26}$$

$$\mathbf{K}_I(\boldsymbol{\theta}) = \tilde{\mathbf{T}}^T[\hat{\bar{\mathbf{K}}}_{bb}^{1}\delta_{10}, \ldots, \hat{\bar{\mathbf{K}}}_{bb}^{N_s}\delta_{N_s0}]\tilde{\mathbf{T}} + \sum_{j=1}^{n_\theta}\tilde{\mathbf{T}}^T[\hat{\bar{\mathbf{K}}}_{bb}^{1}\delta_{1j}, \ldots, \hat{\bar{\mathbf{K}}}_{bb}^{N_s}\delta_{N_sj}]\tilde{\mathbf{T}}h^j(\theta_j) \tag{2.27}$$

The result of the above parametrization is that all the matrices involved in expanding the different matrices that define the transformation matrix $\mathbf{T}_R$ are independent of the values of the vector of model parameters $\boldsymbol{\theta}$. Thus, these matrices are computed and assembled once from the reference model.

2.3.5 Reduced-Order Matrices $\hat{M}_R$ and $\hat{K}_R$

The improved reduced-order mass matrix $\hat{\mathbf{M}}_R$ and stiffness matrix $\hat{\mathbf{K}}_R$ are given in Eqs. (1.50) and (1.51). In view of the parametrization of matrices $\hat{\mathbf{M}}$, $\hat{\mathbf{K}}$, and $\mathbf{T}_D$ in Eqs. (2.13), (2.14), and (2.17), respectively, together with the parametrization of the different matrices involved in the definition of $\mathbf{T}_R$, it is clear that $\hat{\mathbf{M}}_R$ and $\hat{\mathbf{K}}_R$ can be expressed in terms of a set of matrices independent of the vector of system parameters $\boldsymbol{\theta}$. This set of matrices is computed and assembled once from the reference model, thus avoiding the re-assembly of the various matrices involved in defining the reduced-order matrices for different values of the vector of model parameters $\boldsymbol{\theta}$, required by the corresponding simulation processes. The explicit expansion of the enhanced reduced-order mass matrix $\hat{\mathbf{M}}_R$ and stiffness matrix $\hat{\mathbf{K}}_R$ in terms of the model parameters $\boldsymbol{\theta}$ (similar to Eqs. (2.20) and (2.21)) is quite involved, and is not shown here for the sake of simplicity in notation. In what follows, such an expansion is derived for one particular case, for illustration purposes.

2.3.6 Expansion of $\hat{M}_R$ and $\hat{K}_R$ Under Partial Invariant Conditions of T_R

In what follows, it is assumed that the mass matrix is constant and independent of the model parameters $\boldsymbol{\theta}$. As indicated in the previous chapter, the contribution of the residual normal modes is enhancing the dynamic behavior of the internal degrees of freedom at the substructure level. Thus, the use of higher order modes can be considered a secondary type of effect with respect to the internal coordinates. Based on this observation, all the matrices involved in the definition of the transformation matrix $\mathbf{T}_R$ are treated as invariant, that is, independent of the model parameters $\boldsymbol{\theta}$, except for the matrices associated with the interface coordinates, $\mathbf{M}_I$ and $\mathbf{K}_I$. This is clearly an approximation, since some of the matrices included in the definition of $\mathbf{T}_R$, such as $\bar{\mathbf{F}}$ and $\boldsymbol{\Lambda}$ (see Eqs. (2.22) and (2.26)), depend on the model parameters. Note that the matrices $\mathbf{M}_I$, $\tilde{\mathbf{M}}_{ib}$, and $\mathbf{M}_{iI}$ are already independent of $\boldsymbol{\theta}$, due to the invariant assumption of the mass matrix with respect to the model parameters. It is anticipated that the previous approximation will give sufficiently accurate results. The actual validation and applicability of this assumption is left for subsequent chapters with applications. From Eqs. (1.50) and (1.51), it is clear that the parametrizations of $\hat{\mathbf{M}}_R$ and $\hat{\mathbf{K}}_R$ depend on the parametrization of the matrices

$$\hat{\mathbf{M}}_D \ , \ \mathbf{T}_R^T\hat{\mathbf{M}}\mathbf{T}_D \ , \ \mathbf{T}_R^T\hat{\mathbf{M}}\mathbf{T}_R \ , \ \hat{\mathbf{K}}_D \ , \ \mathbf{T}_R^T\hat{\mathbf{K}}\mathbf{T}_D \ , \ \mathbf{T}_R^T\hat{\mathbf{K}}\mathbf{T}_R \tag{2.28}$$

The expansions of the matrices $\hat{\mathbf{M}}_D$ and $\hat{\mathbf{K}}_D$ are given in Eqs. (2.20) and (2.21), respectively, with $g^j(\cdot) = 1$ in this case. If the previous invariant assumptions of $\mathbf{T}_R$ are considered, the matrices $\mathbf{T}_R^T\hat{\mathbf{M}}\mathbf{T}_R$ and $\mathbf{T}_R^T\hat{\mathbf{K}}\mathbf{T}_R$ allow the parametrization [18]

$$\mathbf{T}_R^T\hat{\mathbf{M}}\mathbf{T}_R(\boldsymbol{\theta}) = \begin{bmatrix} \mathbf{T}_{R1}^T[\bar{\mathbf{M}}_{ii}^1, \ldots, \bar{\mathbf{M}}_{ii}^{N_s}]\mathbf{T}_{R1} & \mathbf{T}_{R1}^T[\bar{\mathbf{M}}_{ii}^1, \ldots, \bar{\mathbf{M}}_{ii}^{N_s}]\mathbf{T}_{R2}(\boldsymbol{\theta}) \\ \mathbf{T}_{R2}^T(\boldsymbol{\theta})[\bar{\mathbf{M}}_{ii}^{1^T}, \ldots, \bar{\mathbf{M}}_{ii}^{N_s^T}]\mathbf{T}_{R1} & \mathbf{T}_{R2}^T(\boldsymbol{\theta})[\bar{\mathbf{M}}_{ii}^1, \ldots, \bar{\mathbf{M}}_{ii}^{N_s}]\mathbf{T}_{R2}(\boldsymbol{\theta}) \end{bmatrix} \tag{2.29}$$

and

$$\begin{aligned} \mathbf{T}_R^T\hat{\mathbf{K}}\mathbf{T}_R(\boldsymbol{\theta}) = & \begin{bmatrix} \mathbf{T}_{R1}^T[\bar{\mathbf{K}}_{ii}^1\delta_{10}, \ldots, \bar{\mathbf{K}}_{ii}^{N_s}\delta_{N_s0}]\mathbf{T}_{R1} & \mathbf{T}_{R1}^T[\bar{\mathbf{K}}_{ii}^1\delta_{10}, \ldots, \bar{\mathbf{K}}_{ii}^{N_s}\delta_{N_s0}]\mathbf{T}_{R2}(\boldsymbol{\theta}) \\ \mathbf{T}_{R2}^T(\boldsymbol{\theta})[\bar{\mathbf{K}}_{ii}^{1^T}\delta_{10}, \ldots, \bar{\mathbf{K}}_{ii}^{N_s^T}\delta_{N_s0}]\mathbf{T}_{R1} & \mathbf{T}_{R2}^T(\boldsymbol{\theta})[\bar{\mathbf{K}}_{ii}^1\delta_{10}, \ldots, \bar{\mathbf{K}}_{ii}^{N_s}\delta_{N_s0}]\mathbf{T}_{R2}(\boldsymbol{\theta}) \end{bmatrix} \\ & + \sum_{j=1}^{n_\theta} \begin{bmatrix} \mathbf{T}_{R1}^T[\bar{\mathbf{K}}_{ii}^1\delta_{1j}, \ldots, \bar{\mathbf{K}}_{ii}^{N_s}\delta_{N_sj}]\mathbf{T}_{R1} & \mathbf{T}_{R1}^T[\bar{\mathbf{K}}_{ii}^1\delta_{1j}, \ldots, \bar{\mathbf{K}}_{ii}^{N_s}\delta_{N_sj}]\mathbf{T}_{R2}(\boldsymbol{\theta}) \\ \mathbf{T}_{R2}^T(\boldsymbol{\theta})[\bar{\mathbf{K}}_{ii}^{1^T}\delta_{1j}, \ldots, \bar{\mathbf{K}}_{ii}^{N_s^T}\delta_{N_sj}]\mathbf{T}_{R1} & \mathbf{T}_{R2}^T(\boldsymbol{\theta})[\bar{\mathbf{K}}_{ii}^1\delta_{1j}, \ldots, \bar{\mathbf{K}}_{ii}^{N_s}\delta_{N_sj}]\mathbf{T}_{R2}(\boldsymbol{\theta}) \end{bmatrix} h^j(\theta_j) \end{aligned} \tag{2.30}$$

where $\mathbf{T}_{R1}$ and $\mathbf{T}_{R2}(\boldsymbol{\theta})$ are the nonzero components of the 2×2 partitioned transformation matrix $\mathbf{T}_R$, defined in Eq. (1.47), and given by

$$\begin{aligned} \mathbf{T}_{R1} &= -\bar{\mathbf{F}}\tilde{\mathbf{M}}_{ib}\tilde{\mathbf{T}}(\mathbf{M}_I - \mathbf{M}_{iI}^T\mathbf{M}_{iI})^{-1}\mathbf{M}_{iI}^T\boldsymbol{\Lambda} \\ \mathbf{T}_{R2}(\boldsymbol{\theta}) &= \bar{\mathbf{F}}\tilde{\mathbf{M}}_{ib}\tilde{\mathbf{T}}(\mathbf{M}_I - \mathbf{M}_{iI}^T\mathbf{M}_{iI})^{-1}\mathbf{K}_I(\boldsymbol{\theta}) \end{aligned} \tag{2.31}$$

where the matrices $\bar{\mathbf{F}}$ and $\boldsymbol{\Lambda}$ are evaluated at some nominal value of the model parameters. For example, the nominal value may correspond to the most probable value of the model parameters, however, it is noted that such selection is problem-dependent. The expansion of $\mathbf{K}_I(\boldsymbol{\theta})$ in terms of the model parameters is given in Eq. (2.27).

Similarly, the matrices $\mathbf{T}_R^T\hat{\mathbf{M}}\mathbf{T}_D$ and $\mathbf{T}_R^T\hat{\mathbf{K}}\mathbf{T}_D$ can be expanded as [18]

$$\begin{aligned} \mathbf{T}_R^T\hat{\mathbf{M}}\mathbf{T}_D(\boldsymbol{\theta}) = & \begin{bmatrix} \mathbf{T}_{R1}^T[\bar{\mathbf{M}}_{ii}^1\bar{\boldsymbol{\Phi}}_{id}^1, \ldots, \bar{\mathbf{M}}_{ii}^{N_s}\bar{\boldsymbol{\Phi}}_{id}^{N_s}] & \mathbf{0} \\ \mathbf{T}_{R2}^T(\boldsymbol{\theta})[\bar{\mathbf{M}}_{ii}^1\bar{\boldsymbol{\Phi}}_{id}^1, \ldots, \bar{\mathbf{M}}_{ii}^{N_s}\bar{\boldsymbol{\Phi}}_{id}^{N_s}] & \mathbf{0} \end{bmatrix} \\ & + \begin{bmatrix} \mathbf{0} & \mathbf{T}_{R1}^T[(\bar{\mathbf{M}}_{ii}^1\bar{\boldsymbol{\Psi}}_{ib}^1 + \bar{\mathbf{M}}_{ib}^1), \ldots, (\bar{\mathbf{M}}_{ii}^{N_s}\bar{\boldsymbol{\Psi}}_{ib}^{N_s} + \bar{\mathbf{M}}_{ib}^{N_s})]\tilde{\mathbf{T}} \\ \mathbf{0} & \mathbf{T}_{R2}^T(\boldsymbol{\theta})[(\bar{\mathbf{M}}_{ii}^1\bar{\boldsymbol{\Psi}}_{ib}^1 + \bar{\mathbf{M}}_{ib}^1), \ldots, (\bar{\mathbf{M}}_{ii}^{N_s}\bar{\boldsymbol{\Psi}}_{ib}^{N_s} + \bar{\mathbf{M}}_{ib}^{N_s})]\tilde{\mathbf{T}} \end{bmatrix} \end{aligned} \tag{2.32}$$

and

$$\begin{aligned} \mathbf{T}_R^T\hat{\mathbf{K}}\mathbf{T}_D(\boldsymbol{\theta}) = & \begin{bmatrix} \mathbf{T}_{R1}^T[\bar{\mathbf{K}}_{ii}^1\bar{\boldsymbol{\Phi}}_{id}^1\delta_{10}, \ldots, \bar{\mathbf{K}}_{ii}^{N_s}\bar{\boldsymbol{\Phi}}_{id}^{N_s}\delta_{N_s0}] & \mathbf{0} \\ \mathbf{T}_{R2}^T(\boldsymbol{\theta})[\bar{\mathbf{K}}_{ii}^1\bar{\boldsymbol{\Phi}}_{id}^1\delta_{10}, \ldots, \bar{\mathbf{K}}_{ii}^{N_s}\bar{\boldsymbol{\Phi}}_{id}^{N_s}\delta_{N_s0}] & \mathbf{0} \end{bmatrix} \\ & + \sum_{j=1}^{n_\theta} \begin{bmatrix} \mathbf{T}_{R1}^T[\bar{\mathbf{K}}_{ii}^1\bar{\boldsymbol{\Phi}}_{id}^1\delta_{1j}, \ldots, \bar{\mathbf{K}}_{ii}^{N_s}\bar{\boldsymbol{\Phi}}_{id}^{N_s}\delta_{N_sj}] & \mathbf{0} \\ \mathbf{T}_{R2}^T(\boldsymbol{\theta})[\bar{\mathbf{K}}_{ii}^1\bar{\boldsymbol{\Phi}}_{id}^1\delta_{1j}, \ldots, \bar{\mathbf{K}}_{ii}^{N_s}\bar{\boldsymbol{\Phi}}_{id}^{N_s}\delta_{N_sj}] & \mathbf{0} \end{bmatrix} h^j(\theta_j) \end{aligned} \tag{2.33}$$

It can be seen that, for the particular case considered in this section, all the basic matrices involved in the expansion of the reduced-order matrices $\hat{\mathbf{M}}_R$ and $\hat{\mathbf{K}}_R$ are independent of the model parameters, except for the matrix $\mathbf{K}_I(\boldsymbol{\theta})$, whose expansion is defined in Eq. (2.27).

2.4 Numerical Implementation: Pseudo-Code No. 4

Based on the formulation presented, the following pseudo-code illustrates how the parametrization of reduced-order matrices is constructed. For the enhanced reduced-order matrices case, the invariant assumption of the mass matrix and the transformation matrix $\mathbf{T}_R$ is considered.

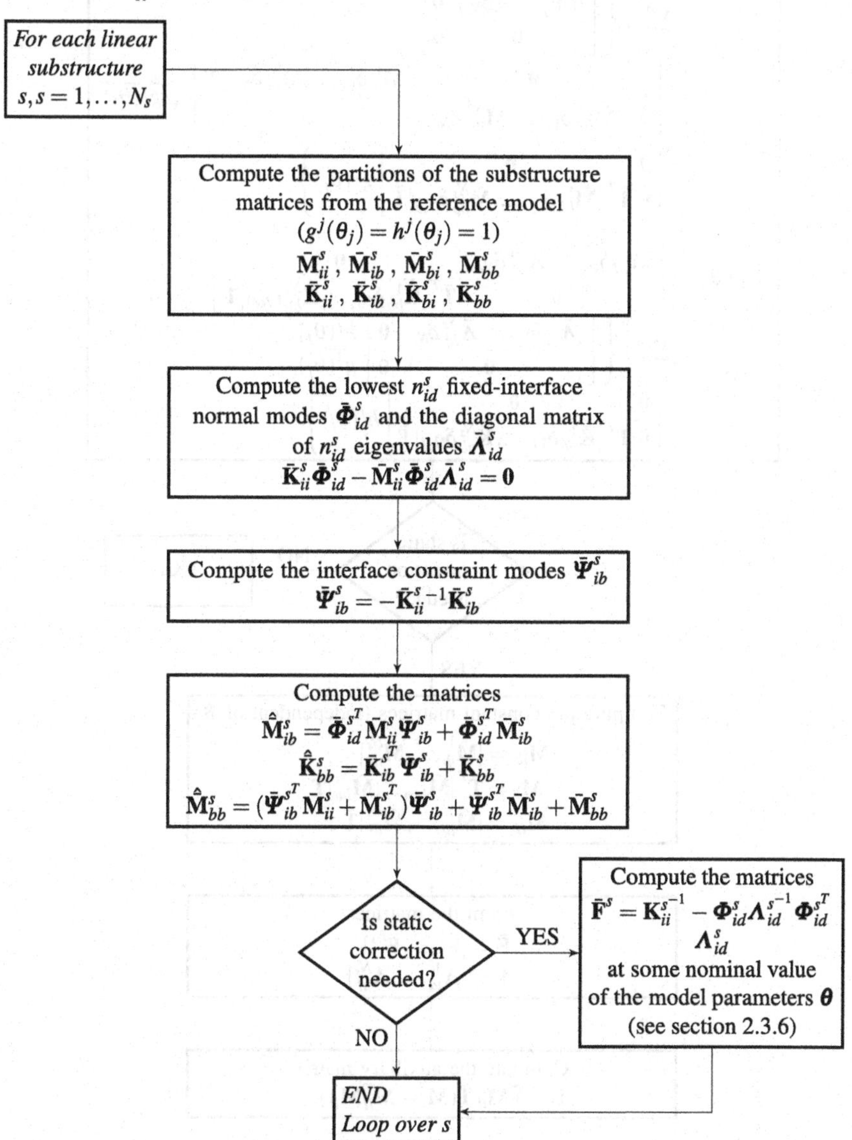

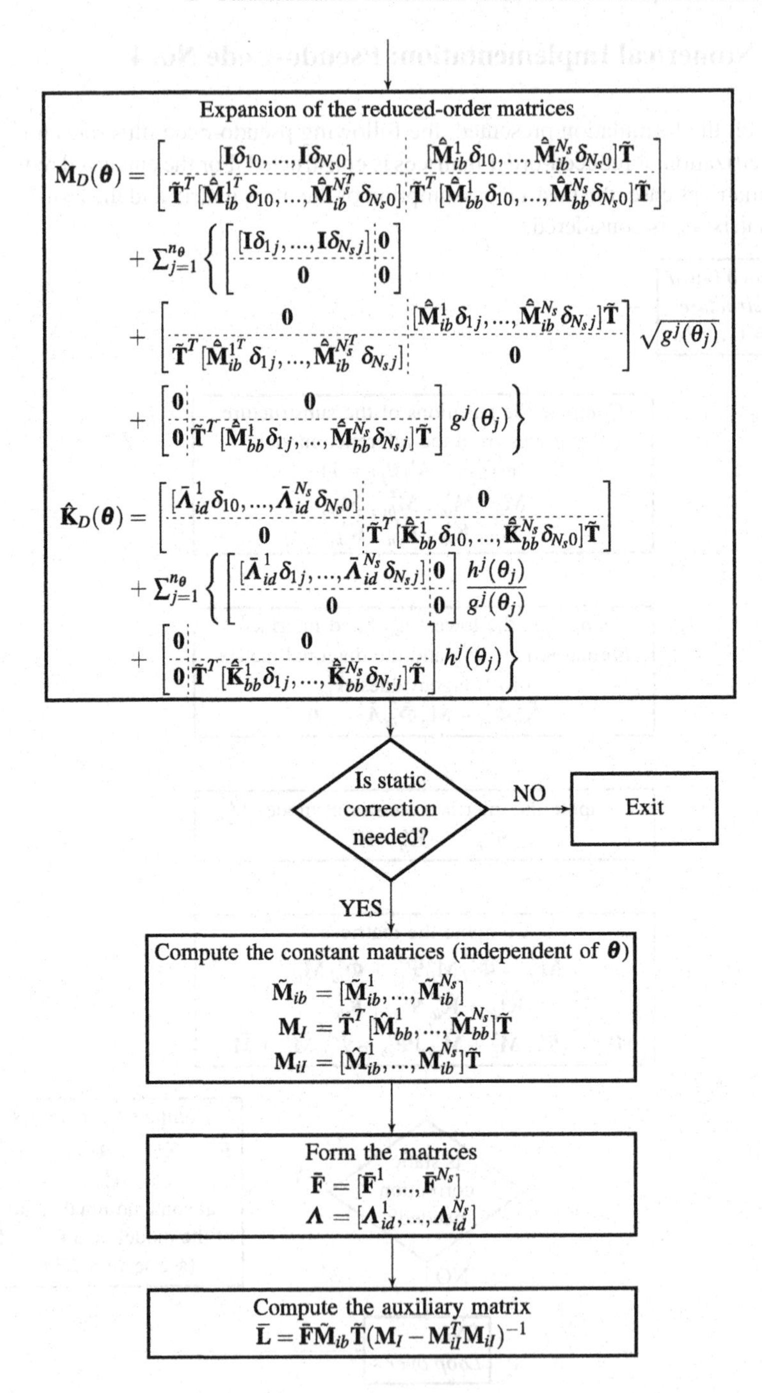
Expansion of the reduced-order matrices
Is static correction needed?
NO
Exit
YES
Compute the constant matrices (independent of θ)
Form the matrices
Compute the auxiliary matrix

Compute the components $\mathbf{T}_{R1}$ and $\mathbf{T}_{R2}$

$$\mathbf{T}_{R1} = -\bar{\mathbf{L}}\mathbf{M}_{il}^{T}\boldsymbol{\Lambda}$$

$$\mathbf{T}_{R2}(\boldsymbol{\theta}) = \bar{\mathbf{L}}\tilde{\mathbf{T}}^{T}[\hat{\mathbf{K}}_{bb}^{1}\delta_{10},\ldots,\hat{\mathbf{K}}_{bb}^{N_s}\delta_{N_s0}]\tilde{\mathbf{T}} + \sum_{j=1}^{n_\theta}\bar{\mathbf{L}}\tilde{\mathbf{T}}^{T}[\hat{\mathbf{K}}_{bb}^{1}\delta_{1j},\ldots,\hat{\mathbf{K}}_{bb}^{N_s}\delta_{N_sj}]\tilde{\mathbf{T}}h^{j}(\theta_j)$$

Expansion of the intermediate matrices $\mathbf{T}_R^T\hat{\mathbf{M}}\mathbf{T}_R$ and $\mathbf{T}_R^T\hat{\mathbf{K}}\mathbf{T}_R$

$$\mathbf{T}_R^T\hat{\mathbf{M}}\mathbf{T}_R(\boldsymbol{\theta}) = \left[\begin{array}{c:c} \mathbf{T}_{R1}^{T}[\bar{\mathbf{M}}_{ii}^{1},\ldots,\bar{\mathbf{M}}_{ii}^{N_s}]\mathbf{T}_{R1} & \mathbf{T}_{R1}^{T}[\bar{\mathbf{M}}_{ii}^{1},\ldots,\bar{\mathbf{M}}_{ii}^{N_s}]\mathbf{T}_{R2}(\boldsymbol{\theta}) \\ \hdashline \mathbf{T}_{R2}^{T}(\boldsymbol{\theta})[\bar{\mathbf{M}}_{ii}^{1^T},\ldots,\bar{\mathbf{M}}_{ii}^{N_s^T}]\mathbf{T}_{R1} & \mathbf{T}_{R2}^{T}(\boldsymbol{\theta})[\bar{\mathbf{M}}_{ii}^{1},\ldots,\bar{\mathbf{M}}_{ii}^{N_s}]\mathbf{T}_{R2}(\boldsymbol{\theta}) \end{array}\right]$$

$$\mathbf{T}_R^T\hat{\mathbf{K}}\mathbf{T}_R(\boldsymbol{\theta}) = \left[\begin{array}{c:c} \mathbf{T}_{R1}^{T}[\bar{\mathbf{K}}_{ii}^{1}\delta_{10},\ldots,\bar{\mathbf{K}}_{ii}^{N_s}\delta_{N_s0}]\mathbf{T}_{R1} & \mathbf{T}_{R1}^{T}[\bar{\mathbf{K}}_{ii}^{1}\delta_{10},\ldots,\bar{\mathbf{K}}_{ii}^{N_s}\delta_{N_s0}]\mathbf{T}_{R2}(\boldsymbol{\theta}) \\ \hdashline \mathbf{T}_{R2}^{T}(\boldsymbol{\theta})[\bar{\mathbf{K}}_{ii}^{1^T}\delta_{10},\ldots,\bar{\mathbf{K}}_{ii}^{N_s^T}\delta_{N_s0}]\mathbf{T}_{R1} & \mathbf{T}_{R2}^{T}(\boldsymbol{\theta})[\bar{\mathbf{K}}_{ii}^{1}\delta_{10},\ldots,\bar{\mathbf{K}}_{ii}^{N_s}\delta_{N_s0}]\mathbf{T}_{R2}(\boldsymbol{\theta}) \end{array}\right]$$

$$+\sum_{j=1}^{n_\theta}\left[\begin{array}{c:c} \mathbf{T}_{R1}^{T}[\bar{\mathbf{K}}_{ii}^{1}\delta_{1j},\ldots,\bar{\mathbf{K}}_{ii}^{N_s}\delta_{N_sj}]\mathbf{T}_{R1} & \mathbf{T}_{R1}^{T}[\bar{\mathbf{K}}_{ii}^{1}\delta_{1j},\ldots,\bar{\mathbf{K}}_{ii}^{N_s}\delta_{N_sj}]\mathbf{T}_{R2}(\boldsymbol{\theta}) \\ \hdashline \mathbf{T}_{R2}^{T}(\boldsymbol{\theta})[\bar{\mathbf{K}}_{ii}^{1^T}\delta_{1j},\ldots,\bar{\mathbf{K}}_{ii}^{N_s^T}\delta_{N_sj}]\mathbf{T}_{R1} & \mathbf{T}_{R2}^{T}(\boldsymbol{\theta})[\bar{\mathbf{K}}_{ii}^{1}\delta_{1j},\ldots,\bar{\mathbf{K}}_{ii}^{N_s}\delta_{N_sj}]\mathbf{T}_{R2}(\boldsymbol{\theta}) \end{array}\right]h^{j}(\theta_j)$$

Expansion of the intermediate matrices $\mathbf{T}_R^T\hat{\mathbf{M}}\mathbf{T}_D$ and $\mathbf{T}_R^T\hat{\mathbf{K}}\mathbf{T}_D$

$$\mathbf{T}_R^T\hat{\mathbf{M}}\mathbf{T}_D(\boldsymbol{\theta}) = \left[\begin{array}{c:c} \mathbf{T}_{R1}^{T}[\bar{\mathbf{M}}_{ii}^{1}\bar{\boldsymbol{\Phi}}_{id}^{1},\ldots,\bar{\mathbf{M}}_{ii}^{N_s}\bar{\boldsymbol{\Phi}}_{id}^{N_s}] & \mathbf{0} \\ \hdashline \mathbf{T}_{R2}^{T}(\boldsymbol{\theta})[\bar{\mathbf{M}}_{ii}^{1}\bar{\boldsymbol{\Phi}}_{id}^{1},\ldots,\bar{\mathbf{M}}_{ii}^{N_s}\bar{\boldsymbol{\Phi}}_{id}^{N_s}] & \mathbf{0} \end{array}\right]$$

$$+\left[\begin{array}{c:c} \mathbf{0} & \mathbf{T}_{R1}^{T}[(\bar{\mathbf{M}}_{ii}^{1}\bar{\boldsymbol{\Psi}}_{ib}^{1}+\bar{\mathbf{M}}_{ib}^{1}),\ldots,(\bar{\mathbf{M}}_{ii}^{N_s}\bar{\boldsymbol{\Psi}}_{ib}^{N_s}+\bar{\mathbf{M}}_{ib}^{N_s})]\tilde{\mathbf{T}} \\ \hdashline \mathbf{0} & \mathbf{T}_{R2}^{T}(\boldsymbol{\theta})[(\bar{\mathbf{M}}_{ii}^{1}\bar{\boldsymbol{\Psi}}_{ib}^{1}+\bar{\mathbf{M}}_{ib}^{1}),\ldots,(\bar{\mathbf{M}}_{ii}^{N_s}\bar{\boldsymbol{\Psi}}_{ib}^{N_s}+\bar{\mathbf{M}}_{ib}^{N_s})]\tilde{\mathbf{T}} \end{array}\right]$$

$$\mathbf{T}_R^T\hat{\mathbf{K}}\mathbf{T}_D(\boldsymbol{\theta}) = \left[\begin{array}{c:c} \mathbf{T}_{R1}^{T}[\bar{\mathbf{K}}_{ii}^{1}\bar{\boldsymbol{\Phi}}_{id}^{1}\delta_{10},\ldots,\bar{\mathbf{K}}_{ii}^{N_s}\bar{\boldsymbol{\Phi}}_{id}^{N_s}\delta_{N_s0}] & \mathbf{0} \\ \hdashline \mathbf{T}_{R2}^{T}(\boldsymbol{\theta})[\bar{\mathbf{K}}_{ii}^{1}\bar{\boldsymbol{\Phi}}_{id}^{1}\delta_{10},\ldots,\bar{\mathbf{K}}_{ii}^{N_s}\bar{\boldsymbol{\Phi}}_{id}^{N_s}\delta_{N_s0}] & \mathbf{0} \end{array}\right]$$

$$+\sum_{j=1}^{n_\theta}\left[\begin{array}{c:c} \mathbf{T}_{R1}^{T}[\bar{\mathbf{K}}_{ii}^{1}\bar{\boldsymbol{\Phi}}_{id}^{1}\delta_{1j},\ldots,\bar{\mathbf{K}}_{ii}^{N_s}\bar{\boldsymbol{\Phi}}_{id}^{N_s}\delta_{N_sj}] & \mathbf{0} \\ \hdashline \mathbf{T}_{R2}^{T}(\boldsymbol{\theta})[\bar{\mathbf{K}}_{ii}^{1}\bar{\boldsymbol{\Phi}}_{id}^{1}\delta_{1j},\ldots,\bar{\mathbf{K}}_{ii}^{N_s}\bar{\boldsymbol{\Phi}}_{id}^{N_s}\delta_{N_sj}] & \mathbf{0} \end{array}\right]h^{j}(\theta_j)$$

Parametrization of the reduced matrices $\hat{\mathbf{M}}_R$ and $\hat{\mathbf{K}}_R$

$$\hat{\mathbf{M}}_R(\boldsymbol{\theta}) = \hat{\mathbf{M}}_D(\boldsymbol{\theta}) + \mathbf{T}_R^T\hat{\mathbf{M}}\mathbf{T}_D(\boldsymbol{\theta}) + (\mathbf{T}_R^T\hat{\mathbf{M}}\mathbf{T}_D)^T(\boldsymbol{\theta}) + \mathbf{T}_R^T\hat{\mathbf{M}}\mathbf{T}_R(\boldsymbol{\theta})$$

$$\hat{\mathbf{K}}_R(\boldsymbol{\theta}) = \hat{\mathbf{K}}_D(\boldsymbol{\theta}) + \mathbf{T}_R^T\hat{\mathbf{K}}\mathbf{T}_D(\boldsymbol{\theta}) + (\mathbf{T}_R^T\hat{\mathbf{K}}\mathbf{T}_D)^T(\boldsymbol{\theta}) + \mathbf{T}_R^T\hat{\mathbf{K}}\mathbf{T}_R(\boldsymbol{\theta})$$

References

1. D. Akçay, H.J.M. Geijselaers, M.H.M. Ellenbroek, A. de Boer, Dynamic substructuring and reanalysis methods in a surrogate-based design optimization environment. Struct. Multidiscip. Optim. (2011)
2. P. Angelikopoulos, C. Papadimitriou, P. Koumoutsakos, Bayesian uncertainty quantification and propagation in molecular dynamics simulations: a high performance computing framework. J. Chem. Phys. **137**, 1441103-1–144103-19 (2012)
3. P. Angelikopoulos, C. Papadimitriou, P. Koumoutsakos, X-TMCMC: adaptive kriging for Bayesian inverse modeling. Comput. Methods Appl. Mech. Eng. **289**, 409–428 (2015)
4. S.K. Au, F.L. Zhang, Fundamental two-stage formulation for Bayesian system identification, part I: general theory. Mech. Syst. Signal Process. **66–67**, 31–42 (2016)

5. I. Behmanesh, B. Moaveni, G. Lombaert, C. Papadimitriou, Hierarchical Bayesian model updating for structural identification. Mech. Syst. Signal Process. **64–65**, 360376 (2015)
6. G. Falsone, N. Impollonia, A new approach for the stochastic analysis of finite element modelled structures with uncertain parameters. Comput. Methods Appl. Mech. Eng. **191**(44), 5067–5085 (2002)
7. O. Giannini, M. Hanss, The component mode transformation method: a fast implementation of fuzzy arithmetic for uncertainty management in structural dynamics. J. Sound Vib. **3311**, 1340–1357 (2008)
8. B. Goller, M. Broggi, A. Calvi, G.I. Schuëller, A stochastic model updating technique for complex aerospace structures. Finite Elem. Anal. Des. **47**, 739–752 (2011)
9. B. Goller, H.J. Pradlwarter, G.I. Schuëller, An interpolation scheme for the approximation of dynamical systems. Comput. Methods Appl. Mech. Eng. **200**, 414–423 (2011)
10. L. Hinke, F. Dohnal, B.R. Mace, T.P. Waters, N.S. Ferguson, Component mode synthesis as a framework for uncertainty analysis. J. Sound Vib. **324**, 161–178 (2009)
11. S.K. Hong, B.I. Epureanu, M.P. Castanier, D. Gorsich, Parametric reduced-order models for prediction the vibration response of complex structures with component damage and uncertainties. J. Sound Vib. **330**(6), 1091–1110 (2011)
12. Y. Huang, J.L. Beck, Hierarchical sparse Bayesian learning for structural health monitoring with incomplete modal data. Int. J. Uncertain. Quantif. **5**(2), 139169 (2015)
13. H.A. Jensen, E. Millas, D. Kusanovic, C. Papadimitriou, Model-reduction techniques for Bayesian finite element model updating using dynamic response data. Comput. Methods Appl. Mech. Eng. **279**, 301–324 (2014)
14. H.A. Jensen, F. Mayorga, C. Papadimitriou, Reliability sensitivity analysis of stochastic finite element models. Comput. Methods Appl. Mech. Eng. **296**, 327–353 (2015)
15. H.A. Jensen, A. Muñoz, C. Papadimitriou, C. Vergara, An enhanced substructure coupling technique for dynamic re-analyses: application to simulation-based problems. Comput. Methods Appl. Mech. Eng. **307**, 215–234 (2016)
16. H.A. Jensen, A. Muñoz, C. Papadimitriou, E. Millas, Model reduction techniques for reliability-based design problems of complex structural systems. Reliab. Eng. Syst. Saf. **149**, 204–217 (2016)
17. H.A. Jensen, C. Esse, V. Araya, C. Papadimitriou, Implementation of an adaptive meta-model for Bayesian finite element model updating in time domain. Reliab. Eng. Syst. Saf. **160**, 174–190 (2017)
18. H.A. Jensen, V. Araya, A. Muñoz, M. Valdebenito, A physical domain-based substructuring as a framework for dynamic modeling and reanalysis of systems. Comput. Methods Appl. Mech. Eng. **326**, 656–678 (2017)
19. A. Kundu, F.A. DiazDelaO, S. Adhikari, M.I. Friswell, A hybrid spectral and metamodeling approach for the stochastic finite element analysis of structural dynamic systems. Comput. Methods Appl. Mech. Eng. **270**, 201–219 (2014)
20. G. Muscolino, A. Sofi, Analysis of structures with random axial stiffness described by imprecise probability density functions. Comput. Struct. **184**, 1–13 (2017)
21. J.B. Nagel, B. Sudret, A unified framework for multilevel uncertainty quantification in Bayesian inverse problems. Probab. Eng. Mech. **43**, 68–84 (2016)
22. C. Papadimitriou, Bayesian uncertainty quantification and propagation in structural dynamics simulations, in *Proceedings of the 9th International Conference on Structural Dynamics, EURODYN2014*, Porto, Portugal, 30 June- 2 July 2014
23. C. Papadimitriou, D.Ch. Papadioti, Component mode synthesis techniques for finite element model updating. Comput. Struct. **126**, 15–28 (2013)
24. M.F. Pellissetti, R.G. Ghanem, Iterative solution of systems of linear equations arising in the context of stochastic finite elements. Adv. Eng. Softw. **31**(8–9), 607–616 (2000)
25. H.J. Pradlwarter, G.I. Schuëller, G.S. Szekely, Random eigenvalue problems for large systems. Comput. Struct. **80**, 2415–2424 (2002)
26. D. Straub, I. Papaioannou, Bayesian updating with structural reliability methods. J. Eng. Mech. **141**(3), 04014134 (2015)

27. J. Wang, L. Katafygiotis, Z. Feng, An efficient simulation method for the first excursion problem of linear structures subjected to stochastic wind loads. Computers and Structures, accepted for publication, 2015
28. K.V. Yuen, S.C. Kuok, Bayesian methods for updating dynamic models. Appl. Mech. Rev. **64**, 010802 (2011)
29. F.L. Zhang, S.K. Au, Fundamental two-stage formulation for Bayesian system identification, part II: application to ambient vibration data. Mech. Syst. Signal Process. **66–67**, 43–61 (2016)
30. E. Zio, N. Pedroni, Monte Carlo simulation-based sensitivity analysis of the model of a thermal-hydraulic passive system. Reliab. Eng. Syst. Saf. **107**, 90–106 (2012)
31. K.M. Zuev, J. Beck, Global optimization using the asymptotically independent Markov sampling method. Comput. Struct. **80**, 2415–2424 (2002)

Chapter 3
Parametrization of Reduced-Order Models Based on Global Interface Reduction

Abstract An interpolation scheme for approximating the interface modes in terms of the model parameters is presented in this chapter. The approximation scheme involves a set of support points in the model parameters space and a number of interpolation coefficients that are determined by the singular value decomposition technique. The approximate interface modes are combined with the parametrization scheme introduced in Chap. 2 to derive the corresponding reduced-order matrices. Pseudo-codes are provided to illustrate how the interface modes are approximated and how the parametrization of the reduced-order matrices is constructed based on interface reduction.

3.1 Meta-Model for Global Interface Modes

The assumption that the substructure matrices depend on only one model parameter is no longer valid for the interface matrices $\mathbf{M}_I$ and $\mathbf{K}_I$, defined in Eqs. (1.45) and (1.49), respectively. In general, these matrices depend on the entire set of model parameters $\boldsymbol{\theta}$ [12, 15]. Therefore, the parametrization scheme presented in the previous chapter can no longer be applied to the interface modes. In other words, a direct interface analysis must be performed for each new sample during the corresponding simulation processes. In this context, different strategies can be considered in order to avoid direct evaluation of the interface quantities for different samples of $\boldsymbol{\theta}$. For example, the interface modes can be considered constant and can be updated every few iterations during the analyses. Another possibility considered in this chapter is to use an interpolation scheme to approximate the interface modes in terms of the model parameters.

H. Jensen and C. Papadimitriou, *Sub-structure Coupling for Dynamic Analysis*,
Lecture Notes in Applied and Computational Mechanics 89,
https://doi.org/10.1007/978-3-030-12819-7_3

3.1.1 Baseline Information

It is assumed that the interface matrices $\mathbf{M}_I$ and $\mathbf{K}_I$, defined in Eqs. (1.45) and (1.49), respectively, have been assembled at L support points in the model parameters space, or $\boldsymbol{\theta}^l, l = 1, \ldots, L$, and the associated eigenvalue problems

$$\mathbf{K}_I(\boldsymbol{\theta}^l)\boldsymbol{\Upsilon}_I(\boldsymbol{\theta}^l) - \mathbf{M}_I(\boldsymbol{\theta}^l)\boldsymbol{\Upsilon}_I(\boldsymbol{\theta}^l)\boldsymbol{\Omega}_I(\boldsymbol{\theta}^l) = \mathbf{0} \ , \ l = 1, \ldots, L \tag{3.1}$$

have been solved. In addition, the nominal solution for $\boldsymbol{\theta}^0$ has also been computed. The support points $\boldsymbol{\theta}^l, l = 1, \ldots, L$ are distributed around the nominal point $\boldsymbol{\theta}^0$. If the definition of the interface matrices and the parametrization of the matrices $\hat{\mathbf{M}}^s_{bb}$ and $\hat{\mathbf{K}}^s_{bb}$, given in Eq. (2.18), are considered, $\mathbf{M}_I$ and $\mathbf{K}_I$ evaluated at the support point $\boldsymbol{\theta}^l$ can be expressed as

$$\mathbf{M}_I(\boldsymbol{\theta}^l) = \tilde{\mathbf{T}}^T[\hat{\hat{\mathbf{M}}}^1_{bb}\delta_{10}, \ldots, \hat{\hat{\mathbf{M}}}^{N_s}_{bb}\delta_{N_s0}]\tilde{\mathbf{T}} + \sum_{j=1}^{n_\theta} \tilde{\mathbf{T}}^T[\hat{\hat{\mathbf{M}}}^1_{bb}\delta_{1j}, \ldots, \hat{\hat{\mathbf{M}}}^{N_s}_{bb}\delta_{N_sj}]\tilde{\mathbf{T}}g^j(\theta^l_j) \tag{3.2}$$

$$\mathbf{K}_I(\boldsymbol{\theta}^l) = \tilde{\mathbf{T}}^T[\hat{\hat{\mathbf{K}}}^1_{bb}\delta_{10}, \ldots, \hat{\hat{\mathbf{K}}}^{N_s}_{bb}\delta_{N_s0}]\tilde{\mathbf{T}} + \sum_{j=1}^{n_\theta} \tilde{\mathbf{T}}^T[\hat{\hat{\mathbf{K}}}^1_{bb}\delta_{1j}, \ldots, \hat{\hat{\mathbf{K}}}^{N_s}_{bb}\delta_{N_sj}]\tilde{\mathbf{T}}h^j(\theta^l_j) \tag{3.3}$$

where θ^l_j is the jth component of the support point $\boldsymbol{\theta}^l$.

3.1.2 Approximation of Interface Modes

To derive an approximation for the n_{IR} kept interface modes, the matrix $\boldsymbol{\Upsilon}_I$ evaluated at a point $\boldsymbol{\theta}^* = (1-\xi_l)\boldsymbol{\theta}^0 + \xi_l\boldsymbol{\theta}^l$, $0 \le \xi_l \le 1$ is first approximated by a linear interpolation as [7]

$$\hat{\boldsymbol{\Upsilon}}_I(\theta^*) = (1-\xi_l)\boldsymbol{\Upsilon}_I(\boldsymbol{\theta}^0) + \xi_l\boldsymbol{\Upsilon}_I(\boldsymbol{\theta}^l) \tag{3.4}$$

If the previous expansion is generalized for the case where a set of L support points $\boldsymbol{\theta}^l, l = 1, \ldots, L$ is available, the matrix $\boldsymbol{\Upsilon}_I$ evaluated at a sample point $\boldsymbol{\theta}^k$ can be approximated by

$$\hat{\boldsymbol{\Upsilon}}_I(\boldsymbol{\theta}^k) = (1 - \sum_{l=1}^{L}\xi^k_l)\boldsymbol{\Upsilon}_I(\boldsymbol{\theta}^0) + \sum_{l=1}^{L}\xi^k_l\boldsymbol{\Upsilon}_I(\boldsymbol{\theta}^l) \tag{3.5}$$

where the coefficient ξ_l^k represents the contribution of the support point $\boldsymbol{\theta}^l$ to the simulation point $\boldsymbol{\theta}^k$. In order to consider only interpolations, the simulation point $\boldsymbol{\theta}^k$ should belong to the n_θ-dimensional convex hull of the support points [1, 3, 17, 18]. The previous expression is not used directly, since the approximated interface modes are not obtained directly from the solution of an eigenvalue problem; instead, they are used as the subspace to span the interface modes. In other words, the approximate eigenvectors $\boldsymbol{\Upsilon}_I(\boldsymbol{\theta}^k)$ are defined as a linear combination of the vectors composing the matrix $\hat{\boldsymbol{\Upsilon}}_I(\boldsymbol{\theta}^k)$, that is,

$$\boldsymbol{\Upsilon}_I(\boldsymbol{\theta}^k) = \hat{\boldsymbol{\Upsilon}}_I(\boldsymbol{\theta}^k)\mathbf{Q}(\boldsymbol{\theta}^k) \tag{3.6}$$

where $\mathbf{Q}(\boldsymbol{\theta}^k) \in R^{n_{IR} \times n_{IR}}$ is an auxiliary transformation matrix. Based on the previous transformation and the definition of the interface eigenvalue problem in (1.54), it can be seen that the auxiliary matrix $\mathbf{Q}(\boldsymbol{\theta}^k)$ can be obtained from the solution of the reduced eigenvalue problem

$$\left[\hat{\boldsymbol{\Upsilon}}_I^T(\boldsymbol{\theta}^k)\mathbf{K}_I(\boldsymbol{\theta}^k)\hat{\boldsymbol{\Upsilon}}_I(\boldsymbol{\theta}^k)\right]\mathbf{Q}(\boldsymbol{\theta}^k) = \left[\hat{\boldsymbol{\Upsilon}}_I^T(\boldsymbol{\theta}^k)\mathbf{M}_I(\boldsymbol{\theta}^k)\hat{\boldsymbol{\Upsilon}}_I(\boldsymbol{\theta}^k)\right]\mathbf{Q}(\boldsymbol{\theta}^k)\boldsymbol{\Omega}_I(\boldsymbol{\theta}^k) \tag{3.7}$$

where the matrices

$$\hat{\boldsymbol{\Upsilon}}_I^T(\boldsymbol{\theta}^k)\mathbf{K}_I(\boldsymbol{\theta}^k)\hat{\boldsymbol{\Upsilon}}_I(\boldsymbol{\theta}^k) \tag{3.8}$$

and

$$\hat{\boldsymbol{\Upsilon}}_I^T(\boldsymbol{\theta}^k)\mathbf{M}_I(\boldsymbol{\theta}^k)\hat{\boldsymbol{\Upsilon}}_I(\boldsymbol{\theta}^k) \tag{3.9}$$

are of dimension equal to $n_{IR} \times n_{IR}$. Note that the interface matrices evaluated at the sample point $\boldsymbol{\theta}^k$ can be written as in Eqs. (3.2) and (3.3), that is,

$$\mathbf{M}_I(\boldsymbol{\theta}^k) = \tilde{\mathbf{T}}^T[\hat{\mathbf{M}}_{bb}^1\delta_{10}, \ldots, \hat{\mathbf{M}}_{bb}^{N_s}\delta_{N_s0}]\tilde{\mathbf{T}} + \sum_{j=1}^{n_\theta}\tilde{\mathbf{T}}^T[\hat{\mathbf{M}}_{bb}^1\delta_{1j}, \ldots, \hat{\mathbf{M}}_{bb}^{N_s}\delta_{N_sj}]\tilde{\mathbf{T}}g^j(\theta_j^k) \tag{3.10}$$

$$\mathbf{K}_I(\boldsymbol{\theta}^k) = \tilde{\mathbf{T}}^T[\hat{\mathbf{K}}_{bb}^1\delta_{10}, \ldots, \hat{\mathbf{K}}_{bb}^{N_s}\delta_{N_s0}]\tilde{\mathbf{T}} + \sum_{j=1}^{n_\theta}\tilde{\mathbf{T}}^T[\hat{\mathbf{K}}_{bb}^1\delta_{1j}, \ldots, \hat{\mathbf{K}}_{bb}^{N_s}\delta_{N_sj}]\tilde{\mathbf{T}}h^j(\theta_j^k) \tag{3.11}$$

where θ_j^k is the j component of the sample point $\boldsymbol{\theta}^k$. The solution of the previous reduced eigenvalue problem, together with Eq. (3.6), provides an approximation for the interface modes at the sample point $\boldsymbol{\theta}^k$. In addition, the reduced eigenvalue problem (3.7) gives an approximation of the eigenvalues $\boldsymbol{\Omega}_I(\boldsymbol{\theta}^k)$.

3.1.3 Determination of Interpolation Coefficients

As indicated before, the interpolation coefficients represent the contribution of the support points to the new simulation point. In order to obtain the interpolation coefficients, the norm of the difference between the support points $\boldsymbol{\theta}^l, l = 1, \ldots, L$ and the simulation point $\boldsymbol{\theta}^k$ is first minimized, that is,

$$\text{Min}_{l=1,\ldots,L} \quad \| \boldsymbol{\theta}^l - \boldsymbol{\theta}^k \| \tag{3.12}$$

If the nearest point to $\boldsymbol{\theta}^k$ is denoted by $\boldsymbol{\theta}^q, q \in \{1, \ldots, L\}$, the corresponding interpolation coefficient ξ_q^k is obtained by projecting $\boldsymbol{\theta}^k - \boldsymbol{\theta}^0$ onto $\boldsymbol{\theta}^q - \boldsymbol{\theta}^0$, which yields

$$\xi_q^k = \frac{(\boldsymbol{\theta}^k - \boldsymbol{\theta}^0)^T \cdot (\boldsymbol{\theta}^q - \boldsymbol{\theta}^0)}{\| (\boldsymbol{\theta}^q - \boldsymbol{\theta}^0) \|^2} \tag{3.13}$$

The remaining part of the vector, or the component perpendicular to $\boldsymbol{\theta}^q - \boldsymbol{\theta}^0$, is given by

$$\mathbf{v}^k = (\boldsymbol{\theta}^k - \boldsymbol{\theta}^0) - \xi_q^k (\boldsymbol{\theta}^q - \boldsymbol{\theta}^0) \tag{3.14}$$

This vector is then represented as a linear combination of the remaining support points $\boldsymbol{\theta}^l, l = 1, \ldots, L, l \neq q$ through

$$\mathbf{v}^k = [\boldsymbol{\theta}^1 - \boldsymbol{\theta}^0, \ldots, \boldsymbol{\theta}^{q-1} - \boldsymbol{\theta}^0, \boldsymbol{\theta}^{q+1} - \boldsymbol{\theta}^0, \ldots, \boldsymbol{\theta}^L - \boldsymbol{\theta}^0] \, \boldsymbol{\tau}^k \tag{3.15}$$

where the components of the vector $\boldsymbol{\tau}^k$ are given by

$$\boldsymbol{\tau}^k = < \tau_1^k, \ldots, \tau_{q-1}^k, \tau_{q+1}^k, \ldots, \tau_L^k >^T \tag{3.16}$$

The coefficients are obtained as the solution to Eq. (3.15), which is solved by using the singular value decomposition (SVD) technique [6, 13, 16]. This technique has the advantage of being applicable to cases of under- and over-determined system of equations. The solution for the coefficients $\boldsymbol{\xi}^k, \xi_l^k, l = 1, \ldots, L$ is then obtained by considering the two parts. This gives

$$\boldsymbol{\xi}^k = < \tau_1^k, \ldots, \tau_{q-1}^k, \xi_q^k, \tau_{q+1}^k, \ldots, \tau_L^k >^T \tag{3.17}$$

The interpolation scheme above guarantees that the approximation is exact in each support point. In fact, if $\boldsymbol{\theta}^k = \boldsymbol{\theta}^l$, where $\boldsymbol{\theta}^l$ is one of the support points, then $\xi_l^k = 1$ and $\mathbf{v}^k = \mathbf{0}$, and therefore $\boldsymbol{\tau}^k = \mathbf{0}$. This result guarantees that $\hat{\boldsymbol{\Upsilon}}_I(\boldsymbol{\theta}^k) = \boldsymbol{\Upsilon}_I(\boldsymbol{\theta}^l)$ and $\mathbf{Q}(\boldsymbol{\theta}^k) = \mathbf{I}$, and therefore

$$\boldsymbol{\Upsilon}_I(\boldsymbol{\theta}^k) = \boldsymbol{\Upsilon}_I(\boldsymbol{\theta}^l) \tag{3.18}$$

It can be seen from the previous formulation that the potential time-consuming step of computing the interface modes for different values of the model parameters only has to be performed for the support points. The accuracy of this global surrogate model can be increased by densifying the region of interest in the model parameter space with additional support points. Alternatively, higher-order schemes, such as the one considered in the next section, can also be used.

3.1.4 Higher-Order Approximations

If the dependence of the interface modes is nonlinear with respect to the model parameters, a linear interpolation scheme may be insufficient for accuracy. To improve the accuracy of the approximations, higher-order interpolation schemes can be used. For a linear interpolation, the approximation is based on the eigensolution of L support points $\boldsymbol{\theta}^l, l = 1, \ldots, L$, and at the nominal value $\boldsymbol{\theta}^0$. For example, in the case of a quadratic interpolation scheme the eigensolutions for the support points $\boldsymbol{\theta}^{(-l)}, l = 1, \ldots, L$ are also needed. These points are defined as the symmetric points with respect to $\boldsymbol{\theta}^0$, that is, $\boldsymbol{\theta}^{(-l)} = \boldsymbol{\theta}^0 - (\boldsymbol{\theta}^l - \boldsymbol{\theta}^0), l = 1, \ldots, L$. Analogous to the linear interpolation, the quadratic approximation of the interface modes evaluated at point $\boldsymbol{\theta}^* = \frac{\xi_l(\xi_l-1)}{2}(\boldsymbol{\theta}^{(-l)}) + (1-\xi_l^2)\boldsymbol{\theta}^0 + \frac{\xi_l(\xi_l+1)}{2}(\boldsymbol{\theta}^l), 1 \leq \xi_l \leq 1$ is given by [7]

$$\hat{\boldsymbol{\Upsilon}}_I(\boldsymbol{\theta}^*) = \frac{1}{2}\xi_l(\xi_l - 1)\boldsymbol{\Upsilon}_I(\boldsymbol{\theta}^{(-l)}) + (1-\xi_l^2)\boldsymbol{\Upsilon}_I(\boldsymbol{\theta}^0) + \frac{1}{2}\xi_l(\xi_l+1)\boldsymbol{\Upsilon}_I(\boldsymbol{\theta}^l) \quad (3.19)$$

If these results are generalized to the case where the set of support points $\{\boldsymbol{\theta}^l, \boldsymbol{\theta}^{(-l)}, l = 1, \ldots, L\}$ is available, the interface modes evaluated at the simulation point $\boldsymbol{\theta}^k$ can be written in the form

$$\hat{\boldsymbol{\Upsilon}}_I(\boldsymbol{\theta}^k) = \sum_{l=1}^{L} \frac{1}{2}\xi_l^k(\xi_l^k - 1)\boldsymbol{\Upsilon}_I(\boldsymbol{\theta}^{(-l)}) + (1 - \sum_{l=1}^{L} \xi_l^{k^2})\boldsymbol{\Upsilon}_I(\boldsymbol{\theta}^0) + \sum_{l=1}^{L} \frac{1}{2}\xi_l^k(\xi_l^k + 1)\boldsymbol{\Upsilon}_I(\boldsymbol{\theta}^l) \quad (3.20)$$

where the coefficient ξ_l^k represents the contribution of the support point $\boldsymbol{\theta}^l$ to the simulation point $\boldsymbol{\theta}^k$. The coefficients $\xi_l^k, l = 1, \ldots, L$ are obtained as described in Sect. 3.1.3, except for the computation of the nearest point to $\boldsymbol{\theta}^k$. In this case, the minimum distance between the simulation point $\boldsymbol{\theta}^k$ and the $2L$ support points, or $\{\boldsymbol{\theta}^l, \boldsymbol{\theta}^{(-l)}, l = 1, \ldots, L\}$, is determined by

$$\mathrm{Min}_{l=\pm 1,\ldots,\pm L} \quad \| \boldsymbol{\theta}^l - \boldsymbol{\theta}^k \| \quad (3.21)$$

If the nearest point $\boldsymbol{\theta}^q$ corresponds to one of the support points in the set $\{\boldsymbol{\theta}^{(-l)}, l = 1, \ldots, L\}$, the corresponding interpolation coefficient ξ_q^k is obtained as

$$\xi_q^k = -\frac{(\boldsymbol{\theta}^k - \boldsymbol{\theta}^0)^T \cdot (\boldsymbol{\theta}^q - \boldsymbol{\theta}^0)}{\| (\boldsymbol{\theta}^q - \boldsymbol{\theta}^0) \|^2} \tag{3.22}$$

The remaining coefficients $\xi_l^k, l = 1 \ldots, L, l \neq q$ are obtained as in Eq. (3.15). As discussed in Sect. 3.1.2, the vectors composing the matrix $\hat{\boldsymbol{\Upsilon}}_I(\boldsymbol{\theta}^k)$ are used as the subspace to span the approximated interface modes $\boldsymbol{\Upsilon}_I(\boldsymbol{\theta}^k)$ (Eq. (3.6)).

3.1.5 *Support Points*

As previously mentioned, the support points are distributed around the nominal point $\boldsymbol{\theta}^0$. The number of support points is problem-dependent and based on several factors, such as the level of accuracy to be reached, the dimension of the uncertain parameter space, and the range of the uncertain model parameters. Different approaches can be considered for choosing the nominal point and the corresponding support points. For example, the nominal point may correspond to the reference model of the structure, or it can be chosen as the mean value of the uncertain model parameters. Then, the support points can be generated by a number of sampling methods, such as random sampling, Latin Hypercube sampling, orthogonal sampling, etc. [2, 4, 8–10, 14, 16]. Alternatively, adaptive schemes can also be devised, in which the nominal and support points are updated during the corresponding simulation processes in order to improve convergence and maintain accuracy [1, 5, 11]. The computational efficiency and accuracy of different schemes for choosing the nominal and support points is left to subsequent chapters with applications.

3.2 Numerical Implementation: Pseudo-Code No. 5

Based on the formulation presented, the following pseudo-code illustrates how the global interface modes are approximated at a sample point $\boldsymbol{\theta}^k$.

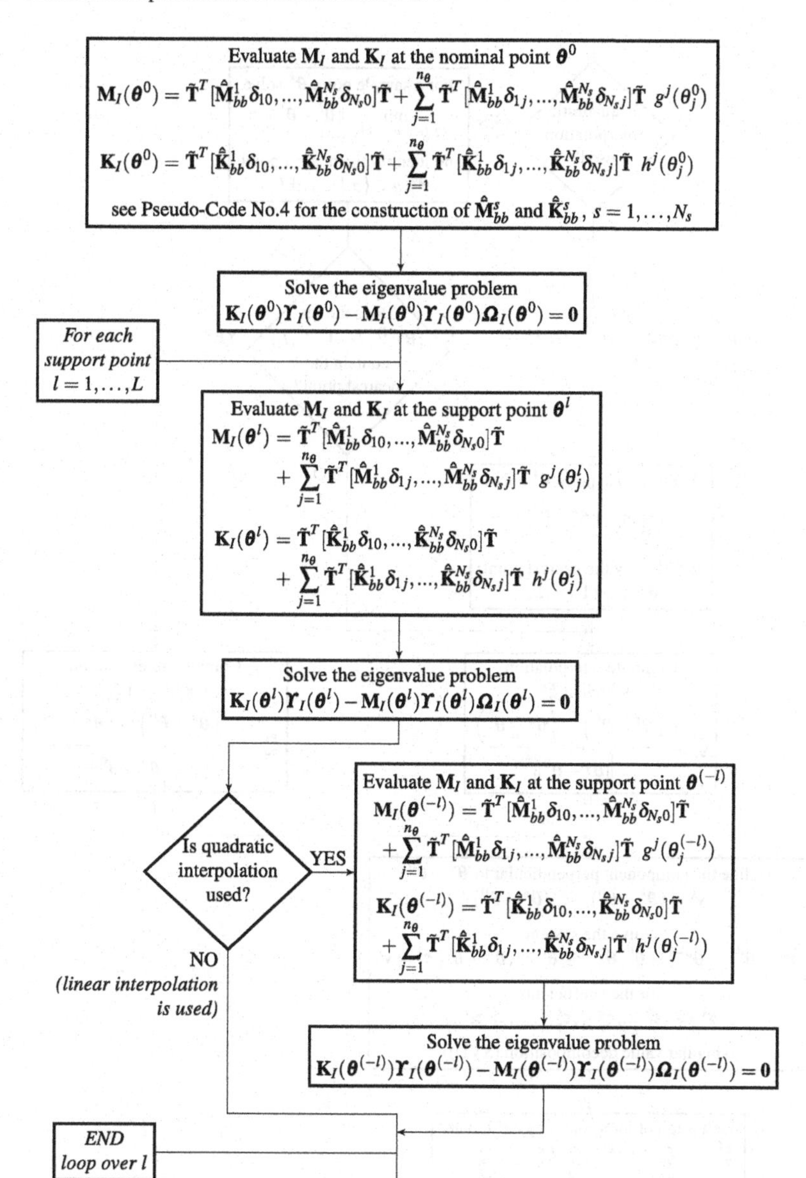

Evaluate $\mathbf{M}_I$ and $\mathbf{K}_I$ at the nominal point $\boldsymbol{\theta}^0$
$\mathbf{M}_I(\boldsymbol{\theta}^0) = \tilde{\mathbf{T}}^T[\hat{\mathbf{M}}_{bb}^1\delta_{10},...,\hat{\mathbf{M}}_{bb}^{N_s}\delta_{N_s0}]\tilde{\mathbf{T}} + \sum_{j=1}^{n_\theta}\tilde{\mathbf{T}}^T[\hat{\mathbf{M}}_{bb}^1\delta_{1j},...,\hat{\mathbf{M}}_{bb}^{N_s}\delta_{N_sj}]\tilde{\mathbf{T}}\ g^j(\theta_j^0)$
$\mathbf{K}_I(\boldsymbol{\theta}^0) = \tilde{\mathbf{T}}^T[\hat{\mathbf{K}}_{bb}^1\delta_{10},...,\hat{\mathbf{K}}_{bb}^{N_s}\delta_{N_s0}]\tilde{\mathbf{T}} + \sum_{j=1}^{n_\theta}\tilde{\mathbf{T}}^T[\hat{\mathbf{K}}_{bb}^1\delta_{1j},...,\hat{\mathbf{K}}_{bb}^{N_s}\delta_{N_sj}]\tilde{\mathbf{T}}\ h^j(\theta_j^0)$
see Pseudo-Code No.4 for the construction of $\hat{\mathbf{M}}_{bb}^s$ and $\hat{\mathbf{K}}_{bb}^s$, $s = 1,\ldots,N_s$
Solve the eigenvalue problem
$\mathbf{K}_I(\boldsymbol{\theta}^0)\boldsymbol{\Upsilon}_I(\boldsymbol{\theta}^0) - \mathbf{M}_I(\boldsymbol{\theta}^0)\boldsymbol{\Upsilon}_I(\boldsymbol{\theta}^0)\boldsymbol{\Omega}_I(\boldsymbol{\theta}^0) = \mathbf{0}$
For each support point $l = 1,\ldots,L$
Evaluate $\mathbf{M}_I$ and $\mathbf{K}_I$ at the support point $\boldsymbol{\theta}^l$
$\mathbf{M}_I(\boldsymbol{\theta}^l) = \tilde{\mathbf{T}}^T[\hat{\mathbf{M}}_{bb}^1\delta_{10},...,\hat{\mathbf{M}}_{bb}^{N_s}\delta_{N_s0}]\tilde{\mathbf{T}} + \sum_{j=1}^{n_\theta}\tilde{\mathbf{T}}^T[\hat{\mathbf{M}}_{bb}^1\delta_{1j},...,\hat{\mathbf{M}}_{bb}^{N_s}\delta_{N_sj}]\tilde{\mathbf{T}}\ g^j(\theta_j^l)$
$\mathbf{K}_I(\boldsymbol{\theta}^l) = \tilde{\mathbf{T}}^T[\hat{\mathbf{K}}_{bb}^1\delta_{10},...,\hat{\mathbf{K}}_{bb}^{N_s}\delta_{N_s0}]\tilde{\mathbf{T}} + \sum_{j=1}^{n_\theta}\tilde{\mathbf{T}}^T[\hat{\mathbf{K}}_{bb}^1\delta_{1j},...,\hat{\mathbf{K}}_{bb}^{N_s}\delta_{N_sj}]\tilde{\mathbf{T}}\ h^j(\theta_j^l)$
Solve the eigenvalue problem
$\mathbf{K}_I(\boldsymbol{\theta}^l)\boldsymbol{\Upsilon}_I(\boldsymbol{\theta}^l) - \mathbf{M}_I(\boldsymbol{\theta}^l)\boldsymbol{\Upsilon}_I(\boldsymbol{\theta}^l)\boldsymbol{\Omega}_I(\boldsymbol{\theta}^l) = \mathbf{0}$
Is quadratic interpolation used?
YES
NO
(linear interpolation is used)
Evaluate $\mathbf{M}_I$ and $\mathbf{K}_I$ at the support point $\boldsymbol{\theta}^{(-l)}$
$\mathbf{M}_I(\boldsymbol{\theta}^{(-l)}) = \tilde{\mathbf{T}}^T[\hat{\mathbf{M}}_{bb}^1\delta_{10},...,\hat{\mathbf{M}}_{bb}^{N_s}\delta_{N_s0}]\tilde{\mathbf{T}} + \sum_{j=1}^{n_\theta}\tilde{\mathbf{T}}^T[\hat{\mathbf{M}}_{bb}^1\delta_{1j},...,\hat{\mathbf{M}}_{bb}^{N_s}\delta_{N_sj}]\tilde{\mathbf{T}}\ g^j(\theta_j^{(-l)})$
$\mathbf{K}_I(\boldsymbol{\theta}^{(-l)}) = \tilde{\mathbf{T}}^T[\hat{\mathbf{K}}_{bb}^1\delta_{10},...,\hat{\mathbf{K}}_{bb}^{N_s}\delta_{N_s0}]\tilde{\mathbf{T}} + \sum_{j=1}^{n_\theta}\tilde{\mathbf{T}}^T[\hat{\mathbf{K}}_{bb}^1\delta_{1j},...,\hat{\mathbf{K}}_{bb}^{N_s}\delta_{N_sj}]\tilde{\mathbf{T}}\ h^j(\theta_j^{(-l)})$
Solve the eigenvalue problem
$\mathbf{K}_I(\boldsymbol{\theta}^{(-l)})\boldsymbol{\Upsilon}_I(\boldsymbol{\theta}^{(-l)}) - \mathbf{M}_I(\boldsymbol{\theta}^{(-l)})\boldsymbol{\Upsilon}_I(\boldsymbol{\theta}^{(-l)})\boldsymbol{\Omega}_I(\boldsymbol{\theta}^{(-l)}) = \mathbf{0}$
END loop over l

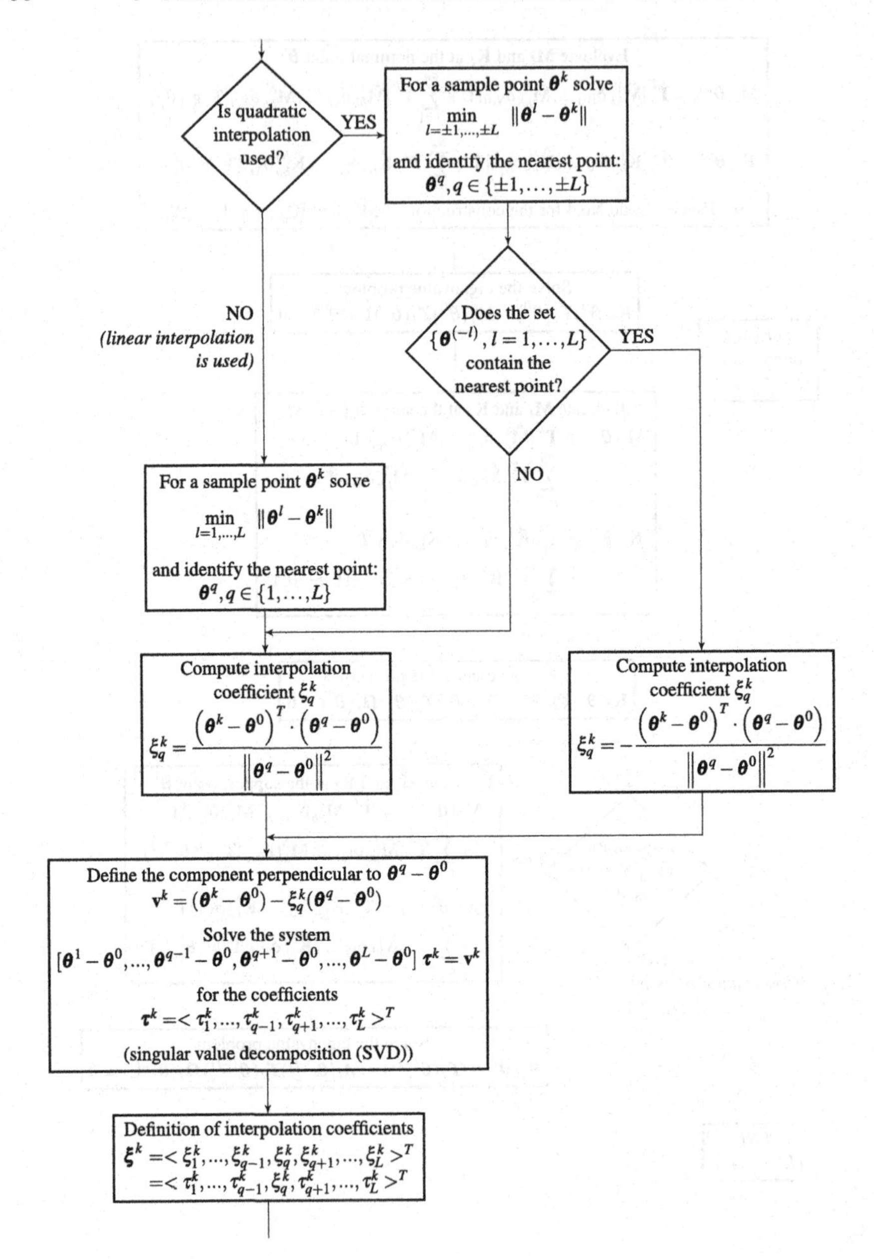
Is quadratic interpolation used?
YES
For a sample point $\boldsymbol{\theta}^k$ solve
$\min_{l=\pm1,\ldots,\pm L} \|\boldsymbol{\theta}^l - \boldsymbol{\theta}^k\|$
and identify the nearest point:
$\boldsymbol{\theta}^q, q \in \{\pm1,\ldots,\pm L\}$
NO
(linear interpolation is used)
Does the set $\{\boldsymbol{\theta}^{(-l)}, l = 1,\ldots,L\}$ contain the nearest point?
YES
NO
For a sample point $\boldsymbol{\theta}^k$ solve
$\min_{l=1,\ldots,L} \|\boldsymbol{\theta}^l - \boldsymbol{\theta}^k\|$
and identify the nearest point:
$\boldsymbol{\theta}^q, q \in \{1,\ldots,L\}$
Compute interpolation coefficient ξ_q^k
$\xi_q^k = \frac{(\boldsymbol{\theta}^k - \boldsymbol{\theta}^0)^T \cdot (\boldsymbol{\theta}^q - \boldsymbol{\theta}^0)}{\|\boldsymbol{\theta}^q - \boldsymbol{\theta}^0\|^2}$
Compute interpolation coefficient ξ_q^k
$\xi_q^k = -\frac{(\boldsymbol{\theta}^k - \boldsymbol{\theta}^0)^T \cdot (\boldsymbol{\theta}^q - \boldsymbol{\theta}^0)}{\|\boldsymbol{\theta}^q - \boldsymbol{\theta}^0\|^2}$
Define the component perpendicular to $\boldsymbol{\theta}^q - \boldsymbol{\theta}^0$
$\mathbf{v}^k = (\boldsymbol{\theta}^k - \boldsymbol{\theta}^0) - \xi_q^k(\boldsymbol{\theta}^q - \boldsymbol{\theta}^0)$
Solve the system
$[\boldsymbol{\theta}^1 - \boldsymbol{\theta}^0, \ldots, \boldsymbol{\theta}^{q-1} - \boldsymbol{\theta}^0, \boldsymbol{\theta}^{q+1} - \boldsymbol{\theta}^0, \ldots, \boldsymbol{\theta}^L - \boldsymbol{\theta}^0]\, \boldsymbol{\tau}^k = \mathbf{v}^k$
for the coefficients
$\boldsymbol{\tau}^k = < \tau_1^k, \ldots, \tau_{q-1}^k, \tau_{q+1}^k, \ldots, \tau_L^k >^T$
(singular value decomposition (SVD))
Definition of interpolation coefficients
$\boldsymbol{\xi}^k = < \xi_1^k, \ldots, \xi_{q-1}^k, \xi_q^k, \xi_{q+1}^k, \ldots, \xi_L^k >^T$
$= < \tau_1^k, \ldots, \tau_{q-1}^k, \xi_q^k, \tau_{q+1}^k, \ldots, \tau_L^k >^T$

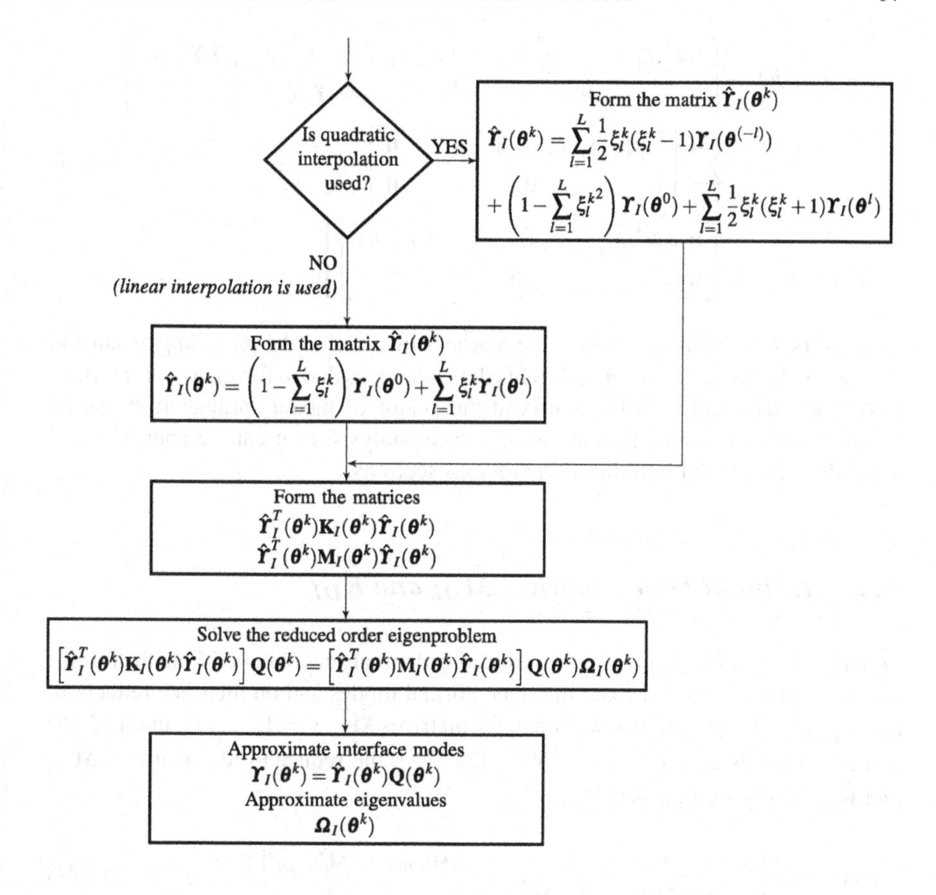

3.3 Reduced-Order Matrices Based on Global Interface Reduction

3.3.1 Transformation Matrix T_{DI}

The transformation matrix $\mathbf{T}_{DI}$ considers the effect of the dominant fixed-interface normal modes and interface reduction, and it is defined in Eq. (1.60). If the parametrization of the fixed-interface normal modes $\boldsymbol{\Phi}_{ii}^s$ given in Eq. (2.7) is considered along with the independence of the interface constraint modes $\bar{\boldsymbol{\Psi}}_{ib}^s$ on the model parameters, the transformation matrix $\mathbf{T}_{DI}$ can be expanded in the form [12]

$$\mathbf{T}_{DI}(\boldsymbol{\theta}) = \begin{bmatrix} [\bar{\boldsymbol{\Phi}}^1_{id}\delta_{10}, \ldots, \bar{\boldsymbol{\Phi}}^{N_s}_{id}\delta_{N_s0}] & [\bar{\boldsymbol{\Psi}}^1_{ib}\delta_{10}, \ldots, \bar{\boldsymbol{\Psi}}^{N_s}_{ib}\delta_{N_s0}]\tilde{\mathbf{T}}\boldsymbol{\Upsilon}_I(\boldsymbol{\theta}) \\ \mathbf{0} & \boldsymbol{\Upsilon}_I(\boldsymbol{\theta}) \end{bmatrix}$$
$$+ \sum_{j=1}^{n_\theta} \left\{ \begin{bmatrix} [\bar{\boldsymbol{\Phi}}^1_{id}\delta_{1j}, \ldots, \bar{\boldsymbol{\Phi}}^{N_s}_{id}\delta_{N_sj}] & \mathbf{0} \\ \mathbf{0} & \mathbf{0} \end{bmatrix} \frac{1}{\sqrt{g^j(\theta_j)}} \right. \tag{3.23}$$
$$\left. + \begin{bmatrix} \mathbf{0} & [\bar{\boldsymbol{\Psi}}^1_{ib}\delta_{1j}, \ldots, \bar{\boldsymbol{\Psi}}^{N_s}_{ib}\delta_{N_sj}]\tilde{\mathbf{T}}\boldsymbol{\Upsilon}_I(\boldsymbol{\theta}) \\ \mathbf{0} & \mathbf{0} \end{bmatrix} \right\}$$

It can be observed that some of the matrices involved in the preceding expansion depend on the model parameters $\boldsymbol{\theta}$. This dependence is due to the matrix of interface modes $\boldsymbol{\Upsilon}_I$. The value of this matrix at the vector of model parameters $\boldsymbol{\theta}$ can be computed directly by performing an interface analysis, or it can be approximated using the meta-model introduced in previous sections.

3.3.2 Reduced-Order Matrices $\hat{\mathbf{M}}_{DI}$ and $\hat{\mathbf{K}}_{DI}$

Equations (1.63) and (1.64) give the reduced-order matrices obtained from the formulation based on dominant fixed-interface normal modes and on interface reduction, respectively. Based on the expansion of the matrices $\hat{\mathbf{M}}^s_{ib}$, $s = 1, \ldots, N_s$ in Eq. (2.18) and the matrices $\boldsymbol{\Lambda}^s_{id}$, $s = 1, \ldots, N_s$ in Eq. (2.8), the reduced-order matrices $\hat{\mathbf{M}}_{DI}$ and $\hat{\mathbf{K}}_{DI}$ can be written as [12]

$$\hat{\mathbf{M}}_{DI}(\boldsymbol{\theta}) = \begin{bmatrix} \mathbf{I} & [\hat{\mathbf{M}}^1_{ib}\delta_{10}, \ldots, \hat{\mathbf{M}}^{N_s}_{ib}\delta_{N_s0}]\tilde{\mathbf{T}}\boldsymbol{\Upsilon}_I(\boldsymbol{\theta}) \\ \boldsymbol{\Upsilon}_I(\boldsymbol{\theta})^T\tilde{\mathbf{T}}^T[\hat{\mathbf{M}}^{1^T}_{ib}\delta_{10}, \ldots, \hat{\mathbf{M}}^{N_s^T}_{ib}\delta_{N_s0}] & \mathbf{I} \end{bmatrix} \tag{3.24}$$
$$+ \sum_{j=1}^{n_\theta} \begin{bmatrix} \mathbf{0} & [\hat{\mathbf{M}}^1_{ib}\delta_{1j}, \ldots, \hat{\mathbf{M}}^{N_s}_{ib}\delta_{N_sj}]\tilde{\mathbf{T}}\boldsymbol{\Upsilon}_I(\boldsymbol{\theta}) \\ \boldsymbol{\Upsilon}_I{}^T(\boldsymbol{\theta})\tilde{\mathbf{T}}^T[\hat{\mathbf{M}}^{1^T}_{ib}\delta_{1j}, \ldots, \hat{\mathbf{M}}^{N_s^T}_{ib}\delta_{N_sj}] & \mathbf{0} \end{bmatrix} \sqrt{g^j(\theta_j)}$$

and

$$\hat{\mathbf{K}}_{DI}(\boldsymbol{\theta}) = \begin{bmatrix} [\bar{\boldsymbol{\Lambda}}^1_{id}\delta_{10}, \ldots, \bar{\boldsymbol{\Lambda}}^{N_s}_{id}\delta_{N_s0}] & \mathbf{0} \\ \mathbf{0} & \mathbf{0} \end{bmatrix} \tag{3.25}$$
$$+ \sum_{j=1}^{n_\theta} \begin{bmatrix} [\bar{\boldsymbol{\Lambda}}^1_{id}\delta_{1j}, \ldots, \bar{\boldsymbol{\Lambda}}^{N_s}_{id}\delta_{N_sj}] & \mathbf{0} \\ \mathbf{0} & \mathbf{0} \end{bmatrix} \frac{h^j(\theta_j)}{g^j(\theta_j)}$$
$$+ \begin{bmatrix} \mathbf{0} & \mathbf{0} \\ \mathbf{0} & \boldsymbol{\Omega}_I(\boldsymbol{\theta}) \end{bmatrix}$$

As in Eq. (3.23), some of the matrices involved in characterizing $\hat{\mathbf{M}}_{DI}$ and $\hat{\mathbf{K}}_{DI}$ depend on the model parameters $\boldsymbol{\theta}$. In this case, the dependence is due to the matrix of interface modes $\boldsymbol{\Upsilon}_I$ and the corresponding matrix of eigenvalues $\boldsymbol{\Omega}_I$. As previously pointed out, these can be either directly computed or approximated using the meta-model introduced in preceding sections.

3.3.3 Transformation Matrix $\mathbf{T}_{RI}$

The transformation matrix that considers the effect of residual fixed-interface normal modes and interface reduction is shown in Eq. (1.72). In terms of the model parameters, the expansion of some of the matrices involved in the definition of $\mathbf{T}_{RI}$, that is, $\bar{\mathbf{F}}$, $\tilde{\mathbf{M}}_{ib}$, and $\boldsymbol{\Lambda}$ are given in Eqs. (2.22), (2.23), and (2.26), respectively. Furthermore, based on the parametrization of the matrices $\hat{\mathbf{M}}^{s}_{ib}, s = 1, \ldots, N_s$ in Eq. (2.18), the matrix $\mathbf{M}_{iIR}$ in Eq. (1.71) can be written as

$$\mathbf{M}_{iIR}(\boldsymbol{\theta}) = [\hat{\mathbf{M}}^{1}_{ib}\delta_{10}, \ldots, \hat{\mathbf{M}}^{N_s}_{ib}\delta_{N_s0}]\tilde{\mathbf{T}}\boldsymbol{\Upsilon}_I(\boldsymbol{\theta}) + \sum_{j=1}^{n_\theta}[\hat{\mathbf{M}}^{1}_{ib}\delta_{1j}, \ldots, \hat{\mathbf{M}}^{N_s}_{ib}\delta_{N_sj}]\tilde{\mathbf{T}}\boldsymbol{\Upsilon}_I(\boldsymbol{\theta})\sqrt{g^j(\theta_j)} \tag{3.26}$$

Thus, the basic matrices included in the transformation matrix $\mathbf{T}_{RI}$ can be expressed in terms of a set of matrices that are independent of the values of the vector of model parameters $\boldsymbol{\theta}$, together with the matrices corresponding to the interface modes and eigenvalues, or $\boldsymbol{\Upsilon}_I(\boldsymbol{\theta})$ and $\boldsymbol{\Omega}_I(\boldsymbol{\theta})$.

3.3.4 Reduced-Order Matrices $\hat{\mathbf{M}}_{RI}$ and $\hat{\mathbf{K}}_{RI}$

The enhanced reduced-order mass matrix $\hat{\mathbf{M}}_{RI}$ and reduced-order stiffness matrix $\hat{\mathbf{K}}_{RI}$ are defined in Eqs. (1.74) and (1.75), respectively. These matrices consider interface reduction and the contribution of residual fixed-interface normal modes. If the expansion of the matrices $\hat{\mathbf{M}}$, $\hat{\mathbf{K}}$, and $\mathbf{T}_{DI}$ in Eqs. (2.13), (2.14) and (3.23), respectively, are considered along with the characterization of the different matrices involved in the definition of $\mathbf{T}_{RI}$, $\hat{\mathbf{M}}_{RI}$ and $\hat{\mathbf{K}}_{RI}$ can be written in terms of a set of matrices that are independent of the values of the vector of model parameters $\boldsymbol{\theta}$, the parametrization functions $h^j(\theta_j)$ and $g^j(\theta_j)$, and the interface matrices $\boldsymbol{\Upsilon}_I(\boldsymbol{\theta})$ and $\boldsymbol{\Omega}_I(\boldsymbol{\theta})$. The specific expansion of these matrices in terms of the model parameters is somewhat complex, and is thus not presented here for the sake of simplicity in notation. For illustration purposes, the expansion of the enhanced reduced-order matrices $\hat{\mathbf{M}}_{RI}$ and $\hat{\mathbf{K}}_{RI}$ is provided in the next section for one specific case.

3.3.5 Expansion of $\hat{\mathbf{M}}_{RI}$ and $\hat{\mathbf{K}}_{RI}$ Under Global Invariant Conditions of $\mathbf{T}_{RI}$

In the following, it is assumed that the mass matrix is independent of the model parameters $\boldsymbol{\theta}$. In addition, all the matrices involved in the definition of the transformation matrix $\mathbf{T}_{RI}$ are treated as invariant and are evaluated at some nominal value of the model parameters. Note that in Sect. 2.3.6, the matrices involved in the definition of $\mathbf{T}_R$ that are associated with the interface coordinates were considered dependent on the model parameters in the analysis. However, in this case, the dependence of the interface matrices on the model parameters is taken into account in the transformation matrix $\mathbf{T}_{DI}$, not in the components of $\mathbf{T}_{RI}$. The actual validation of the global invariant assumption of $\mathbf{T}_{RI}$ is left for subsequent chapters with applications. It can be observed from the definition of the enhanced reduced-order matrices in Eqs. (1.74) and (1.75) that the parametrization of these matrices depends on the expansion of the matrices

$$\hat{\mathbf{M}}_{DI}\ ,\ \mathbf{T}_{RI}^T\hat{\mathbf{M}}\mathbf{T}_{DI}\ ,\ \mathbf{T}_{RI}^T\hat{\mathbf{M}}\mathbf{T}_{RI}\ ,\ \hat{\mathbf{K}}_{DI}\ ,\ \mathbf{T}_{RI}^T\hat{\mathbf{K}}\mathbf{T}_{DI}\ ,\ \mathbf{T}_{RI}^T\hat{\mathbf{K}}\mathbf{T}_{RI} \tag{3.27}$$

The parametrization of $\hat{\mathbf{M}}_{DI}$ and $\hat{\mathbf{K}}_{DI}$ is given in Eqs. (3.24) and (3.25), respectively, with $g^j(\cdot) = 1$. Based on the previous invariant assumptions, the matrices $\mathbf{T}_{RI}^T\hat{\mathbf{M}}\mathbf{T}_{RI}$ and $\mathbf{T}_{RI}^T\hat{\mathbf{K}}\mathbf{T}_{RI}$ can be expanded as [12]

$$\mathbf{T}_{RI}^T\hat{\mathbf{M}}\mathbf{T}_{RI} = \begin{bmatrix} \mathbf{T}_{RI1}^T[\bar{\mathbf{M}}_{ii}^1, \ldots, \bar{\mathbf{M}}_{ii}^{N_s}]\mathbf{T}_{RI1} & \mathbf{T}_{RI1}^T[\bar{\mathbf{M}}_{ii}^1, \ldots, \bar{\mathbf{M}}_{ii}^{N_s}]\mathbf{T}_{RI2} \\ \mathbf{T}_{RI2}^T[\bar{\mathbf{M}}_{ii}^{1^T}, \ldots, \bar{\mathbf{M}}_{ii}^{N_s^T}]\mathbf{T}_{RI1} & \mathbf{T}_{RI2}^T[\bar{\mathbf{M}}_{ii}^1, \ldots, \bar{\mathbf{M}}_{ii}^{N_s}]\mathbf{T}_{RI2} \end{bmatrix} \tag{3.28}$$

and

$$\begin{aligned} \mathbf{T}_{RI}^T\hat{\mathbf{K}}\mathbf{T}_{RI}(\boldsymbol{\theta}) = & \begin{bmatrix} \mathbf{T}_{RI1}^T[\bar{\mathbf{K}}_{ii}^1\delta_{10}, \ldots, \bar{\mathbf{K}}_{ii}^{N_s}\delta_{N_s0}]\mathbf{T}_{RI1} & \mathbf{T}_{RI1}^T[\bar{\mathbf{K}}_{ii}^1\delta_{10}, \ldots, \bar{\mathbf{K}}_{ii}^{N_s}\delta_{N_s0}]\mathbf{T}_{RI2} \\ \mathbf{T}_{RI2}^T[\bar{\mathbf{K}}_{ii}^{1^T}\delta_{10}, \ldots, \bar{\mathbf{K}}_{ii}^{N_s^T}\delta_{N_s0}]\mathbf{T}_{RI1} & \mathbf{T}_{RI2}^T[\bar{\mathbf{K}}_{ii}^1\delta_{10}, \ldots, \bar{\mathbf{K}}_{ii}^{N_s}\delta_{N_s0}]\mathbf{T}_{RI2} \end{bmatrix} \\ & + \sum_{j=1}^{n_\theta} \begin{bmatrix} \mathbf{T}_{RI1}^T[\bar{\mathbf{K}}_{ii}^1\delta_{1j}, \ldots, \bar{\mathbf{K}}_{ii}^{N_s}\delta_{N_sj}]\mathbf{T}_{RI1} & \mathbf{T}_{RI1}^T[\bar{\mathbf{K}}_{ii}^1\delta_{1j}, \ldots, \bar{\mathbf{K}}_{ii}^{N_s}\delta_{N_sj}]\mathbf{T}_{RI2} \\ \mathbf{T}_{RI2}^T[\bar{\mathbf{K}}_{ii}^{1^T}\delta_{1j}, \ldots, \bar{\mathbf{K}}_{ii}^{N_s^T}\delta_{N_sj}]\mathbf{T}_{RI1} & \mathbf{T}_{RI2}^T[\bar{\mathbf{K}}_{ii}^1\delta_{1j}, \ldots, \bar{\mathbf{K}}_{ii}^{N_s}\delta_{N_sj}]\mathbf{T}_{RI2} \end{bmatrix} h^j(\theta_j) \end{aligned} \tag{3.29}$$

where $\mathbf{T}_{RI1}$ and $\mathbf{T}_{RI2}$ are the nonzero components of the 2×2 partitioned transformation matrix $\mathbf{T}_{RI}$ defined in Eq. (1.72), given by

$$\begin{aligned} \mathbf{T}_{RI1} &= -\bar{\mathbf{F}}\tilde{\mathbf{M}}_{ib}\tilde{\mathbf{T}}\boldsymbol{\Upsilon}_I(\mathbf{I}_I - \mathbf{M}_{iIR}^T\mathbf{M}_{iIR})^{-1}\mathbf{M}_{iIR}^T\boldsymbol{\Lambda} \\ \mathbf{T}_{RI2} &= \bar{\mathbf{F}}\tilde{\mathbf{M}}_{ib}\tilde{\mathbf{T}}\boldsymbol{\Upsilon}_I(\mathbf{I}_I - \mathbf{M}_{iIR}^T\mathbf{M}_{iIR})^{-1}\boldsymbol{\Omega}_I \end{aligned} \tag{3.30}$$

As previously mentioned, these matrices are evaluated at some nominal value of the model parameters.

Likewise, the matrices $\mathbf{T}_{RI}^T\hat{\mathbf{M}}\mathbf{T}_{DI}$ and $\mathbf{T}_{RI}^T\hat{\mathbf{K}}\mathbf{T}_{DI}$ can be written as [12]

$$\mathbf{T}_{RI}^T\hat{\mathbf{M}}\mathbf{T}_{DI}(\boldsymbol{\theta}) = \begin{bmatrix} \mathbf{T}_{RI1}^T[\bar{\mathbf{M}}_{ii}^1\bar{\boldsymbol{\Phi}}_{id}^1, \ldots, \bar{\mathbf{M}}_{ii}^{N_s}\bar{\boldsymbol{\Phi}}_{id}^{N_s}] & \mathbf{0} \\ \mathbf{T}_{RI2}^T[\bar{\mathbf{M}}_{ii}^1\bar{\boldsymbol{\Phi}}_{id}^1, \ldots, \bar{\mathbf{M}}_{ii}^{N_s}\bar{\boldsymbol{\Phi}}_{id}^{N_s}] & \mathbf{0} \end{bmatrix} + \begin{bmatrix} \mathbf{0} & \mathbf{T}_{RI1}^T[(\bar{\mathbf{M}}_{ii}^1\bar{\boldsymbol{\Psi}}_{ib}^1 + \bar{\mathbf{M}}_{ib}^1), \ldots, (\bar{\mathbf{M}}_{ii}^{N_s}\bar{\boldsymbol{\Psi}}_{ib}^{N_s} + \bar{\mathbf{M}}_{ib}^{N_s})]\tilde{\mathbf{T}}\boldsymbol{\Upsilon}_I(\boldsymbol{\theta}) \\ \mathbf{0} & \mathbf{T}_{RI2}^T[(\bar{\mathbf{M}}_{ii}^1\bar{\boldsymbol{\Psi}}_{ib}^1 + \bar{\mathbf{M}}_{ib}^1), \ldots, (\bar{\mathbf{M}}_{ii}^{N_s}\bar{\boldsymbol{\Psi}}_{ib}^{N_s} + \bar{\mathbf{M}}_{ib}^{N_s})]\tilde{\mathbf{T}}\boldsymbol{\Upsilon}_I(\boldsymbol{\theta}) \end{bmatrix} \tag{3.31}$$

and

$$\mathbf{T}_{RI}^T\hat{\mathbf{K}}\mathbf{T}_{DI}(\boldsymbol{\theta}) = \begin{bmatrix} \mathbf{T}_{RI1}^T[\bar{\mathbf{K}}_{ii}^1\bar{\boldsymbol{\Phi}}_{id}^1\delta_{10}, \ldots, \bar{\mathbf{K}}_{ii}^{N_s}\bar{\boldsymbol{\Phi}}_{id}^{N_s}\delta_{N_s0}] & \mathbf{0} \\ \mathbf{T}_{RI2}^T[\bar{\mathbf{K}}_{ii}^1\bar{\boldsymbol{\Phi}}_{id}^1\delta_{10}, \ldots, \bar{\mathbf{K}}_{ii}^{N_s}\bar{\boldsymbol{\Phi}}_{id}^{N_s}\delta_{N_s0}] & \mathbf{0} \end{bmatrix} + \sum_{j=1}^{n_\theta} \begin{bmatrix} \mathbf{T}_{RI1}^T[\bar{\mathbf{K}}_{ii}^1\bar{\boldsymbol{\Phi}}_{id}^1\delta_{1j}, \ldots, \bar{\mathbf{K}}_{ii}^{N_s}\bar{\boldsymbol{\Phi}}_{id}^{N_s}\delta_{N_sj}] & \mathbf{0} \\ \mathbf{T}_{RI2}^T[\bar{\mathbf{K}}_{ii}^1\bar{\boldsymbol{\Phi}}_{id}^1\delta_{1j}, \ldots, \bar{\mathbf{K}}_{ii}^{N_s}\bar{\boldsymbol{\Phi}}_{id}^{N_s}\delta_{N_sj}] & \mathbf{0} \end{bmatrix} h^j(\theta_j) \tag{3.32}$$

It should be noted that most of the matrices involved in the expansion of the reduced-order mass matrix $\hat{\mathbf{M}}_{RI}$ and the reduced-order stiffness matrix $\hat{\mathbf{K}}_{RI}$ are independent of the model parameters, except the matrix of interface modes $\boldsymbol{\Upsilon}_I(\boldsymbol{\theta})$. As previously mentioned, this matrix can be directly computed, or it can be approximated by using the proposed meta-model introduced in previous sections.

3.4 Numerical Implementation: Pseudo-Code No. 6

Based on the formulation presented, the following pseudo-code illustrates how the parametrization of the reduced-order matrices based on interface reduction is constructed. For the enhanced reduced-order matrices case, the invariant assumption of the mass matrix and the transformation matrix $\mathbf{T}_{RI}$ is considered.

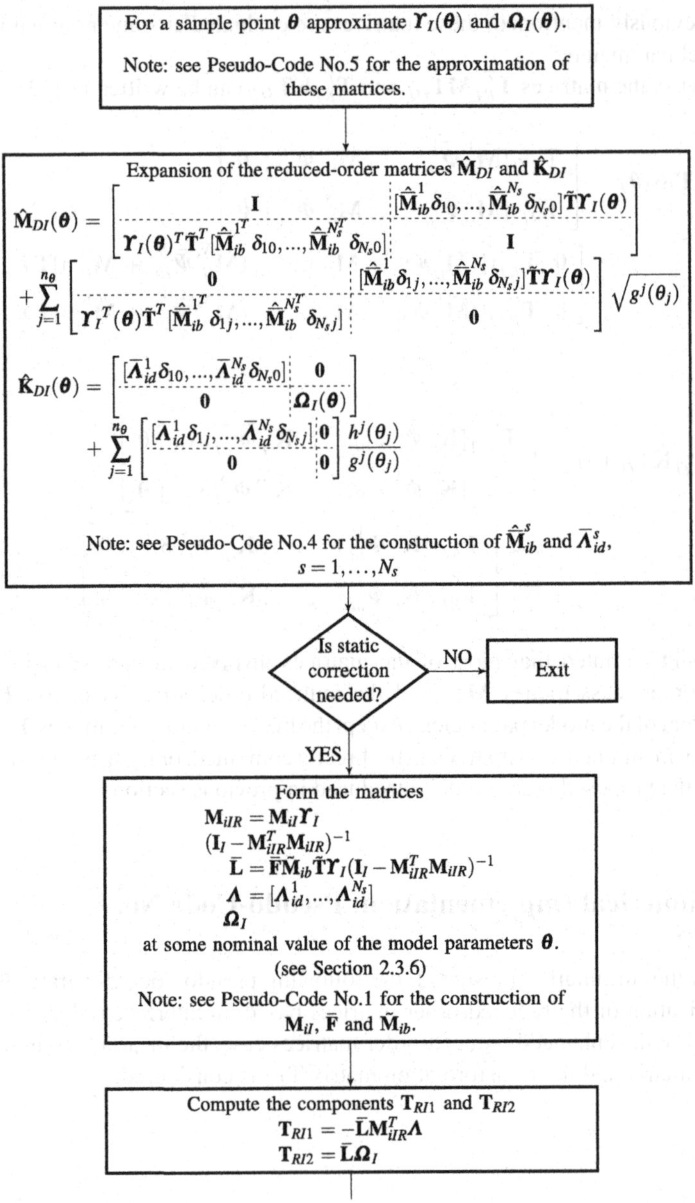

For a sample point $\boldsymbol{\theta}$ approximate $\boldsymbol{\Upsilon}_I(\boldsymbol{\theta})$ and $\boldsymbol{\Omega}_I(\boldsymbol{\theta})$.

Note: see Pseudo-Code No.5 for the approximation of these matrices.

↓

Expansion of the reduced-order matrices $\hat{\mathbf{M}}_{DI}$ and $\hat{\mathbf{K}}_{DI}$

$$\hat{\mathbf{M}}_{DI}(\boldsymbol{\theta}) = \begin{bmatrix} \mathbf{I} & [\hat{\overline{\mathbf{M}}}_{ib}^{1}\delta_{10},\ldots,\hat{\overline{\mathbf{M}}}_{ib}^{N_s}\delta_{N_s0}]\tilde{\mathbf{T}}\boldsymbol{\Upsilon}_I(\boldsymbol{\theta}) \\ \boldsymbol{\Upsilon}_I(\boldsymbol{\theta})^T\tilde{\mathbf{T}}^T[\hat{\overline{\mathbf{M}}}_{ib}^{1^T}\delta_{10},\ldots,\hat{\overline{\mathbf{M}}}_{ib}^{N_s^T}\delta_{N_s0}] & \mathbf{I} \end{bmatrix} + \sum_{j=1}^{n_\theta} \begin{bmatrix} \mathbf{0} & [\hat{\overline{\mathbf{M}}}_{ib}^{1}\delta_{1j},\ldots,\hat{\overline{\mathbf{M}}}_{ib}^{N_s}\delta_{N_sj}]\tilde{\mathbf{T}}\boldsymbol{\Upsilon}_I(\boldsymbol{\theta}) \\ \boldsymbol{\Upsilon}_I^T(\boldsymbol{\theta})\tilde{\mathbf{T}}^T[\hat{\overline{\mathbf{M}}}_{ib}^{1^T}\delta_{1j},\ldots,\hat{\overline{\mathbf{M}}}_{ib}^{N_s^T}\delta_{N_sj}] & \mathbf{0} \end{bmatrix} \sqrt{g^j(\theta_j)}$$

$$\hat{\mathbf{K}}_{DI}(\boldsymbol{\theta}) = \begin{bmatrix} [\overline{\boldsymbol{\Lambda}}_{id}^{1}\delta_{10},\ldots,\overline{\boldsymbol{\Lambda}}_{id}^{N_s}\delta_{N_s0}] & \mathbf{0} \\ \mathbf{0} & \boldsymbol{\Omega}_I(\boldsymbol{\theta}) \end{bmatrix} + \sum_{j=1}^{n_\theta} \begin{bmatrix} [\overline{\boldsymbol{\Lambda}}_{id}^{1}\delta_{1j},\ldots,\overline{\boldsymbol{\Lambda}}_{id}^{N_s}\delta_{N_sj}] & \mathbf{0} \\ \mathbf{0} & \mathbf{0} \end{bmatrix} \frac{h^j(\theta_j)}{g^j(\theta_j)}$$

Note: see Pseudo-Code No.4 for the construction of $\hat{\overline{\mathbf{M}}}_{ib}^{s}$ and $\overline{\boldsymbol{\Lambda}}_{id}^{s}$, $s = 1,\ldots,N_s$

↓

Is static correction needed?

NO → Exit

YES ↓

Form the matrices

$\mathbf{M}_{iIR} = \mathbf{M}_{iI}\boldsymbol{\Upsilon}_I$

$(\mathbf{I}_I - \mathbf{M}_{iIR}^T\mathbf{M}_{iIR})^{-1}$

$\overline{\mathbf{L}} = \overline{\mathbf{F}}\tilde{\mathbf{M}}_{ib}\tilde{\mathbf{T}}\boldsymbol{\Upsilon}_I(\mathbf{I}_I - \mathbf{M}_{iIR}^T\mathbf{M}_{iIR})^{-1}$

$\boldsymbol{\Lambda} = [\boldsymbol{\Lambda}_{id}^{1},\ldots,\boldsymbol{\Lambda}_{id}^{N_s}]$

$\boldsymbol{\Omega}_I$

at some nominal value of the model parameters $\boldsymbol{\theta}$. (see Section 2.3.6)

Note: see Pseudo-Code No.1 for the construction of $\mathbf{M}_{iI}$, $\overline{\mathbf{F}}$ and $\tilde{\mathbf{M}}_{ib}$.

↓

Compute the components $\mathbf{T}_{RI1}$ and $\mathbf{T}_{RI2}$

$\mathbf{T}_{RI1} = -\overline{\mathbf{L}}\mathbf{M}_{iIR}^T\boldsymbol{\Lambda}$

$\mathbf{T}_{RI2} = \overline{\mathbf{L}}\boldsymbol{\Omega}_I$

Expansion of the intermediate matrices $\mathbf{T}_{RI}^T\hat{\mathbf{M}}\mathbf{T}_{RI}$ and $\mathbf{T}_{RI}^T\hat{\mathbf{K}}\mathbf{T}_{RI}$

$$\mathbf{T}_{RI}^T\hat{\mathbf{M}}\mathbf{T}_{RI} = \begin{bmatrix} \mathbf{T}_{RI1}^T[\overline{\mathbf{M}}_{ii}^1,\ldots,\overline{\mathbf{M}}_{ii}^{N_s}]\mathbf{T}_{RI1} & \mathbf{T}_{RI1}^T[\overline{\mathbf{M}}_{ii}^1,\ldots,\overline{\mathbf{M}}_{ii}^{N_s}]\mathbf{T}_{RI2} \\ \mathbf{T}_{RI2}^T[\overline{\mathbf{M}}_{ii}^{1^T},\ldots,\overline{\mathbf{M}}_{ii}^{N_s^T}]\mathbf{T}_{RI1} & \mathbf{T}_{RI2}^T[\overline{\mathbf{M}}_{ii}^1,\ldots,\overline{\mathbf{M}}_{ii}^{N_s}]\mathbf{T}_{RI2} \end{bmatrix}$$

$$\mathbf{T}_{RI}^T\hat{\mathbf{K}}\mathbf{T}_{RI}(\boldsymbol{\theta}) = \begin{bmatrix} \mathbf{T}_{RI1}^T[\overline{\mathbf{K}}_{ii}^1\delta_{10},\ldots,\overline{\mathbf{K}}_{ii}^{N_s}\delta_{N_s0}]\mathbf{T}_{RI1} & \mathbf{T}_{R1}^T[\overline{\mathbf{K}}_{ii}^1\delta_{10},\ldots,\overline{\mathbf{K}}_{ii}^{N_s}\delta_{N_s0}]\mathbf{T}_{RI2} \\ \mathbf{T}_{RI2}^T[\overline{\mathbf{K}}_{ii}^{1^T}\delta_{10},\ldots,\overline{\mathbf{K}}_{ii}^{N_s^T}\delta_{N_s0}]\mathbf{T}_{RI1} & \mathbf{T}_{R2}^T[\overline{\mathbf{K}}_{ii}^1\delta_{10},\ldots,\overline{\mathbf{K}}_{ii}^{N_s}\delta_{N_s0}]\mathbf{T}_{RI2} \end{bmatrix}$$
$$+ \sum_{j=1}^{n_\theta} \begin{bmatrix} \mathbf{T}_{RI1}^T[\overline{\mathbf{K}}_{ii}^1\delta_{1j},\ldots,\overline{\mathbf{K}}_{ii}^{N_s}\delta_{N_sj}]\mathbf{T}_{RI1} & \mathbf{T}_{RI1}^T[\overline{\mathbf{K}}_{ii}^1\delta_{1j},\ldots,\overline{\mathbf{K}}_{ii}^{N_s}\delta_{N_sj}]\mathbf{T}_{RI2} \\ \mathbf{T}_{RI2}^T[\overline{\mathbf{K}}_{ii}^{1^T}\delta_{1j},\ldots,\overline{\mathbf{K}}_{ii}^{N_s^T}\delta_{N_sj}]\mathbf{T}_{RI1} & \mathbf{T}_{RI2}^T[\overline{\mathbf{K}}_{ii}^1\delta_{1j},\ldots,\overline{\mathbf{K}}_{ii}^{N_s}\delta_{N_sj}]\mathbf{T}_{RI2} \end{bmatrix} h^j(\theta_j)$$

Expansion of the intermediate matrices $\mathbf{T}_{RI}^T\hat{\mathbf{M}}\mathbf{T}_{DI}$ and $\mathbf{T}_{RI}^T\hat{\mathbf{K}}\mathbf{T}_{DI}$

$$\mathbf{T}_{RI}^T\hat{\mathbf{M}}\mathbf{T}_{DI}(\boldsymbol{\theta}) = \begin{bmatrix} \mathbf{T}_{RI1}^T[\overline{\mathbf{M}}_{ii}^1\overline{\boldsymbol{\Phi}}_{id}^1,..,\overline{\mathbf{M}}_{ii}^{N_s}\overline{\boldsymbol{\Phi}}_{id}^{N_s}] & \mathbf{0} \\ \mathbf{T}_{RI2}^T[\overline{\mathbf{M}}_{ii}^1\overline{\boldsymbol{\Phi}}_{id}^1,..,\overline{\mathbf{M}}_{ii}^{N_s}\overline{\boldsymbol{\Phi}}_{id}^{N_s}] & \mathbf{0} \end{bmatrix}$$
$$+ \begin{bmatrix} \mathbf{0} & \mathbf{T}_{RI1}^T[(\overline{\mathbf{M}}_{ii}^1\overline{\boldsymbol{\Psi}}_{ib}^1+\overline{\mathbf{M}}_{ib}^1),..,(\overline{\mathbf{M}}_{ii}^{N_s}\overline{\boldsymbol{\Psi}}_{ib}^{N_s}+\overline{\mathbf{M}}_{ib}^{N_s})]\tilde{\mathbf{T}}\boldsymbol{\Upsilon}_I(\boldsymbol{\theta}) \\ \mathbf{0} & \mathbf{T}_{RI2}^T[(\overline{\mathbf{M}}_{ii}^1\overline{\boldsymbol{\Psi}}_{ib}^1+\overline{\mathbf{M}}_{ib}^1),..,(\overline{\mathbf{M}}_{ii}^{N_s}\overline{\boldsymbol{\Psi}}_{ib}^{N_s}+\overline{\mathbf{M}}_{ib}^{N_s})]\tilde{\mathbf{T}}\boldsymbol{\Upsilon}_I(\boldsymbol{\theta}) \end{bmatrix}$$

$$\mathbf{T}_{RI}^T\hat{\mathbf{K}}\mathbf{T}_{DI}(\boldsymbol{\theta}) = \begin{bmatrix} \mathbf{T}_{RI1}^T[\overline{\mathbf{K}}_{ii}^1\overline{\boldsymbol{\Phi}}_{id}^1\delta_{10},..,\overline{\mathbf{K}}_{ii}^{N_s}\overline{\boldsymbol{\Phi}}_{id}^{N_s}\delta_{N_s0}] & \mathbf{0} \\ \mathbf{T}_{RI2}^T[\overline{\mathbf{K}}_{ii}^1\overline{\boldsymbol{\Phi}}_{id}^1\delta_{10},..,\overline{\mathbf{K}}_{ii}^{N_s}\overline{\boldsymbol{\Phi}}_{id}^{N_s}\delta_{N_s0}] & \mathbf{0} \end{bmatrix}$$
$$+ \sum_{j=1}^{n_\theta} \begin{bmatrix} \mathbf{T}_{RI1}^T[\overline{\mathbf{K}}_{ii}^1\overline{\boldsymbol{\Phi}}_{id}^1\delta_{1j},..,\overline{\mathbf{K}}_{ii}^{N_s}\overline{\boldsymbol{\Phi}}_{id}^{N_s}\delta_{N_sj}] & \mathbf{0} \\ \mathbf{T}_{RI2}^T[\overline{\mathbf{K}}_{ii}^1\overline{\boldsymbol{\Phi}}_{id}^1\delta_{1j},..,\overline{\mathbf{K}}_{ii}^{N_s}\overline{\boldsymbol{\Phi}}_{id}^{N_s}\delta_{N_sj}] & \mathbf{0} \end{bmatrix} h^j(\theta_j)$$

Parametrization of the reduced matrices $\hat{\mathbf{M}}_{RI}$ and $\hat{\mathbf{K}}_{RI}$

$$\hat{\mathbf{M}}_{RI}(\boldsymbol{\theta}) = \hat{\mathbf{M}}_{DI}(\boldsymbol{\theta}) + \mathbf{T}_{RI}^T\hat{\mathbf{M}}\mathbf{T}_{DI}(\boldsymbol{\theta}) + (\mathbf{T}_{RI}^T\hat{\mathbf{M}}\mathbf{T}_{DI})^T(\boldsymbol{\theta}) + \mathbf{T}_{RI}^T\hat{\mathbf{M}}\mathbf{T}_{RI}$$
$$\hat{\mathbf{K}}_{RI}(\boldsymbol{\theta}) = \hat{\mathbf{K}}_{DI}(\boldsymbol{\theta}) + \mathbf{T}_{RI}^T\hat{\mathbf{K}}\mathbf{T}_{DI}(\boldsymbol{\theta}) + (\mathbf{T}_{RI}^T\hat{\mathbf{K}}\mathbf{T}_{DI})^T(\boldsymbol{\theta}) + \mathbf{T}_{RI}^T\hat{\mathbf{K}}\mathbf{T}_{RI}(\boldsymbol{\theta})$$

3.5 Treatment of Local Interface Modes

The treatment of local interface modes is similar to that considered for global interface modes; that is, the same interpolation scheme introduced in previous sections can be used for local interface modes. In fact, according to Eq. (1.76), the partition matrices $\mathbf{K}_{Ill}$ and $\mathbf{M}_{Ill}$ of the interface matrices $\mathbf{K}_I$ and $\mathbf{M}_I$, associated with the physical coordinates at the interface l, or $\mathbf{u}_I^l(t)$, satisfy the eigenvalue problem

$$\mathbf{K}_{Ill}\boldsymbol{\Upsilon}_{Ill} - \mathbf{M}_{Ill}\boldsymbol{\Upsilon}_{Ill}\boldsymbol{\Omega}_{Ill} = \mathbf{0} \ , \quad l = 1, \ldots, N_l \tag{3.33}$$

where $\boldsymbol{\Upsilon}_{Ill}$ contains the selected local interface modes and $\boldsymbol{\Omega}_{Ill}$ is the diagonal matrix that contains the corresponding eigenvalues. The interpolation schemes presented in the previous sections can be applied directly to the local interface modes $\boldsymbol{\Upsilon}_{Ill}, l = 1, \ldots, N_l$. Note that, in this case, the support points $\boldsymbol{\theta}^l, l = 1, \ldots, L$ and the simulation point $\boldsymbol{\theta}^k$ involve only the model parameters associated with the corresponding interface. The number of model parameters related to a given interface is generally much smaller than the total number of model parameters n_θ. Following the procedure outlined above, the corresponding transformation matrices $\mathbf{T}_{DIL}(\boldsymbol{\theta})$ and $\mathbf{T}_{RIL}(\boldsymbol{\theta})$ in Eqs. (1.82) and (1.87), respectively, and the reduced-order matrices $\hat{\mathbf{M}}_{DIL}(\boldsymbol{\theta}), \hat{\mathbf{K}}_{DIL}(\boldsymbol{\theta}), \hat{\mathbf{M}}_{RIL}(\boldsymbol{\theta})$, and $\hat{\mathbf{K}}_{RIL}(\boldsymbol{\theta})$ in Eqs. (1.85), (1.86), (1.93), and (1.94), respectively, can be defined accordingly.

As indicated in Sect. 3.1, alternative strategies can be considered in order to avoid the direct evaluation of the interface quantities for different samples of the model parameters $\boldsymbol{\theta}$. For example, the local interface modes $\boldsymbol{\Upsilon}_{Ill}, l = 1, \ldots, N_l$ can be considered constant, independent of the model parameters. Clearly, this choice is critical to getting accurate results with the least number of interface modes over the region of variation of the model parameters associated with the interface l. In this regard, in order to improve convergence and to maintain accuracy, the local interface modes can be updated every few iterations during the simulation processes. The computational efficiency and accuracy of these simple alternative strategies is problem-dependent.

3.6 Final Remarks

The efficiency of the parametrization scheme presented in Chap. 2 is based on the assumption that the substructure matrices depend only on one model parameter. In general, the fixed-interface normal modes and interface constraint modes must be recomputed in each new sample point $\boldsymbol{\theta}^k$. When the substructure matrices depend on two or more parameters, a representation similar to Eqs. (2.20) and (2.21) is no longer applicable, and the reduced substructure matrices of the reduced-order model should be re-assembled from the substructure stiffness and mass matrices for new values of the vector of model parameters. If the repeated computation is confined to a small number of substructures, significant amounts of time, effort, and cost may still be saved, since the estimation of the fixed-interface modes and the interface constraint modes for most of the substructures do not need to be repeated during the corresponding simulation processes.

In general, interpolation schemes like the ones considered in this chapter can also be used to approximate the fixed-interface normal modes and the interface constraint modes. In other words, normal and constraint modes can be approximated at different values of the model parameters, in terms of the corresponding modes of a family of models defined at a number of support points. Similarly, the proposed interpolation scheme can also be used to approximate other quantities involved in the construction of reduced-order models, such as the transformation matrices that account for the contribution of the residual fixed-interface normal modes. However, the use of approximation schemes for the above quantities is not pursued in this monograph.

References

1. P. Angelikopoulos, C. Papadimitriou, P. Koumoutsakos, X-TMCMC: adaptive kriging for Bayesian inverse modeling. Comput. Methods Appl. Mech. Eng. **289**, 409–428 (2015)
2. T.M. Barry, Recommendations on the testing and use of pseudo-random number generators used in Monte Carlo analysis for risk assesment. Risk. Anal. **16**(1), 93–105 (1996)
3. M. Drugan, D. Thierens, Recombination operators and selection strategies for evolutionary Markov chain Monte Carlo algorithms. Evol. Intell. **3**(2), 79–109 (2010)
4. G.S. Fishman, *Monte Carlo: Concepts, Algorithms and Applications* (Springer, New York, 1996)
5. D.G. Giovanis, I. Papaioannou, D. Straub, V. Papadopoulos, Bayesian updating with subset simulation using artificial neural netwoks. Comput. Methods Appl. Mech. Eng. **319**, 124–145 (2017)
6. G.H. Golub, C. Reinsch, Singular value decomposition and least squares solutions. Numer. Math. **14**(5), 403–420 (1970)
7. B. Goller, H.J. Pradlwarter, G.I. Schuëller, An interpolation scheme for the approximation of dynamical systems. Comput. Methods Appl. Mech. Eng. **200**, 414–423 (2011)
8. J.C. Helton, F.J. Davis, Latin hypercube sampling and the propagation of uncertainty in analyses of complex systems. Reliab. Eng. Syst. Saf. **81**, 23–69 (2003)
9. R.L. Iman, Uncertainty and sensitivity analysis for computer modeling applications, in *Reliability Techcnology*, ed. by T.A. Cruse. The Winter Annual Meeting of the American Society of Mechanical Engineers, vol. 28 (American Society of Mechanical Engineers, Aerospace Division, New York, 1992), pp. 153–168
10. R.L. Iman, W.J. Conover, A distribution-free approach to inducing rank correlation among input variables. Commun. Stat. Simul. Comput. **B11**(3), 311–334 (1982)
11. H.A. Jensen, C. Esse, V. Araya, C. Papadimitriou, Implementation of an adaptive meta-model for Bayesian finite element model updating in time domain. Reliab. Eng. Syst. Saf. **160**, 174–190 (2017)
12. H.A. Jensen, V. Araya, A. Muñoz, M. Valdebenito, A physical domain-based substructuring as a framework for dynamic modeling and reanalysis of systems. Comput. Methods Appl. Mech. Eng. **326**, 656–678 (2017)
13. V.C. Klema, A.J. Laub, The singular value decomposition: its computation and some applications. IEEE Trans. Autom. Control. **AC-25**(2), 163–176 (1980)
14. P. L'Ecuyer, Random number generation, in *Handbook of Simulation: Principles, Methodology, Advances, Application and Practice*, ed. by J. Banks (Wiley, New York, 1998), pp. 93–137
15. C. Papadimitriou, D.Ch. Papadioti, Component mode synthesis techniques for finite element model updating. Comput. Struct. **126**, 15–28 (2013)
16. W.H. Press, S.A. Teukolsky, W.T. Vetterling, B.P. Flannery, *Numerical Recipes. The Art of Scientific Computing* (Cambridge University Press, New York, 2007)
17. R. Seidel, Convex hull computations (Chap. 19), in *Handbook of Discrete and Computational Geometry*, ed. by J.E. Goodman, J. O'Rourke (CRC Press, Boca Raton, 1997), pp. 361–375
18. S.S. Skiena, Convex hull, *The Algorithm Design Manual* (Springer, New York, 1997), pp. 351–354

Part II
Application to Reliability Problems

Chapter 4
Reliability Analysis of Dynamical Systems

Abstract The use of reduced-order models in the context of reliability analysis of dynamical systems under stochastic excitation is explored in this chapter. A stochastic excitation model based on a point-source model is introduced, and it is used for the generation of ground motions. The corresponding reliability analysis represents a high-dimensional reliability problem whose solution is carried out by an advanced simulation technique. Two application problems are considered in order to evaluate the effectiveness of the proposed model reduction technique. The first example consists of a two-dimensional frame structure, while the second example considers an involved nonlinear finite element building model. The results show that an important reduction in computational effort can be achieved without compromising the accuracy of the reliability estimates.

4.1 Motivation

Reliability analysis allows the possibility of accounting for the unavoidable effects of uncertainty over the performance of a structure. In this context, the level of safety of a structure can be measured in terms of the reliability, which is a metric of plausibility that the structure fulfills certain performance requirements during its lifetime. The complement of the reliability is the probability of failure, that is, the probability that a structure violates prescribed performance criteria. Thus, reliability can be incorporated as one of the performance criteria in the analysis and design of structures to explicitly address the effects of uncertainty [24, 25, 29, 33, 42, 49, 58]. In this framework, it is assumed that the external force vector $\mathbf{f}(t)$ (see Eq. (1.1)) is modeled as a non-stationary stochastic process and characterized by a random variable vector $\mathbf{z} \in \Omega_{\mathbf{z}} \subset R^{n_z}$. This vector is defined in terms of a probability density function $p(\mathbf{z})$. Furthermore, consider a vector $\boldsymbol{\theta} \in \Omega_{\boldsymbol{\theta}} \subset R^{n_\theta}$ of uncertain model parameters. These parameters are characterized in a probabilistic manner by means of a joint probability density function $q(\boldsymbol{\theta})$. It is noted that alternative approaches for modeling uncertainties do exist, as well. For example, methodologies based on non-traditional uncertainty models can be very useful in a number of cases [8, 12, 27, 47, 48]. How-

H. Jensen and C. Papadimitriou, *Sub-structure Coupling for Dynamic Analysis*, Lecture Notes in Applied and Computational Mechanics 89,
https://doi.org/10.1007/978-3-030-12819-7_4

ever, the focus here is on probabilistic approaches. The performance of the structural system due to the excitation is characterized by means of n_r responses of interest

$$r_i(t, \mathbf{z}, \boldsymbol{\theta}) \ , \quad i = 1, \ldots, n_r, \ t \in [0, T] \tag{4.1}$$

where T is the duration of the excitation. Clearly, the aforementioned responses r_i are functions of time t (due to the dynamic nature of the loading), and functions of the system parameter vector $\boldsymbol{\theta}$ and random variable vector $\mathbf{z}$. The response functions $r_i(t, \mathbf{z}, \boldsymbol{\theta})$, $i = 1, \ldots, n_r$, are obtained from the solution of the equation of motion that characterizes the structural model, i.e., Eq. (1.1).

4.2 Reliability Problem Formulation

For structural systems under stochastic excitation, the probability that design conditions are satisfied within a particular reference period provides a useful reliability measure. Such a measure is referred to as the first excursion probability and quantifies the plausibility of the occurrence of unacceptable behavior (failure) of the structural system [63, 68]. Then, first excursion probabilities are used to characterize the level of safety of a structure. Specifically, this probability measures the chances that the uncertain responses exceed prescribed thresholds in magnitude within a specified time interval. Then, a failure event $F(\mathbf{z}, \boldsymbol{\theta})$ can be defined in terms of the so-called normalized demand function $d(\mathbf{z}, \boldsymbol{\theta})$ as [5]

$$F(\mathbf{z}, \boldsymbol{\theta}) = \{d(\mathbf{z}, \boldsymbol{\theta}) > 1\} \tag{4.2}$$

where this function is defined as the maximum of the quotient between the structural responses of interest and their corresponding threshold levels, that is,

$$d(\mathbf{z}, \boldsymbol{\theta}) = \max_{i=1,\ldots,n_r} \left(\max_{t \in [0,T]} \left(\left| \frac{r_i(t, \mathbf{z}, \boldsymbol{\theta})}{r_i^*} \right| \right) \right) \tag{4.3}$$

where r_i^*, $i = 1, \ldots, n_r$, are the acceptable threshold levels of the corresponding responses of interest r_i, $i = 1, \ldots, n_r$. Note that the quotient $r_i(t, \mathbf{z}, \boldsymbol{\theta},)/r_i^*$ can be interpreted as a demand to capacity ratio, as it compares the value of the response $r_i(t, \mathbf{z}, \boldsymbol{\theta})$ with its maximum allowable value r_i^*. It is noted that the concept of failure event does not necessarily imply collapse. In fact, the failure event may refer to, for example, partial damage states or unacceptable system performance.

The probability of occurrence of the failure event F, P_F, can be expressed in terms of the probability integral in the form

$$P_F = \int_{d(\mathbf{z},\boldsymbol{\theta})>1} p(\mathbf{z})\, q(\boldsymbol{\theta})\, d\mathbf{z}\, d\boldsymbol{\theta} \tag{4.4}$$

or in terms of the indicator function $I_F(\mathbf{z}, \boldsymbol{\theta})$ as

$$P_F = \int_{\mathbf{z} \in \Omega_{\mathbf{z}}, \boldsymbol{\theta} \in \Omega_{\boldsymbol{\theta}}} I_F(\mathbf{z}, \boldsymbol{\theta})\ p(\mathbf{z})\ q(\boldsymbol{\theta})\ d\mathbf{z}\ d\boldsymbol{\theta} \tag{4.5}$$

where the indicator function is equal to 1, in the case that the normalized demand function is equal or larger than 1 and 0 otherwise. In general, the probability integral involves a large number of random variables (hundreds or thousands) in the context of dynamical systems under stochastic excitation [5, 38, 40, 56] (see Sect. 4.5). Therefore, this integral represents a high-dimensional reliability problem whose numerical evaluation is extremely demanding from a numerical point of view [17, 21, 50].

4.3 Reliability Estimation

4.3.1 General Remarks

As previously pointed out, the probability integral represents a high-dimensional reliability problem. In addition, the normalized demand function that characterizes the failure event F is usually not explicitly known but must be computed pointwise by applying suitable deterministic numerical techniques, such as finite element analyses. Then, it is essential to minimize the number of such function evaluations. Finally, the probability of failure of a system properly designed is, in general, very small ($P_F \sim 10^{-6}$–10^{-2}). In other words, failure is a rare event. It is also apparent that methods based on numerical integration or standard reliability methods are not suitable for estimating the high-dimensional probability integral. This difficulty favors the application of simulation techniques in order to estimate the probability of failure. In this regard, it is well known that direct Monte Carlo is theoretically applicable for evaluating P_F, but it is inefficient in estimating small probabilities because it requires a very large number of samples (dynamic analyses) to achieve an acceptable level of accuracy [28, 59]. Based on the above conditions, it is clear that the reliability problem is computationally very challenging. Therefore, the estimation of the system reliability has to rely on advanced simulation techniques to limit, to the greatest extent possible, the number of dynamic analyses. Several advanced stochastic simulation methods have been recently developed to cope with this type of problems. Examples of these algorithms include subset simulation [5, 6, 70], line sampling [40], auxiliary domain method [39], horseracing simulation [69], and subset simulation based on hidden variables [7]. Among these algorithms, subset simulation is used in the present implementation due to its generality and flexibility. The generality of the method is due to the fact that it is not based on any geometrical assumption about the topology of the failure domain. Moreover, validation calculations have shown that subset simulation can be applied efficiently to a wide range of complex reliability problems [6, 18, 19, 34, 37, 62]. Even though this is a well known technique in the reliability engineering research community, some of the key aspects of subset simulation are reviewed in this section for completeness.

Based on the above conditions, it is clear that the reliability problem is computationally very challenging.

4.3.2 Basic Ideas

The conceptual idea of subset simulation is to decompose the failure event F into a sequence of nested failure events

$$F = F_m \subset F_{m-1} \subset \cdots \subset F_1 \tag{4.6}$$

so that

$$F = \cap_{k=1}^{m} F_k \tag{4.7}$$

By definition of conditional probability, the probability of failure can be written as

$$P(F) = P(F_m) = P(\cap_{k=1}^{m} F_k) = P(F_1) \prod_{k=1}^{m-1} P(F_{k+1}/F_k) \tag{4.8}$$

In other words, the probability of failure is expressed as a product of $P(F_1)$ and the conditional probabilities $\{P(F_{k+1}/F_k), k = 1, \ldots, m-1\}$. It is seen that, even if $P(F)$ is small, by choosing m and $F_k, k = 1, \ldots, m-1$, appropriately, the conditional probabilities can still be made sufficiently large, and, therefore, can be efficiently evaluated by direct simulation because the failure events are more frequent. The subsets $F_1, F_2, \ldots, F_{m-1}$ are called intermediate failure events. For actual implementation, the intermediate failure events are adaptively chosen using information from simulated samples in order to correspond to some specific values of conditional failure probabilities. To be more specific, the sequence of intermediate failure events is defined as

$$F_k = \{d(\mathbf{z}, \boldsymbol{\theta}) > \delta_k\} \ , \ k = 1, \ldots, m \tag{4.9}$$

where $0 < \delta_1 < \cdots < \delta_{m-1} < 1 = \delta_m$ is a sequence of intermediate threshold values. Note that the failure event $F_m = F$ is defined as $\{F_m = d(\mathbf{z}, \boldsymbol{\theta}) > \delta_m = 1\}$. During subset simulation, the threshold values $\delta_1, \ldots, \delta_{m-1}$ are adaptively selected, so that the conditional failure probabilities are set equal to a pre-established value, for example, p_0. This parameter is called the conditional failure probability. Validation calculations have shown that choosing any value of p_0 between 0.1 and 0.3 will lead to similar efficiency as long as subset simulation is properly implemented [70]. Then, it is seen that the demand function values $\delta_1, \ldots, \delta_{m-1}$ at the specified probability levels are estimated during the subset simulation. In this manner, subset simulation generates samples whose demand function values correspond to specific (pre-established) probability levels. Therefore, the unconditional as well as all conditional failure probabilities are automatically equal to p_0, except for the conditional failure probability in the last step of subset simulation, that is, $P(F_m/F_{m-1})$.

4.3.3 Failure Probability Estimator

The previous result implies that the probability of failure can be expressed in the form

$$P_F = p_0^{m-1} \int_{\mathbf{z}\in\Omega_{\mathbf{z}},\boldsymbol{\theta}\in\Omega_{\boldsymbol{\theta}}} I_F(\mathbf{z},\boldsymbol{\theta}) p(\mathbf{z}|F_{m-1})\, q(\boldsymbol{\theta}|F_{m-1})\, d\mathbf{z}\, d\boldsymbol{\theta} \tag{4.10}$$

where $p(\mathbf{z}|F_{m-1})$ and $q(\boldsymbol{\theta}|F_{m-1})$ are the conditional distributions of the random variable vector $\mathbf{z}$ and uncertain system parameters $\boldsymbol{\theta}$ conditional to the failure event F_{m-1}, respectively. Note that the integral in the above equation corresponds to the expected value of the indicator function with respect to the conditional distributions $p(\mathbf{z}|F_{m-1})$ and $q(\boldsymbol{\theta}|F_{m-1})$. Thus, the probability of failure can also be written as

$$P_F = p_0^{m-1} E_{p(\mathbf{z}|F_{m-1}),q(\boldsymbol{\theta}|F_{m-1})} \left[I_F(\mathbf{z},\boldsymbol{\theta})\right] \tag{4.11}$$

where $E_{p(\mathbf{z}|F_{m-1}),q(\boldsymbol{\theta}|F_{m-1})}[\,\cdot\,]$ is the expectation operator. The probability of failure is then estimated as

$$P_F \approx p_0^{m-1} \frac{1}{N_m} \sum_{i=1}^{N_m} I_F(\mathbf{z}_{m-1,i},\boldsymbol{\theta}_{m-1,i}) \tag{4.12}$$

where $\{(\mathbf{z}_{m-1,i},\boldsymbol{\theta}_{m-1,i}), i = 1,\ldots,N_m\}$ is the set of samples generated at the last stage of subset simulation (conditional level $m-1$).

For actual implementation of subset simulation, it is assumed without much loss of generality that the components of $\mathbf{z}$ are independent, that is,

$$p(\mathbf{z}) = \Pi_{j=1}^{n_{\mathbf{z}}} p_j(z_j) \tag{4.13}$$

where for every j, $p_j(\cdot)$ is a one-dimensional probability density function for z_j. Similarly, the uncertain system parameters $\boldsymbol{\theta}$ are also assumed to be independent and, therefore, the joint probability density function $q(\boldsymbol{\theta})$ takes the form

$$q(\boldsymbol{\theta}) = \Pi_{j=1}^{n_{\theta}} q_j(\theta_j) \tag{4.14}$$

where $q_j(\theta_j)$ represents the probability density function of the basic system parameter θ_j. It is noted that this assumption is not a limitation for a number of cases of interest. However, the estimation of posterior robust failure probability integrals is not covered by the assumption of independence (see Sect. 7.3.4 to address this situation).

4.4 Numerical Implementation

4.4.1 Basic Implementation

Based on the previous conceptual ideas, the basic implementation of subset simulation is as follows.

(1) Generate N_1 samples $\{(\mathbf{z}_{0,i}, \boldsymbol{\theta}_{0,i}), i = 1, \ldots, N_1\}$ by direct Monte Carlo according to the probability density functions $p(\mathbf{z})$ and $q(\boldsymbol{\theta})$, respectively (the subscript 0 denotes that the samples correspond to the unconditional level (level 0)). Set $k = 1$.
(2) Evaluate the normalized demand function to obtain $\{d(\mathbf{z}_{k-1,i}, \boldsymbol{\theta}_{k-1,i}), i = 1, \ldots, N_k\}$. Arrange these values in an increasing order.
(3) Identify the $[(1 - p_0)N_k + 1]$th largest value of the set $\{d(\mathbf{z}_{k-1,i}, \boldsymbol{\theta}_{k-1,i}), i = 1, \ldots, N_k\}$. In the case that this value is equal or larger than 1, set $m = k$, $\delta_m = 1$ and go to step 7. Otherwise, set the intermediate threshold value δ_k equal to the aforementioned $[(1 - p_0)N_k + 1]$th largest value of the set $\{d(\mathbf{z}_{k-1,i}, \boldsymbol{\theta}_{k-1,i}), i = 1, \ldots, N_k\}$.
(4) The kth intermediate failure event is defined as $F_k = \{d(\mathbf{z}, \boldsymbol{\theta}) \geq \delta_k\}$.
(5) The sampling estimate for $P(F_k)$ if $(k = 1)$ or $P(F_k/F_{k-1})$ if $(k > 1)$ is equal to p_0 by construction, where p_0 and N_k are chosen such that $p_0 N_k$ is an integer number.
(6) By construction, there are $p_0 N_k$ samples among $\{(\mathbf{z}_{k-1,i}, \boldsymbol{\theta}_{k-1,i}), i = 1, \ldots, N_k\}$ whose demand function value is equal or greater than δ_k. Starting from each of these conditional samples, Markov chain Monte Carlo simulation is used to generate an additional $(N_{k+1} - p_0 N_k)$ conditional samples that lie in F_k, making a total of N_{k+1} conditional samples $\{(\mathbf{z}_{k,i}, \boldsymbol{\theta}_{k,i}), i = 1, \ldots, N_{k+1}\}$ at level k. The Markov chain samples are drawn by using the modified Metropolis algorithm [5, 45]. Return to step 2 with $k = k + 1$.
(7) The conditional failure probability $P(F_m/F_{m-1})$ is estimated directly by $P(F_m/F_{m-1}) = N_F/N_m$ where N_F is the number of samples that lie in the target failure event F_m. The failure probability is estimated as

$$P_F \approx p_0^{m-1} \frac{1}{N_m} \sum_{i=1}^{N_m} I_{F_m}(\mathbf{z}_{m-1,i}, \boldsymbol{\theta}_{m-1,i}) \tag{4.15}$$

where $\{(\mathbf{z}_{m-1,i}, \boldsymbol{\theta}_{m-1,i}), i = 1, \ldots, N_m\}$ is the set of samples generated at the last stage of subset simulation (conditional level $m - 1$).

For a more detailed implementation of the approach, the reader is referred to [5, 70].

4.4.2 Implementation Issues

The numerical implementation of subset simulation can be improved by considering the parallelization of some independent parts of the algorithm. The highest computational efforts are associated with the dynamic analysis of the structural system. Then, parallelization strategies that exploit the parallelism of those parts of the code where the dynamic analysis is performed can be implemented [2, 55]. The unconditional level of subset simulation (level 0) can be scheduled completely in parallel, since the samples are independent. At higher conditional levels, Markov chains need to be generated. Samples forming a Markov chain depend on the previous samples, which implies inherent dependence and then excludes parallelization. However, the chains themselves are independent from each other, which means that the generation of different chains can be concurrently performed. Thus, a number of chains can be simultaneously run, taking advantage of available parallelization techniques. Additionally, low-level parallelism can also be considered to accelerate the individual model runs (dynamic analysis), improving the numerical implementation even more [13, 66].

4.5 Stochastic Model for Excitation

4.5.1 General Description

Depending on the particular application and the available information, different stochastic excitation models can be used. For example, in the area of seismic engineering, filtered Gaussian white noise-based processes, models based on power spectra, record-based models, point source-based models, multiple point source-based models, and models based on large or small sub-events are usually used [4, 14, 20, 22, 46, 51–53, 57, 60, 67]. In particular, a stochastic point source-based model is used in the present formulation to simulate ground motions. The model is characterized by a series of seismicity parameters, such as the moment magnitude M and the epicentral distance r [4, 14]. The methodology, which was initially developed for generating synthetic ground motions, has been reinterpreted to form a stochastic model for earthquake excitation [36, 65]. According to this approach, high-frequency and low-frequency (pulse) components of the ground motion are independently generated and then combined to form an acceleration time history. The stochastic model represents a practical tool for the description of far and near-field ground motions. It establishes a direct link between the knowledge about the characteristics of the seismic hazard in the structural site and future ground motions. For completeness, some of the basic aspects of the model are presented in this section.

4.5.2 High-Frequency Components

The time history for a specific event of magnitude M and epicentral distance r with high-frequency components of the ground motion is obtained by several steps. First, a discrete white noise sequence is generated as $\mathbf{w}^T = <\sqrt{1/\Delta t}\, w_j>$, $j = 1, \ldots, n_T$, where w_j, $j = 1, \ldots, n_T$, are independent, identically distributed standard Gaussian random variables, Δt is the sampling interval, and n_T is the number of time instants equal to the duration of the excitation T divided by the sampling interval. The white noise sequence is then modulated by an envelope function $e(t, M, r)$, such as the one suggested in [61], at the discrete time instants (see Sect. 4.6.5). Discrete Fourier transform is applied to the modulated white noise sequence. The resulting spectrum is multiplied by a ground motion spectrum (or radiation spectrum) $A(f, M, r)$, after which discrete inverse Fourier transform is applied to transform the sequence back to the time domain to yield the desired ground acceleration time history. The envelope function is the major factor affecting the duration of simulated ground motions for a given moment magnitude M and epicentral distance r. Furthermore, the ground motion spectrum contains information on the physics of the earthquake process as well as other geophysical parameters, such as radiation pattern, density, shear wave velocity in the vicinity of the source, corner frequencies, local site conditions, etc. Details of the procedure as well as the characterization of the envelope function and the ground acceleration spectrum can be found in [1, 4, 14, 15, 61, 65].

4.5.3 Pulse Components

The description of the time history with low-frequency components is based on a simple analytical model developed in [44]. According to the model, the pulse component related to near-field motions is described through a velocity pulse $v(t)$ as

$$v(t) = \frac{A_p}{2}[1 + \cos(\frac{2\pi f_p}{\gamma_p}(t - t_p))] \cos(2\pi f_p(t - t_p) + \nu_p) \ , \ t \in (t_p - \frac{\gamma_p}{2 f_p}, t_p + \frac{\gamma_p}{2 f_p}) \tag{4.16}$$

where A_p, f_p, ν_p, γ_p, and t_p describe the amplitude, prevailing frequency, phase angle, number of half cycles, and time shift, respectively. Outside the time interval, the velocity pulse is equal to zero. Some of the pulse parameters, such as the amplitude and frequency, can be linked to the moment magnitude M and epicentral distance r of the seismic event [16]. The rest of the pulse parameters are considered as independent model parameters, and they have been calibrated by tuning the analytical expression of the velocity pulse to a wide range of recorded near-field ground motions [44].

4.5.4 Synthesis of Near-Field Ground Motions

The synthesis of near-field ground motions is obtained by combining the high- and low-frequency components through the following steps. First, an acceleration time history with high-frequency components and a pulse ground acceleration are generated. The Fourier transforms of these synthetic acceleration time histories are then calculated. Next, the Fourier amplitude spectrum of the synthetic time history with low-frequency components is subtracted from the Fourier amplitude spectrum of the synthetic time history with high-frequency components. A synthetic acceleration time history is constructed, so that its Fourier amplitude spectrum is equal to the difference of the Fourier amplitude spectra calculated before, and its phase coincides with the phase of the Fourier transform of the synthetic time history with high-frequency components. Finally, the time history generated in the previous steps is superimposed to the acceleration time history corresponding to the velocity pulse [44]. For illustration purposes, Fig. 4.1 shows a synthetic near-field ground motion sample corresponding to the envelope function and radiation spectrum presented in Fig. 4.2 and with near-field pulse parameters $A_p = 27.11$ (cm/s), $f_p = 0.53$ (Hz), $\nu_p = 0.0$ (rad), and $\gamma_p = 1.8$. The existence of the near-field pulse is evident when looking at the velocity time history of the ground motion. It is noted that considering a sampling interval equal to $\Delta t = 0.01$ s, the discrete white noise sequence has more than 1,500 components. In other words, the vector of uncertain parameters $\mathbf{w}$ has more than 1,500 elements in this case.

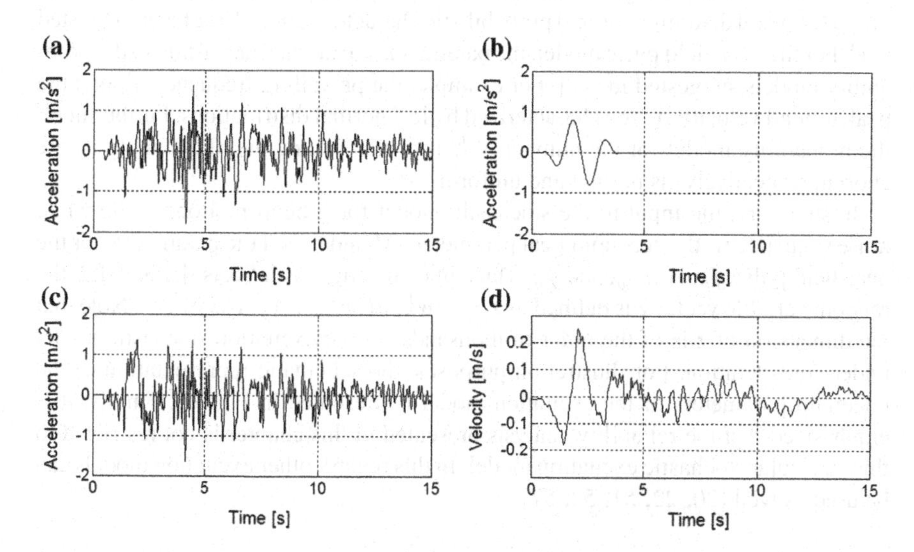

Fig. 4.1 Acceleration time history sample. **a** High-frequency components. **b** Near-field pulse acceleration. **c** Final ground motion (acceleration time history). **d** Final ground motion (velocity time history)

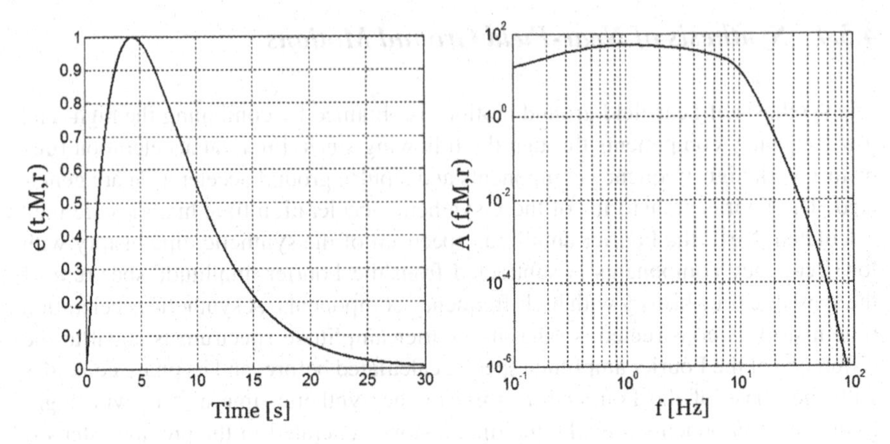

Fig. 4.2 Envelope function $e(t, M, r)$ and radiation spectrum $A(f, M, r)$ for $M = 7.0$ and $r = 20$ km

4.5.5 Seismicity Model

The probabilistic model for the seismic hazard at the structural site is finally complemented by assigning a probability density function to some of the model parameters. In the context of this formulation, the epicentral distance r for the earthquake events is assumed to follow a log-normal distribution. With respect to the moment magnitude M, several deterministic and probabilistic characterizations have been suggested [41]. For the near-field pulse model, the parameters are defined according to the probability models suggested in [44]. For example, the prevailing frequency f_p and the peak ground velocity A_p are characterized by log-normal distributions. Furthermore, the probability model for the number of half cycles γ_p and the phase angle ν_p are chosen, respectively, as normal and uniform.

In summary, the input to the stochastic model for ground motions is the white noise sequence $\mathbf{w}$, the seismological parameters M and r, and the parameters for the near-field pulse f_p, A_p, ν_p, and γ_p. Thus, in connection with Sects. 4.1 and 4.2, the random variable vector $\mathbf{z}$ is defined as $\mathbf{z} =< \mathbf{w}^T, M, r, f_p, A_p, \nu_p, \gamma_p >^T$. Note that the dimension of $\mathbf{z}$ is of the order of thousands for the excitation stochastic model under consideration. For illustration purposes, the schematic representation of the uncertain parameters of the excitation model is presented in Fig. 4.3. Finally, it is emphasized that the reliability analysis presented in this chapter is not restricted to this particular stochastic excitation model. In this regard, other excitation models can be used as well [20, 22, 51, 52, 57].

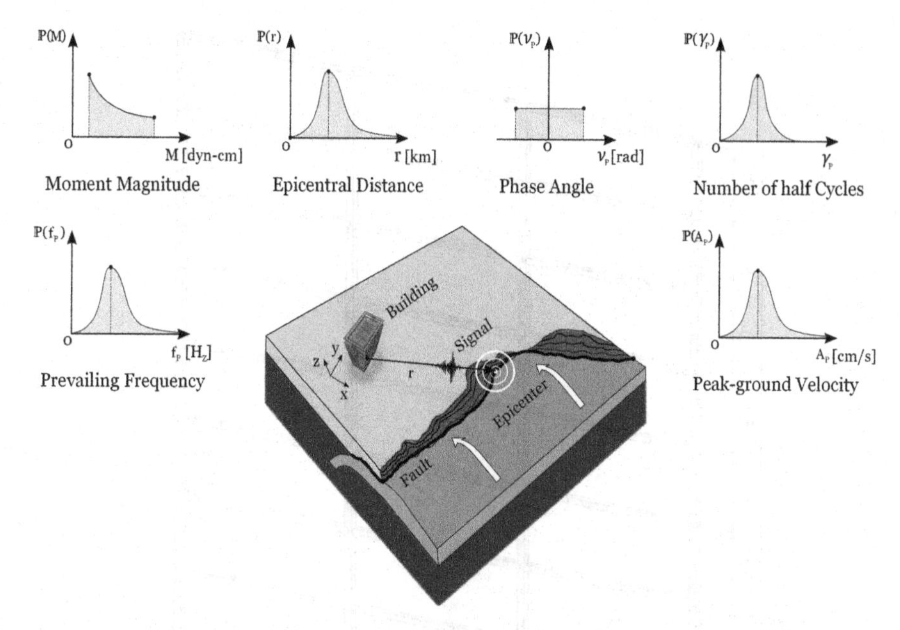

Fig. 4.3 Uncertain seismological and near-fault pulse parameters

4.6 Application Problem No. 1

The objective of this application problem is to evaluate the performance and effectiveness of the proposed model reduction technique for the reliability analysis of a two-dimensional frame structure. Different reduced-order models are considered, including models based on fixed-interface normal modes with and without interface reduction.

4.6.1 Model Description and Substructures Characterization

The model, shown in Fig. 4.4, consists of a three-span two-dimensional eight-story frame structure, and it can be considered as one of the moment-resisting frames of a building model.

The structural model has a total length of 30 m and a constant floor height of 5 m, leading to a total height of 40 m. The finite element model comprises 160 two-dimensional beam elements of square cross section with 140 nodes and a total of 408 degrees of freedom. The dimension of the square cross section of the beam elements is equal to 0.4 m. The axial deformation of these elements is neglected with respect to their bending deformation. The basic material properties of the beam and column elements are given by the Young's modulus $E = 2.0 \times 10^{10}$ N/m^2 and mass density

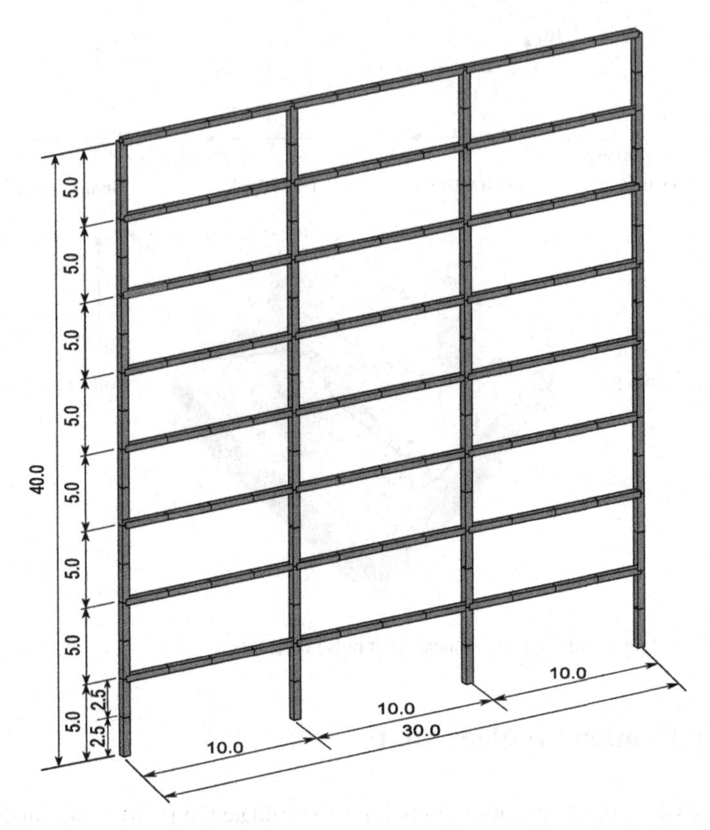

Fig. 4.4 Three-span two-dimensional eight-story frame structure. Application problem No. 1

$\rho = 2{,}500\,\text{kg/m}^3$. The structural model is subdivided into 16 substructures as shown in Fig. 4.5. Substructures $S_i, i = 1, \ldots, 8$, are composed of the column elements of the different floors, while substructures $S_i, i = 9, \ldots, 16$, correspond to the beam elements of the different floors. With this subdivision, there are eight interfaces in the model. The total number of internal degrees of freedom is equal to 312, while 96 degrees of freedom are present at the interfaces.

4.6.2 *Reduced-Order Model Based on Dominant Fixed-Interface Normal Modes*

Two models with a reduced number of fixed-interface normal modes are considered to evaluate the effect of dominant normal modes on the accuracy of the reduced-order model spectral properties. The first model (Model-1) considers the minimum number of fixed-interface normal modes at each substructure, while the second model

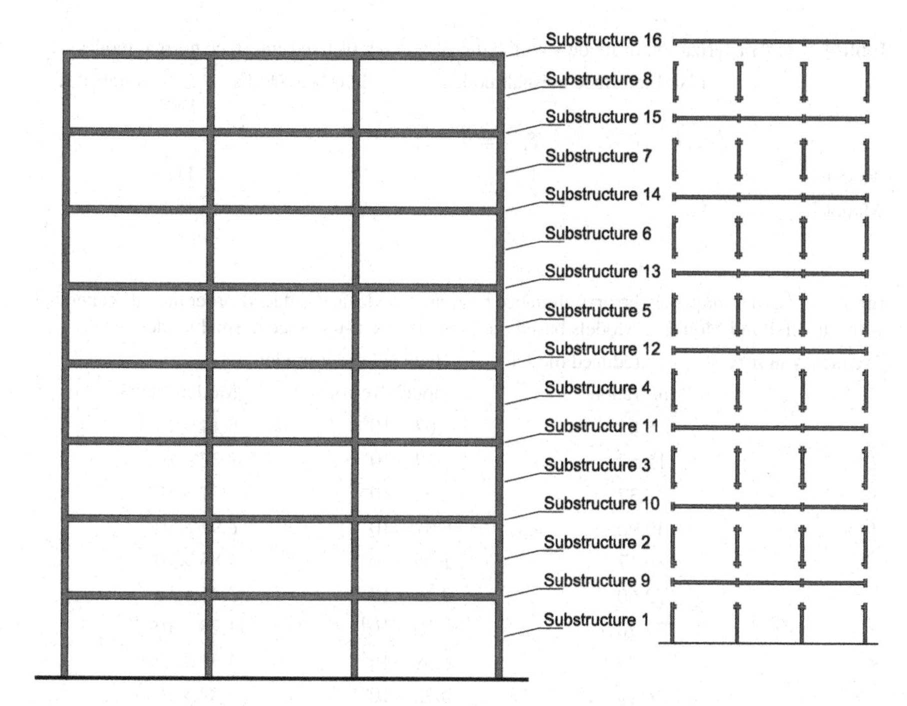

Fig. 4.5 Substructures of the finite element model. Application problem No. 1

(Model-2) includes all fixed-interface normal modes with frequencies inside a target frequency bandwidth. More specifically, Model-1 is characterized by the first fixed-interface normal mode of each substructure (with the lowest frequency). For each substructure of Model-2, all fixed-interface normal modes that have frequency ω such that $\omega \leq \alpha\omega_c$ are retained, with α being a multiplication factor and ω_c being a cut-off frequency that is taken equal to 87.66 rad/s (10th modal frequency of the unreduced reference model). The multiplication factor is selected to be 5 for substructures $S_i, i = 1, \ldots, 8$, and 2 for substructures $S_i, i = 9, \ldots, 16$. The difference between the multiplication factors is due to the fact that spectral properties of substructures $S_i, i = 1, \ldots, 8$ are quite different from substructures $S_i, i = 9, \ldots, 16$, as the lowest frequencies corresponding to substructures 1 to 8 are substantially higher than the lowest frequencies of substructures 9 to 16. The selected multiplication factors define a frequency bandwidth that contains the most important frequencies of each substructure.

With this selection of multiplication factors, four fixed-interface normal modes are kept for each substructure $S_i, i = 1, \ldots, 8$, and three fixed-interface normal modes for each substructure $S_i, i = 9, \ldots, 16$. Table 4.1 characterizes the two models in terms of the number of fixed-interface normal modes of each substructure, total number of interface degrees of freedom, and total number of degrees of freedom. In summary, only 16 generalized coordinates corresponding to the dominant fixed-

Table 4.1 Characterization of models with reduced number of fixed-interface normal modes

	Fixed interface normal modes		Interface DOFs	Total number of DOFs
	$S_i, i = 1, \ldots, 8$	$S_i, i = 9, \ldots, 16$		
Model-1	1	1	96	112
Model-2	4	3	96	152

Table 4.2 Modal frequency error: unreduced reference model and reduced-order models generated from Model-1 and Model-2. Models based on dominant fixed-interface normal modes

Frequency number	Unreduced model	Reduced-order model	
	ω (rad/s)	Model-1(error) (%)	Model-2(error) (%)
1	5.03	1.07×10^{-3}	5.14×10^{-6}
2	15.57	8.69×10^{-3}	2.32×10^{-4}
3	27.37	2.52×10^{-2}	1.77×10^{-3}
4	40.90	6.87×10^{-2}	6.97×10^{-3}
5	56.17	1.25×10^{-1}	1.85×10^{-2}
6	72.60	2.26×10^{-1}	3.45×10^{-2}
7	77.26	9.20×10^{0}	1.28×10^{-2}
8	82.57	6.08×10^{0}	1.31×10^{-2}
9	86.15	3.32×10^{0}	1.32×10^{-2}
10	87.66	3.85×10^{0}	3.78×10^{-2}

interface normal modes are retained for all substructures in Model-1, while 56 generalized coordinates are considered in Model-2. The dimension of the corresponding reduced-order models represents a 72% and 62% reduction with respect to the unreduced model, respectively.

Table 4.2 shows the errors between the modal frequencies using the unreduced reference finite element model and the modal frequencies computed using the reduced-order models generated from Model-1 and Model-2. The reduced-order models are based on dominant fixed-interface normal modes. It is seen that the errors are quite small for the reduced-order model generated from Model-2. The errors for the lowest 10 modes fall below 0.05%. For Model-1, an increase in the errors is observed for modes 7–10, with a range of relative errors between 3% and 10%.

The corresponding matrices of MAC-values between the first 10 modal vectors computed from the unreduced finite element model and from the reduced-order models are shown in terms of a 3-D representation in Figs. 4.6 and 4.7, respectively. It is seen that, for Model-2, the values at the diagonal terms are practically one and zero at the off-diagonal terms. Thus, the modal vectors are consistent for both models. Contrarily, some of the diagonal terms are less than one, while some of the off-diagonal terms exhibit values greater than zero for Model-1. Thus, the reduced-order model generated from Model-1 is not able to accurately characterize the higher

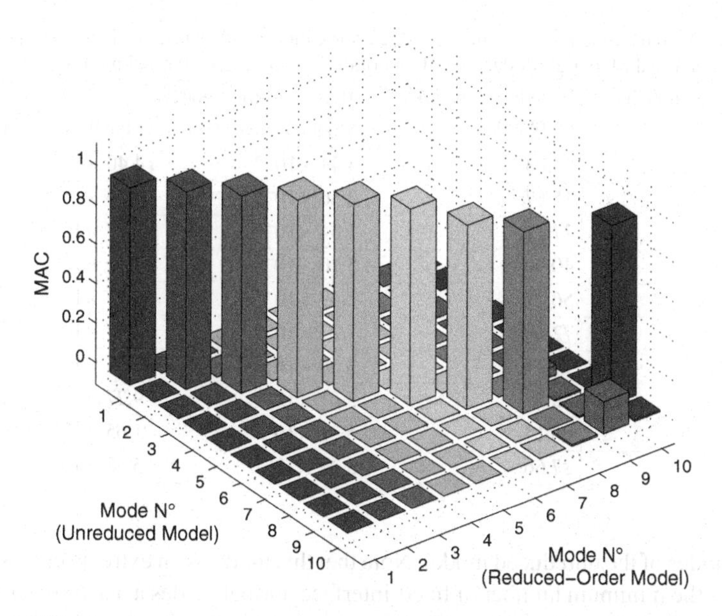

Fig. 4.6 MAC-values between the mode shapes computed from the unreduced finite element model and from the reduced-order model based on dominant normal modes. Reduced-order model generated from Model-1

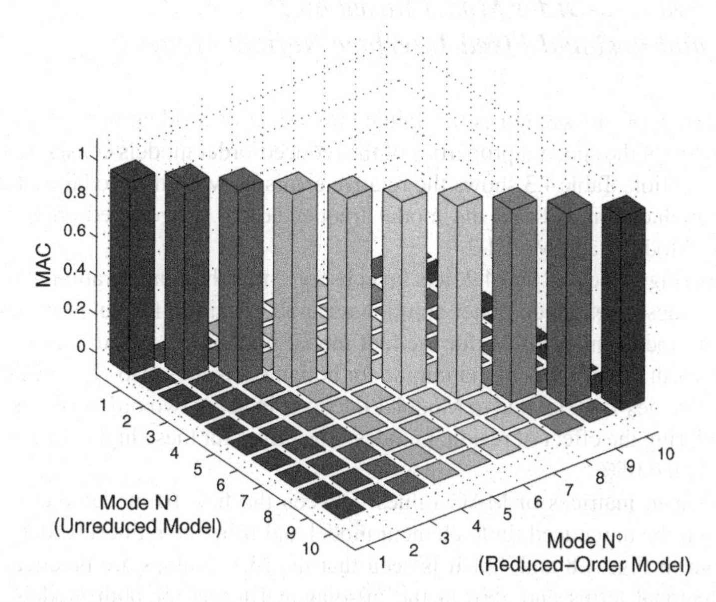

Fig. 4.7 MAC-values between the mode shapes computed from the unreduced finite element model and from the reduced-order model based on dominant normal modes. Reduced-order model generated from Model-2

Table 4.3 Modal frequency error: unreduced reference model and reduced-order models generated from Model-1 and Model-2. Models based on dominant and residual normal modes

Frequency number	Unreduced model ω (rad/s)	Reduced-order model Model-1(error) (%)	Reduced-order model Model-2(error) (%)
1	5.03	4.44×10^{-8}	4.40×10^{-8}
2	15.57	2.56×10^{-8}	7.24×10^{-9}
3	27.37	1.64×10^{-6}	1.07×10^{-6}
4	40.90	5.10×10^{-5}	3.41×10^{-5}
5	56.17	2.49×10^{-4}	1.63×10^{-4}
6	72.60	1.23×10^{-3}	5.37×10^{-4}
7	77.26	3.68×10^{-2}	1.91×10^{-4}
8	82.57	5.67×10^{-2}	9.27×10^{-5}
9	86.15	5.34×10^{-2}	1.65×10^{-4}
10	87.66	5.32×10^{-2}	5.32×10^{-3}

order modes of the unreduced model. Note that this model is an extreme case, since it includes the minimum number of fixed-interface normal modes at each substructure.

4.6.3 Reduced-Order Model Based on Dominant and Residual Fixed-Interface Normal Modes

The objective of this section is to evaluate the effect of residual normal modes on the accuracy of the spectral properties of the reduced-order models considered in the previous section. Table 4.3 shows the relative errors between the modal frequencies of the unreduced model and the modal frequencies of the reduced-order models related to Model-1 and Model-2.

Comparing Tables 4.2 and 4.3, it is first observed that the consideration of residual normal modes gives much better solution accuracy than the formulation based on dominant modes only. In fact, for the first modal frequencies, the difference in the errors is about three orders of magnitude for both models. It is also observed that the errors for modes 7–10 related to Model-1 decrease in about two orders of magnitude by considering the effect of residual modes. The errors for these higher order modes are less than 0.06%.

The related matrices of MAC-values between the first 10 modal vectors computed from the unreduced finite element model and from the reduced-order models are shown in Figs. 4.8 and 4.9. It is seen that the MAC-values are practically one at the diagonal terms and zero at the off-diagonal terms for both models. Thus, the reduced-order model generated from Model-1 is consistent with the unreduced model if the residual normal modes are considered in the formulation. Recall that Model-1 is an utmost case where the minimum number of fixed-interface modes is considered. As previously pointed out, this reduced-order model is not consistent

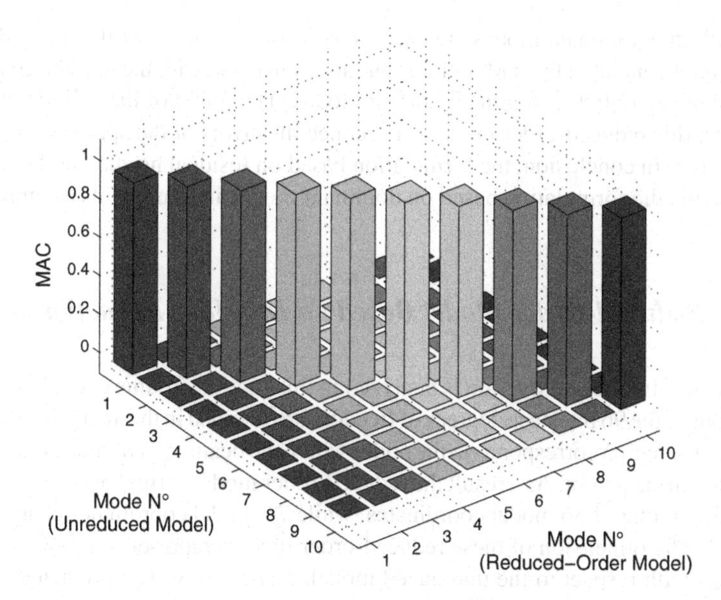

Fig. 4.8 MAC-values between the mode shapes computed from the unreduced finite element model and from the reduced-order model based on dominant and residual normal modes. Reduced-order model generated from Model-1

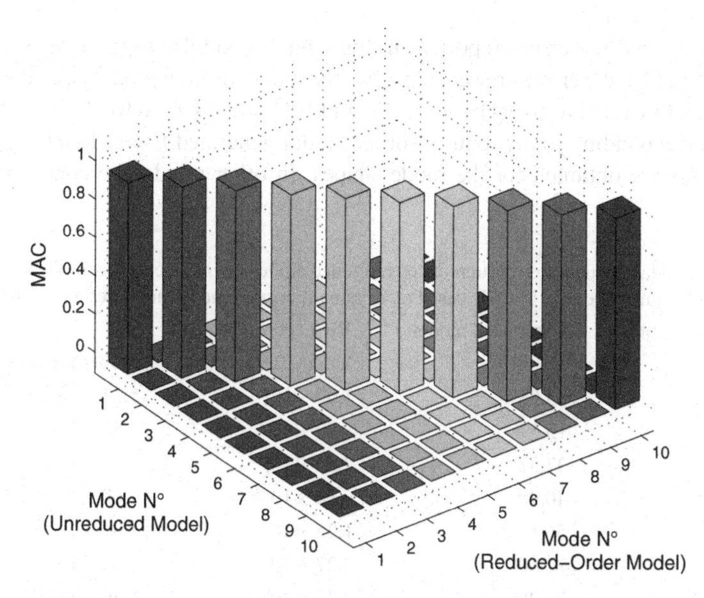

Fig. 4.9 MAC-values between the mode shapes computed from the unreduced finite element model and from the reduced-order model based on dominant and residual normal modes. Reduced-order model generated from Model-2

when only the dominant modes are taken into account. Also, note that the reduced-order model generated from Model-2 is already consistent with the unreduced model by considering only the dominant normal modes. The effect of the residual normal modes on this reduced-order model is to reduce the errors of the spectral properties even further. In conclusion, the formulation based on residual normal modes greatly outperforms the formulation based on dominant modes in terms of its accuracy.

4.6.4 Reduced-Order Model Based on Interface Reduction

The effect of interface reduction is analyzed in this section. To this end, 20 interface modes out of the 96 interface degrees of freedom are retained in the analysis. Note that the interface region corresponds to the nodes where the beam and column elements are connected at each floor. As a result the reduced-order model corresponding to Model-1 includes a total of 36 modal coordinates, while 76 modal coordinates characterize Model-2. The dimension of these reduced-order models represents a 91% and 81% reduction with respect to the unreduced model, respectively. The predicted natural frequencies resulting from both reduced-order models are presented in Table 4.4, and they are compared with the frequencies computed from the unreduced model as a reference. The reduced-order models are based on dominant normal modes and interface reduction.

It is seen that the errors reported in this table are similar to the ones shown in Table 4.2. In fact, the errors are very small for the reduced-order model generated from Model-2, while relative errors between 3% and 10% are observed for the higher-order modes corresponding to the reduced-order model generated from Model-1. Similar conclusions are obtained for the mode shapes. In other words, the contribution of

Table 4.4 Modal frequency error: unreduced reference model and reduced-order models generated from Model-1 and Model-2. Models based on dominant normal modes and interface reduction

Frequency number	Unreduced model ω (rad/s)	Reduced-order model	
		Model-1(error) (%)	Model-2(error) (%)
1	5.03	1.07×10^{-3}	5.14×10^{-6}
2	15.57	8.69×10^{-3}	2.32×10^{-4}
3	27.37	2.52×10^{-2}	1.77×10^{-3}
4	40.90	6.88×10^{-2}	7.00×10^{-3}
5	56.17	1.25×10^{-1}	1.86×10^{-2}
6	72.60	2.27×10^{-1}	3.48×10^{-2}
7	77.26	9.45×10^{0}	1.30×10^{-2}
8	82.57	6.15×10^{0}	1.32×10^{-2}
9	86.15	3.34×10^{0}	1.34×10^{-2}
10	87.66	4.10×10^{0}	4.26×10^{-2}

Table 4.5 Modal frequency error: unreduced reference model and reduced-order models generated from Model-1 and Model-2. Models based on dominant and residual modes, and interface reduction

Frequency number	Unreduced model ω (rad/s)	Reduced-order model	
		Model-1(error) (%)	Model-2(error) (%)
1	5.03	4.44×10^{-8}	4.40×10^{-8}
2	15.57	2.56×10^{-8}	7.24×10^{-9}
3	27.37	1.64×10^{-6}	1.07×10^{-6}
4	40.90	5.10×10^{-5}	3.41×10^{-5}
5	56.17	2.49×10^{-4}	1.63×10^{-4}
6	72.60	1.23×10^{-3}	5.36×10^{-4}
7	77.26	3.67×10^{-2}	1.90×10^{-4}
8	82.57	5.65×10^{-2}	9.27×10^{-5}
9	86.15	5.33×10^{-2}	1.64×10^{-4}
10	87.66	5.30×10^{-2}	5.30×10^{-3}

the first 20 interface modes seems to be adequate in the sense that the accuracy of the reduced-order models remains invariant with this number of interface modes, as the selected interface modes are able to capture the relevant deformation at the interfaces. Validation calculations show that lower interface modes (lower than the 20th interface mode) cannot be neglected for this model. Note that a small number of interface degrees of freedom are present at the interfaces, and, therefore, the number of retained interface modes cannot be too small.

The effect of residual normal modes on the reduced-order models that consider interface reduction is similar to the one observed in the previous section. That is, the errors of the spectral properties are significantly reduced. This effect can be seen in Table 4.5. Note that the errors are virtually the same to the ones reported in Table 4.3.

The matrices of MAC-values between the first 10 modal vectors computed from the unreduced finite element model and from the reduced-order models based on dominant and residual normal modes and interface reduction are shown in Figs. 4.10 and 4.11. Clearly, the reduced-order models are consistent with the unreduced model.

To get more insight into the interface modes, the first two characteristic constraint modes are shown in Figs. 4.12 and 4.13. Recall that these modes are obtained by transforming the interface modes $\boldsymbol{\Upsilon}_I$ into finite element coordinates as indicated in Sect. 1.6.2. The characteristic constraint modes $\boldsymbol{\Upsilon}_{CC}$ provide the principal modes of deformation for the interface, since they capture some characteristic physical motion in the interface region. It is seen that the first characteristic constraint mode captures much of the interface-induced motion seen in the first global mode, whereas the second characteristic constraint mode resembles the second global mode. Thus, the importance of considering an adequate number of interface modes in constructing the reduced-order model is evident.

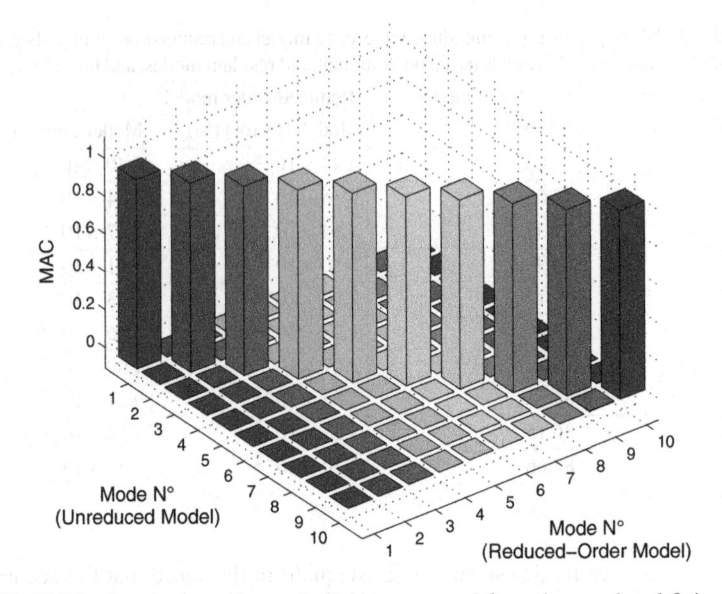

Fig. 4.10 MAC-values between the mode shapes computed from the unreduced finite element model and from the reduced-order model based on dominant and residual normal modes and interface reduction. Reduced-order model generated from Model-1

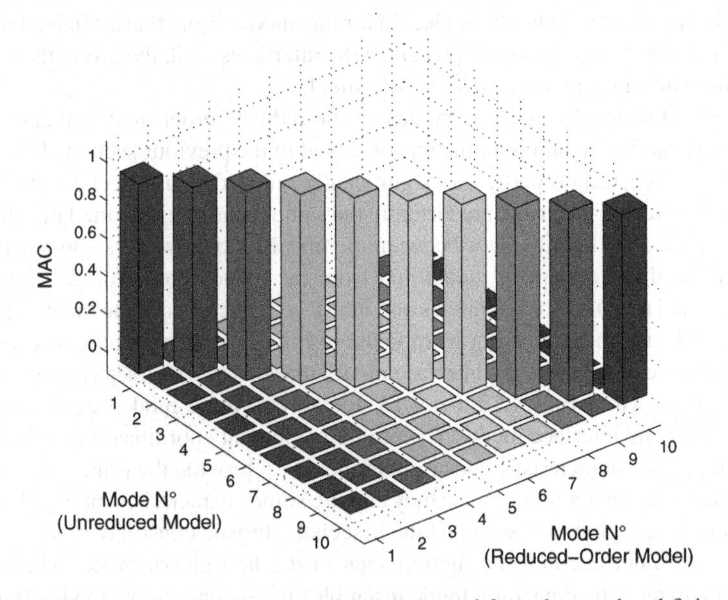

Fig. 4.11 MAC-values between the mode shapes computed from the unreduced finite element model and from the reduced-order model based on dominant and residual normal modes and interface reduction. Reduced-order model generated from Model-2

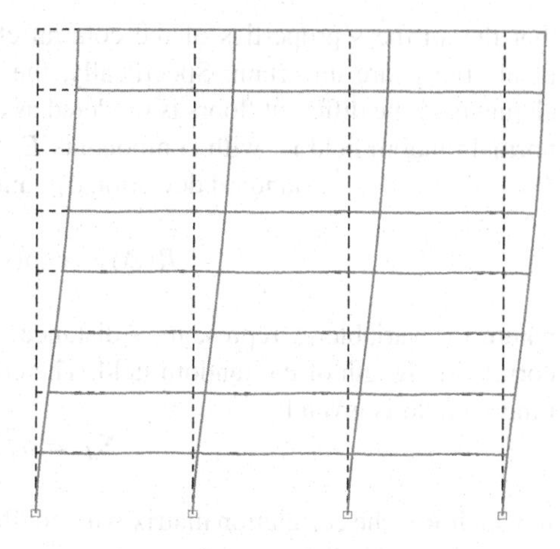

Fig. 4.12 First characteristic constraint mode

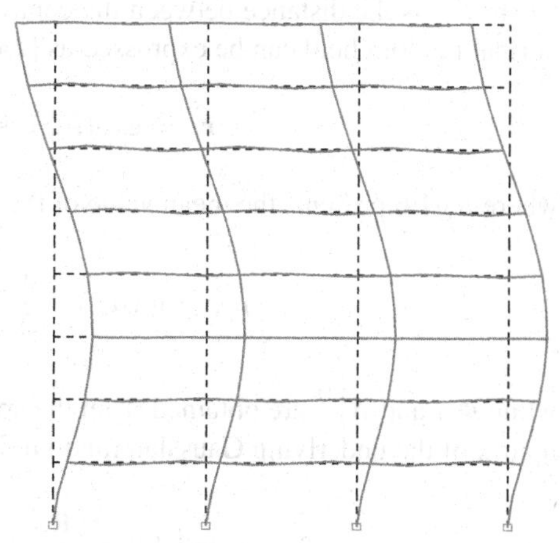

Fig. 4.13 Second characteristic constraint mode

4.6.5 Reliability Problem

To control serviceability, the performance of the structure is characterized in terms of the probability of occurrence of a failure event related to the maximum relative displacement between the top of the model and the ground, or $\delta(t, \mathbf{z}, \boldsymbol{\theta})$. Mathematically, the failure event $F(\mathbf{z}, \boldsymbol{\theta})$ is defined as $F(\mathbf{z}, \boldsymbol{\theta}) = \{d(\mathbf{z}, \boldsymbol{\theta}) \geq 1\}$ where the demand function is given by

$$d(\mathbf{z}, \boldsymbol{\theta}) = \max_{t \in [0,T]} \left(\left| \frac{\delta(t, \mathbf{z}, \boldsymbol{\theta})}{\delta^*} \right| \right) \tag{4.17}$$

where δ^* is the acceptable threshold of the maximum relative displacement of the eighth floor with respect to the ground. Of course, additional responses can be considered in the definition of the failure event. Recall that in the previous expressions, $\boldsymbol{\theta}$ represents the vector of uncertain system parameters. In this regard, it is assumed that the stiffness properties of the column elements, represented by the modulus of elasticity, are uncertain. Specifically, the modulus of elasticity of the column elements of the different floors is modeled as a discrete homogeneous isotropic log-normal random field $\mathbf{r}_E$ with components $E_i, i = 1, \ldots, 8$, mean value $\mu_E \mathbf{1}$ where $\mathbf{1} =< 1, \ldots, 1 >^T$, standard deviation σ_E, and correlation function

$$R(\Delta) = \exp(-\alpha \Delta^2) \tag{4.18}$$

where the variable Δ represents a distance and the parameter α is related to the correlation length of the random field. The corresponding covariance matrix of the random field is given by

$$\boldsymbol{\Sigma}_E = \sigma_E^2 \mathbf{R} \tag{4.19}$$

in which $\mathbf{R}$ is the correlation matrix with coefficients $R_{ij} = R(\Delta_{ij}), i, j = 1, \ldots, 8$, where Δ_{ij} is the distance between the centroid of the i and j floors. Then, the log-normal random field can be expressed as [23, 30, 32, 64]

$$\mathbf{r}_E = \exp(\mu_N \mathbf{1} + \boldsymbol{\Phi}_N \boldsymbol{\Lambda}_N^{1/2} \mathbf{y}) \tag{4.20}$$

where $\mu_N \mathbf{1}$ represents the mean value of the underlying Gaussian random field with

$$\mu_N = \ln(\mu_E) - \frac{1}{2} \ln\left(1 + \frac{\sigma_E^2}{\mu_E^2}\right), \tag{4.21}$$

while $\boldsymbol{\Phi}_N$ and $\boldsymbol{\Lambda}_N^{1/2}$ are obtained from the spectral decomposition of the covariance matrix of the underlying Gaussian random field $\boldsymbol{\Sigma}_N$, with coefficients

$$\boldsymbol{\Sigma}_{Nij} = \ln\left(1 + \frac{\sigma_E^2 \mathbf{R}_{ij}}{\mu_E^2}\right) \quad , \quad i, j = 1, \ldots, 8 \, , \tag{4.22}$$

and $\mathbf{y}$ is a vector of independent standard normal random variables. The mean value and standard deviation of the log-normal random field are set equal to $\mu_E = 2.0 \times 10^{10}\,\text{N/m}^2$ and $\sigma_E = 3.0 \times 10^9\,\text{N/m}^2$, respectively. Thus, the corresponding coefficient of variation of the random field is equal to 15%. A mildly correlated random field is considered by selecting an appropriate value of α. The corresponding correlation function is shown in Fig. 4.14.

The model is excited horizontally by a ground acceleration modeled as indicated in Sect. 4.5. The moment magnitude and epicentral distance are taken as $M = 7.0$ and $r = 25$ km, respectively. The near-field pulse parameters are fixed at their nominal values as suggested in [44], i.e., $A_p = 27.11$ (cm/s), $f_p = 0.53$ (Hz), $\nu_p = 0.0$ (rad), and $\gamma_p = 1.8$. The envelope function to be used is given by [61]

$$e(t, M, r) = a_1 \left(\frac{t}{2T}\right)^{a_2} \cdot \exp\left(-a_3 \cdot \frac{t}{2T}\right) \tag{4.23}$$

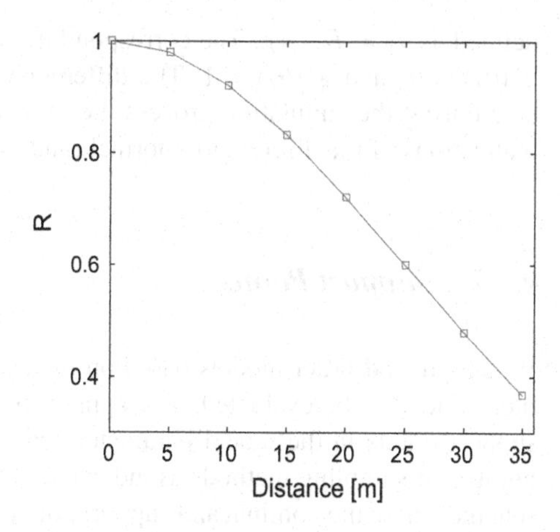

Fig. 4.14 Correlation function of the random field. First application problem

where T corresponds to the duration of the ground motion and the parameters a_1, a_2 and a_3 are defined as

$$a_1 = \left(\frac{e}{\lambda}\right)^{a_2} \ , \ a_2 = \frac{-\lambda \ln(\eta)}{1 + \lambda \cdot (\ln(\lambda) - 1)} \ , \ a_3 = \frac{a_2}{\lambda} \tag{4.24}$$

with parameter values equal to $\lambda = 0.2$ and $\eta = 0.05$. The sampling interval and the duration of the excitation are taken equal to $\Delta t = 0.01$ s and $T = 30$ s, respectively. Thus, the characterization of the stochastic excitation involves more than 3,000 uncertain parameters in this case (white noise sequence). Clearly, the corresponding reliability problem is a high-dimensional problem.

4.6.6 Remarks on the Use of Reduced-Order Models

It is noted that even though subset simulation is an effective advanced simulation technique, the reliability analysis can be computationally very demanding due to the large number of dynamic analyses required during the simulation process (evaluation of the indicator function). Thus, the repetitive generation of reduced-order models for different values of the uncertain model parameters $\boldsymbol{\theta}$ can be computationally expensive due to the substantial computational overhead that arises at the substructure level. To cope with this difficulty, reduced-order models together with the parametrization schemes introduced in Chaps. 2 and 3 are used to estimate the system reliability. With respect to Chap. 2 and based on the previous characterization of the uncertain parameter, it is clear that substructures $S_j, j = 1, \ldots, 8$, depend on the model parameters related to the modulus of elasticity, while substructures $S_j, j = 9, \ldots, 16$, are independent of the model parameters. For implementation purposes, the model parameters associated with substructures $S_j, j = 1, \ldots, 8$, are defined as $\theta_j = E_j/\mu_E$. The corresponding parametrization functions are given by $h^j(\theta_j) = \theta_j$ and $g^j(\theta_j) = 1$. The different values that the model parameters may take during the simulation process, i.e., subset simulation, correspond to different realizations of the discrete log-normal random field.

4.6.7 Support Points

When reduced-order models based on interface reduction are considered, interface modes need to be evaluated. The approximation of these modes involves a set of support points in the model parameter space. These points can be generated by a number of sampling methods as indicated in Sect. 3.1.5. In this section, an adaptive scheme where the nominal and support points are updated during the different stages of subset simulation is introduced. The basic idea is to use support points lying in the vicinity of the intermediate failure domains in order to increase the accuracy of the approximate interface modes.

The selected support points at a given stage of subset simulation are Latin Hypercube samples from a normal distribution whose definition is based on samples from the previous stage. Specifically, at stage k of subset simulation, $N_s = p_0 N$ conditional samples that lie in F_k ($\{\boldsymbol{\theta}_{k-1,i}, i = 1, \ldots, N_s\}$) are obtained. Based on these samples, the sample mean $\bar{\boldsymbol{\theta}}_k$ and the sample covariance matrix Σ_k are computed as

$$\bar{\boldsymbol{\theta}}_k = \frac{1}{N_s} \sum_{i=1}^{N_s} \boldsymbol{\theta}_{k-1,i} \tag{4.25}$$

and

$$\Sigma_k = \frac{1}{N_s} \sum_{i=1}^{N_s} [(\boldsymbol{\theta}_{k-1,i} - \bar{\boldsymbol{\theta}}_k)(\boldsymbol{\theta}_{k-1,i} - \bar{\boldsymbol{\theta}}_k)^T] \tag{4.26}$$

Then, the support points to be used during stage k of subset simulation are generated from the normal distribution $N(\bar{\boldsymbol{\theta}}_k, \beta_k \Sigma_k)$, where β_k is a user-selected parameter scaling the covariance matrix Σ_k. Such a parameter is problem-dependent. Additional conditional samples can also be used for the purpose of defining the sample mean and covariance matrix. In this case, conditional samples can be simulated from the available N_s samples by the Modified Metropolis algorithm [5]. The complete set of conditional samples is then used to characterize the normal distribution from which the support points are generated. The support points generated by the proposed adaptive scheme spread over the important region of failure in the uncertain parameter space for the examples that are considered in this section. To control the accuracy of the global surrogate model, the support points correspond to direct evaluation of the interface modes. In this manner, the propagation of error that occurs in previous stages is avoided. In addition, to consider only interpolations, the point at which the reduced-order model needs to be recomputed, $\boldsymbol{\theta}^*$, should belong to the n_θ-dimensional convex hull of the support points [3, 11]. If this condition is not satisfied, a direct evaluation of the interface modes is required for updating the reduced-order model. In the case of complex failure domains, i.e., when the failure samples are distributed in disjoint sets, a cluster analysis can be performed in each stage of subset simulation, to order the samples into clusters [26]. In this way, support points can be generated for each cluster. The choice concerning which of the set of support points are used for a given sample is based on its distance with respect to the center of each cluster [43]. The use of cluster analyses is not necessary for the numerical examples considered in this chapter.

Alternatively, the support points can be defined in terms of the Markov chains generated from the conditional samples at each stage of subset simulation. As previously pointed out, at each stage of subset simulation, a number of conditional samples that lie in F_k are already available. Starting from these samples, additional samples are simulated through Markov chain Monte Carlo simulation using an adaptive conditional sampling algorithm [31]. In each adaptation step, a number of seeds are chosen at random from the available conditional samples. After running the algorithm for a number of adaptation steps, a set of support points, to be used during the current stage of subset simulation, can be obtained.

4.6.8 Reliability Results

Figure 4.15 shows the probability of failure in terms of the threshold by using the unreduced model and several reduced-order models generated from Model-2. Three reduced-order models are considered in the figure, namely: reduced-order model based on dominant fixed-interface normal modes; model based on dominant and

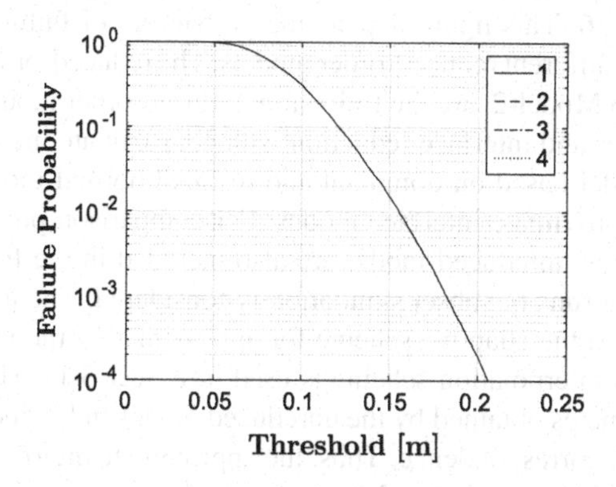

Fig. 4.15 Probability of failure in terms of the threshold level. 1: unreduced model. 2: reduced-order model based on dominant fixed-interface normal modes. 3: reduced-order model based on dominant and residual fixed-interface normal modes. 4: reduced-order model based on dominant and residual fixed-interface normal modes and interface reduction

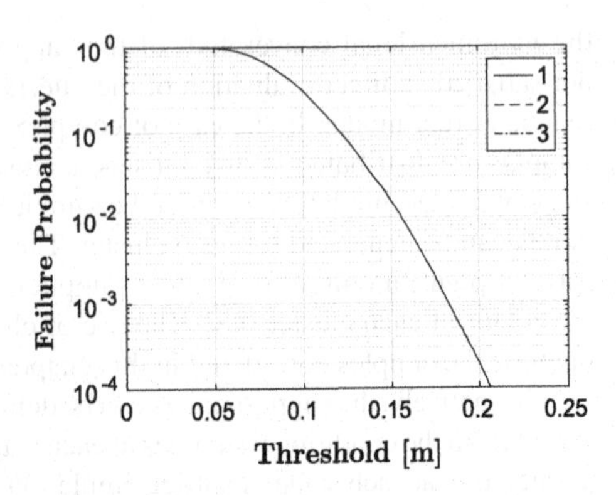

Fig. 4.16 Probability of failure in terms of the threshold level. 1: unreduced model. 2: reduced-order model based on dominant fixed-interface normal modes and approximate interface modes. 3: reduced-order model based on dominant and residual fixed-interface normal modes and approximate interface modes

residual fixed-interface normal modes; and model based on dominant and residual fixed-interface normal modes and interface reduction. In the case of interface reduction, no approximations are considered for the interface modes. In other words, they are directly evaluated during the simulation process. However, partial invariant conditions are assumed for the transformation matrix, which accounts for the contribution of the residual fixed-interface normal modes (see Sects. 2.3.6 and 3.3.5). The curves in the figure correspond to an average of five independent runs of subset simulation. The figure illustrates the whole trend of the probability of failure in terms of different thresholds, not only for one target value. It is observed that the system reliability obtained from the unreduced model coincides with the one obtained from the reduced-order models for all range of thresholds, even for low failure probabilities, i.e. 10^{-4}. Note that in this case, the reduced-order model based on dominant fixed-interface normal modes is adequate in the context of the reliability problem under consideration.

The effect of approximate interface modes on the accuracy of the reliability results is shown in Fig. 4.16. This figure depicts the probability of failure in terms of the threshold by using different reduced-order models. The reduced-order models, which are generated from Model-2, are the following: reduced-order model based on dominant normal modes and interface reduction with approximate interface modes; and reduced-order model based on dominant and residual normal modes and interface reduction with approximate interface modes. For comparison purposes, the results corresponding to the unreduced model are also included in the figure. An average of five independent runs of subset simulation is considered. The number of support points considered in the adaptive scheme for approximating the interface modes is 36, where a linear interpolation scheme is used (see Sect. 3.1). The comparison of the reliability estimates obtained by the unreduced model and reduced-order models shows an excellent correspondence. Thus, the approximate interface modes are able to accurately predict the response of the system and, consequently, its reliability.

Table 4.6 Speedup attained for different models. First application problem

Model	Speedup
Unreduced	1
Reduced-order-model-1	4
Reduced-order-model-2	2
Reduced-order-model-3	2
Reduced-order-model-4	3
Reduced-order-model-5	3

Table 4.7 Description of reduced-order models. First application problem

Model	Description
Reduced-order-model-1	Reduced-order model based on dominant normal modes
Reduced-order-model-2	Reduced-order model based on dominant normal modes and interface modes
Reduced-order-model-3	Reduced-order model based on dominant and residual normal modes and interface modes
Reduced-order-model-4	Reduced-order model based on dominant normal modes and approximate interface modes
Reduced-order-model-5	Reduced-order model based on dominant and residual normal modes and approximate interface modes

4.6.9 Computational Cost

The computational effort involved in the reliability analysis is shown in Table 4.6. Specifically, this table shows the speedup (round to the nearest integer) achieved by different reduced-order models, which are described in Table 4.7. In this context, the speedup is the ratio of the execution time by using the unreduced model and the execution time by using a reduced-order model.

The speedups reported in the table are based on the implementation of the reliability analysis in a four-core computer unit (Intel Core i7 processor). The actual procedure is carried out by using a homemade code based on a Matlab C++ platform. First, it is noted that a speedup equal to 4 is obtained by using the reduced-order model based on dominant normal modes. This value reduces to 2 when interface and residual normal modes are considered. This is mainly due to the update process of the interface modes and the consideration of the residual normal modes during the simulation process. However, when approximate interface modes are considered, the corresponding speedups increase to 3. Thus, the effect of considering approximate interface modes is also positive in terms of the numerical implementation of the reliability analysis.

Based on the previous results, it is seen that the use of reduced-order models for estimating the reliability of the system is rather effective. In fact, a reduction in computational effort by a factor between 2 and 4 is achieved without compromising the accuracy of the reliability estimates. It is expected that a more significant effect will be obtained for more involved finite element models (see next Application Problem).

4.7 Application Problem No. 2

The objective of this example is to explore the effectiveness of reduced-order models based on interface reduction. In particular, an involved nonlinear finite element building model is considered.

4.7.1 Structural Model

The three-dimensional finite element building model shown in Fig. 4.17 is considered as the second application problem. The application involves a 55-story building model with a total height of 190 m. The plan view and the dimensions of a typical floor are shown in Fig. 4.18. The building has a reinforced concrete core of shear walls and a reinforced concrete perimeter moment-resisting frame as shown in Fig. 4.18. The columns of the perimeter have a circular cross section. The floors and walls are modeled by shell elements of different thicknesses. Additionally, beam and column elements are used in the finite element model, which has 89,000 degrees of freedom. Material properties are given by the Young's modulus $E = 2.45 \times 10^{10}\,\text{N/m}^2$, mass density $\rho = 2{,}500\,\text{kg/m}^3$, and Poisson's ratio $\mu = 0.3$. Finally, 5% of critical damping is added to the model.

For an improved performance, the structural system is reinforced with a total of 45 nonlinear vibration control devices placed in two different configurations, i.e., longitudinal (x) and transverse (y) directions. A typical configuration of the vibration control devices, at the floors where they are located, is shown in Fig. 4.19. Each longitudinal device consists of brace and plate elements where a series of metallic U-shaped flexural plates (UFP's) are located between the plates, as shown in Fig. 4.20 [35]. On the other hand, each transverse device consists of concrete walls where the UFP's are located between them, as illustrated in Fig. 4.20.

Each UFP exhibits a one-dimensional hysteretic type of nonlinearity modeled by the restoring force law

$$f_{NL}(t) = \alpha\, k_e\, \delta(t) + (1-\alpha)\, k_e U^y\, z(t) \tag{4.27}$$

where k_e is the pre-yield stiffness, U^y is the yield displacement, α is the factor that defines the extent to which the restoring force is linear, $z(t)$ is a dimensionless hysteretic variable, and $\delta(t)$ is the relative displacement between the upper and lower

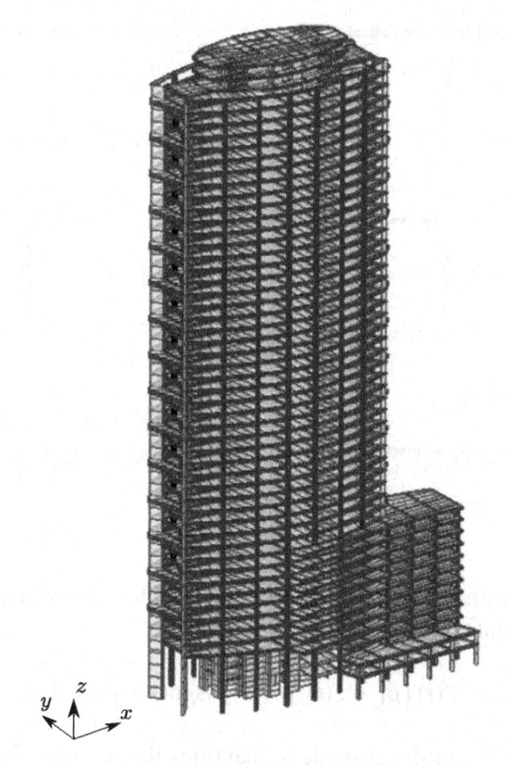

Fig. 4.17 Three-dimensional finite element building model. Example No. 2

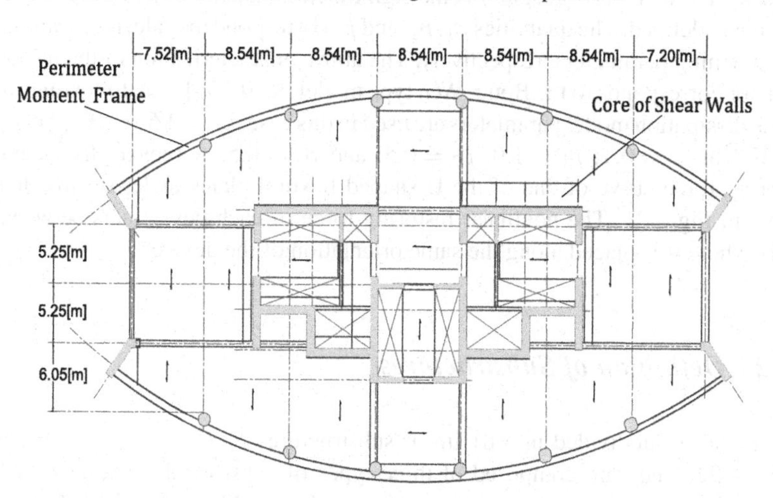

Fig. 4.18 Typical floor plan of the 55-story building model

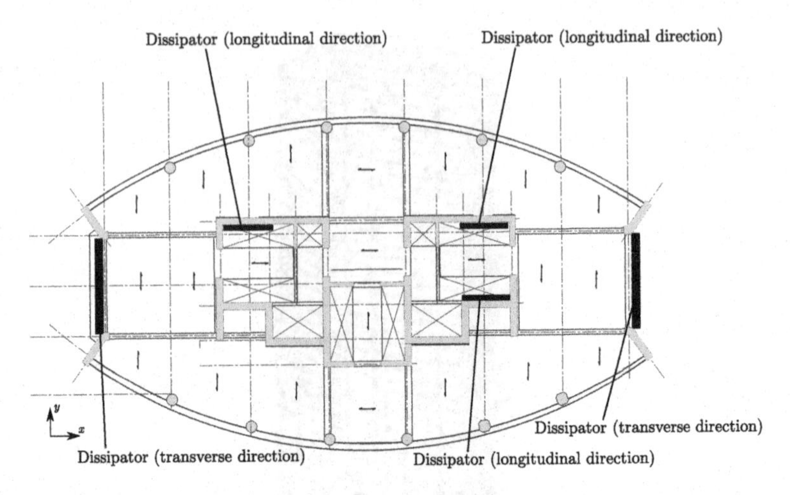

Fig. 4.19 Typical configuration of vibration control devices

surfaces of the flexural plates. The hysteretic variable $z(t)$ satisfies the first-order nonlinear differential equation

$$\dot{z}(t) = \dot{\delta}(t)\left[\beta_1 - z(t)^2[\beta_2 + \beta_3 \text{sgn}(z(t)\dot{\delta}(t))]\right]/U^y \tag{4.28}$$

where β_1, β_2 and β_3 are dimensionless quantities that characterize the properties of the hysteretic behavior, $\text{sgn}(\cdot)$ is the sign function, and all other terms have been previously defined. The quantities β_1, β_2, and β_3 correspond to scale, loop fatness and loop pinching parameters, respectively. The above characterization of the hysteretic behavior corresponds to the Bouc–Wen type model [9, 10, 54]. The following values for the dissipation model parameters are used in this case: $k_e = 2.5 \times 10^6$ N/m; $U^y = 5 \times 10^{-3}$m; $\alpha = 0.1$; $\beta_1 = 1.0$; $\beta_2 = 0.5$; and $\beta_3 = 0.5$. A typical displacement-restoring force curve of one of the U-shaped flexural plates under seismic load is shown in Fig. 4.21. The nonlinear restoring force of each device acts between the floors where it is placed along the same orientation of the device.

4.7.2 *Definition of Substructures*

The model is subdivided into 81 linear substructures $S_i, i = 1, \ldots, 81$, as shown in Fig. 4.22. They are composed of three types of substructures, namely: core of shear walls located between two floors ($S_i, i = 1, \ldots, 27$); slabs of different floors ($S_i, i = 28, \ldots, 54$); and circular columns of the perimeter frame located between two floors and the corresponding slab of the intermediate floor ($S_i, i = 55, \ldots, 81$). Figure 4.23 depicts a typical substructure of each type. In addition, there are 45 nonlinear substructures comprised by the nonlinear vibration control devices defined

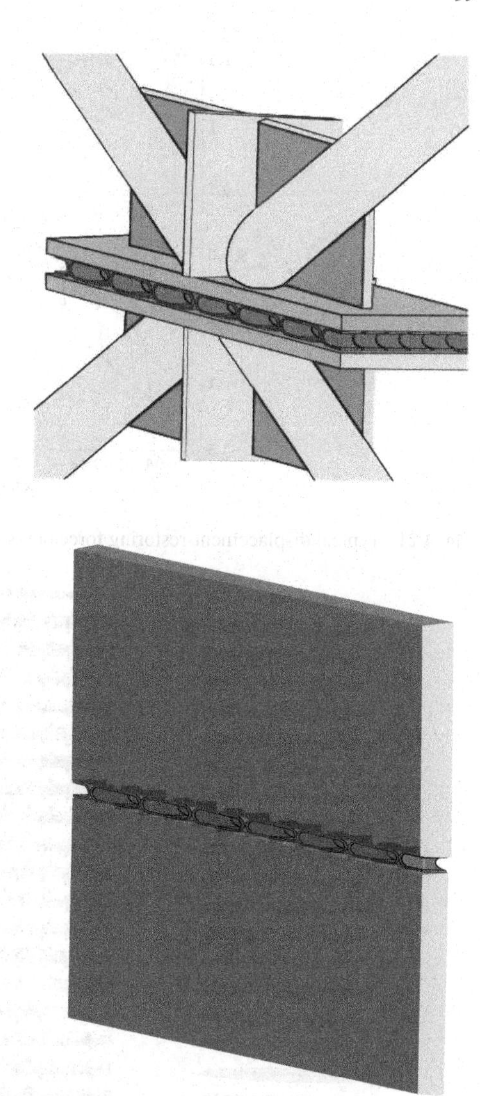

Fig. 4.20 Upper figure: Model of vibration control device in the longitudinal direction. Lower figure: Model of vibration control device in the transverse direction

in the previous section. With this subdivision, the total number of internal degrees of freedom is equal to 65,300, while 23,700 degrees of freedom are present at the interfaces. A small number of fixed-interface normal modes is selected for the model. In particular, a model characterized by only 252 fixed-interface normal modes is considered. In addition, 100 interface modes, which represent about 0.5% of the total number of interface degrees of freedom, are used in the model. Thus, the total number of generalized coordinates of the reduced-order model represents more than 99% reduction with respect to the unreduced finite element model.

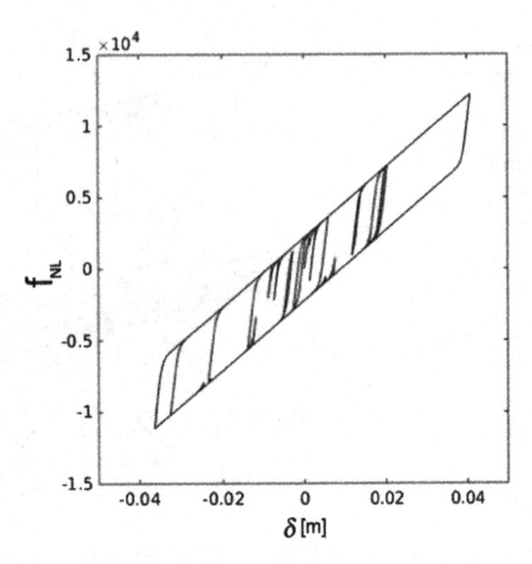

Fig. 4.21 Typical displacement-restoring force curve of one of the U-shaped flexural plates

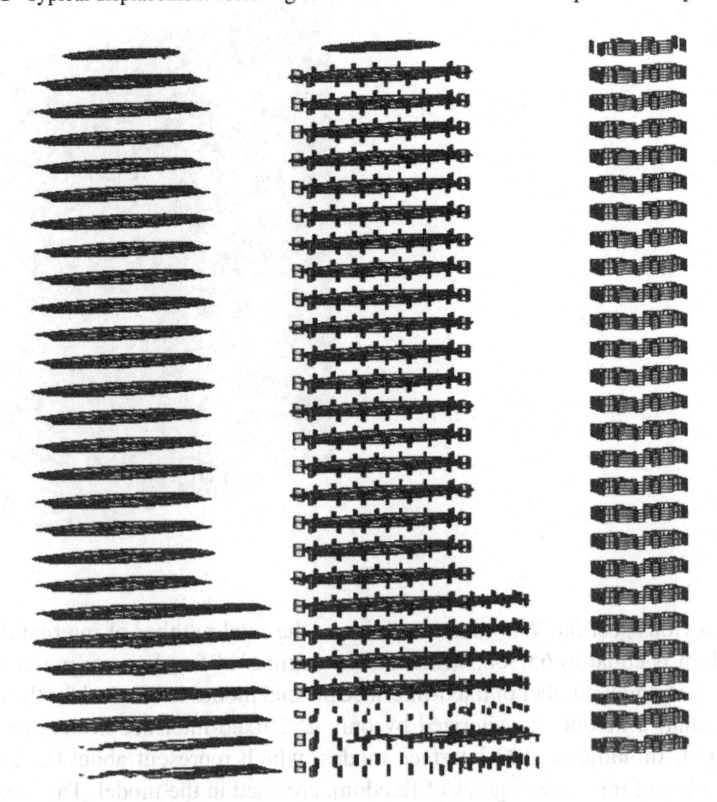

Fig. 4.22 Substructures of the finite element model. Application problem No. 2

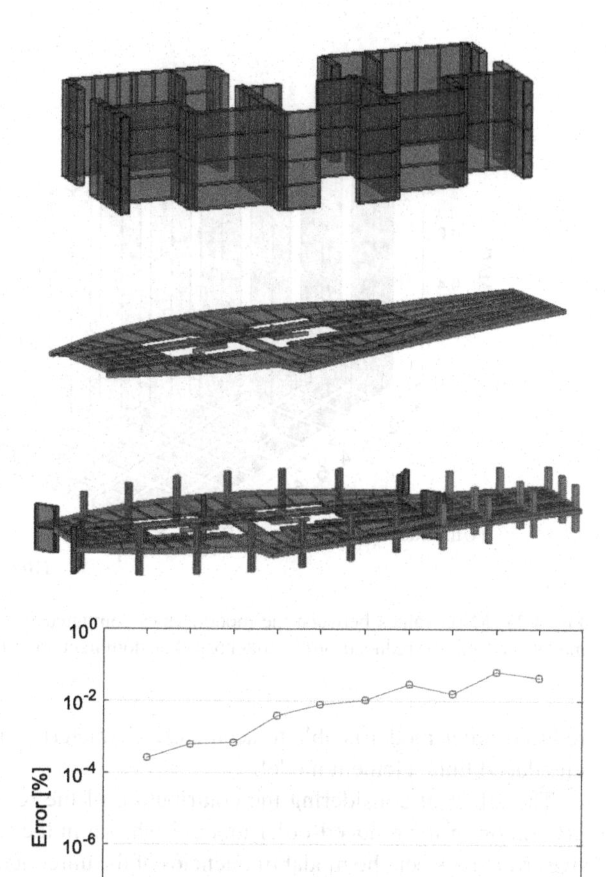

Fig. 4.23 Typical substructure of each type (shear wall, slab, perimeter moment frame and slab)

Fig. 4.24 Relative frequency errors between the modal frequencies of the full finite element model and of the reduced-order model based on dominant normal modes and interface reduction

Figure 4.24 shows the relative errors between the modal frequencies of the unreduced finite element model and the modal frequencies of the reduced-order model based on dominant modes and interface reduction. The first 10 modes are considered for reference purposes. The corresponding MAC-values between the first 10 modal vectors computed from the unreduced finite element model and from the reduced-order model are shown in Fig. 4.25. It is seen that the errors for the modal frequencies are quite small. The accuracy of the results is also seen for the modal vectors. In fact, the values at the diagonal terms of the matrix of MAC-values are one, while the off-diagonal terms are zero. Consequently, the mode shapes of the reduced-order model are consistent with the mode shapes from the unreduced model. Thus, the

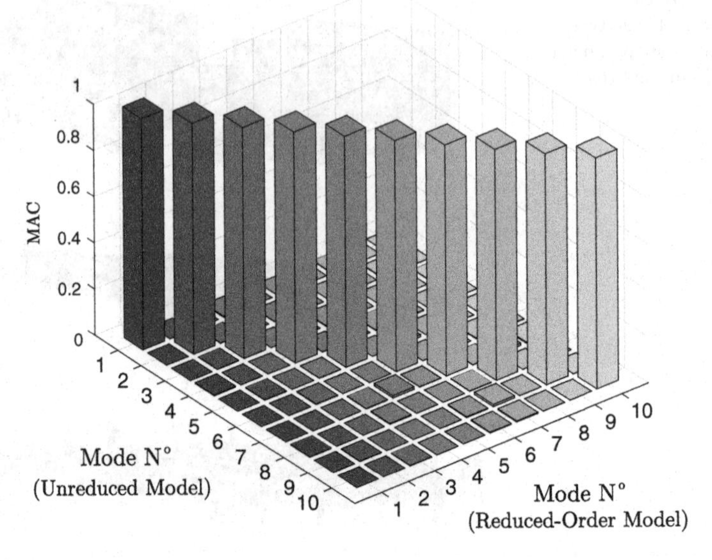

Fig. 4.25 MAC-values between the mode shapes computed from the unreduced finite element model and from the reduced-order model based on dominant normal modes and interface reduction

reduced-order model is able to accurately characterize the important modes of the unreduced finite element model.

The effect of considering the contribution of the residual normal modes in the generation of the reduced-order model is shown in the following figures. The relative errors between the modal frequencies of the unreduced finite element model and the modal frequencies of the reduced-order model are shown in Fig. 4.26. The corresponding matrix of MAC-values between the first 10 modal vectors computed from the unreduced finite element model and from the reduced-order model is shown in Fig. 4.27. The effect of considering the residual normal modes in the analysis is evident. The difference in the errors for the modal frequencies is more than four orders of magnitude with respect to the ones obtained from the reduced-order model based on dominant modes only. Thus, the contribution of the residual normal modes significantly enhances the accuracy of the reduced-order model. In addition, the matrix of MAC-values indicates that both models are consistent. It is important to stress that the construction of the reduced-order model is carried out offline, that is, before the reliability analysis takes place. Thus, this process is independent of the reliability analysis, which can be computationally quite demanding.

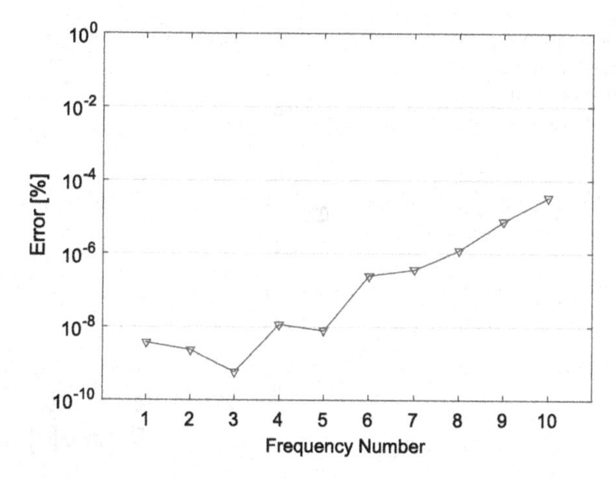

Fig. 4.26 Relative frequency errors between the modal frequencies of the full finite element model and of the reduced-order model. Reduced-order model based on dominant and residual normal modes and interface reduction

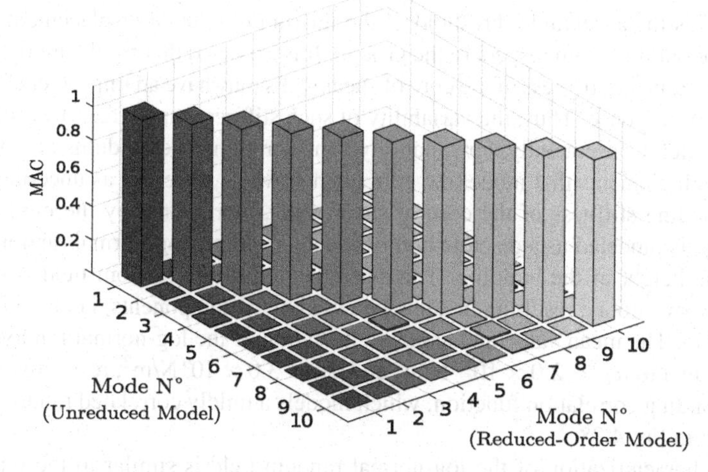

Fig. 4.27 MAC-values between the mode shapes computed from the unreduced finite element model and from the reduced-order model. Reduced-order model based on dominant and residual normal modes and interface reduction

4.7.3 System Reliability

The failure event is formulated as a first excursion problem during the time of analysis as indicated in Sect. 4.2. For illustration purposes, the structural response to be controlled is the displacement at the top of the building. Thus, the corresponding demand function is characterized as

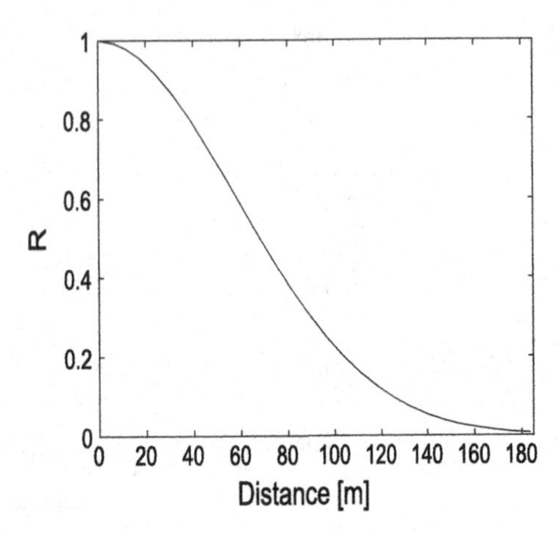

Fig. 4.28 Correlation function of the random field. Second application problem

$$d(\mathbf{z}, \boldsymbol{\theta}) = \max_{t \in [0,T]} \left(\left| \frac{\delta(t, \mathbf{z}, \boldsymbol{\theta})}{\delta^*} \right| \right) \tag{4.29}$$

where δ^* is the acceptable threshold of the maximum relative displacement of the top of the building with respect to the ground. It is expected that for the model under consideration, the stiffness of the core of shear walls may have an important effect on the system response. Thus, the variability of such stiffness may affect the reliability of the model. Consequently, for reliability considerations, the modulus of elasticity of the shell elements that model the core of shear walls is treated as uncertain. The corresponding stiffness of the core of shear walls, represented by the modulus of elasticity, is modeled as a discrete homogeneous isotropic log-normal random field along the height of the building. The discretization of the random field is carried out every two floors, resulting in a discrete field of 27 components, i.e. $\mathbf{r}_E$, $(E_i, i = 1, \ldots, 27)$. The mean value and standard deviation of the log-normal random field are set equal to $\mu_E = 2.0 \times 10^{10}\,\text{N/m}^2$ and $\sigma_E = 3.0 \times 10^9\,\text{N/m}^2$, respectively. The corresponding correlation function, which models a mildly correlated random field, is shown in Fig. 4.28.

The characterization of the log-normal random field is similar to the one considered in Sect. 4.6.5. Based on the previous definition of the substructures and the characterization of the uncertainty, it is clear that the substructures related to the core of shear walls $(S_j, j = 1, \ldots, 27)$ depend on the model parameters associated with the modulus of elasticity. For implementation purposes, the model parameters related to substructures S_j, $j = 1, \ldots, 27$, are defined as $\theta_j = E_j/\mu_E$. The related parametrization functions are given by $h^j(\theta_j) = \theta_j$ and $g^j(\theta_j) = 1$. The other substructures are independent of the model parameters. The different values that the model parameters may assume during the simulation process correspond to different realizations of the discrete log-normal random field. The same excitation used in the previous example is considered in the present application. Note that the characterization of the stochastic excitation involves more than 3,000 random variables. This number of random variables plus the 27 model parameters indicate that the reliability estimation constitutes a high reliability problem. Due to the dimension and complexity of the finite element model at hand, it is expected that the use of model reduction techniques and parametrization schemes will have an important effect on the computational cost of the reliability analysis. Such an effect is illustrated in Sect. 4.7.5.

4.7.4 Results

The probability of failure in terms of the threshold by using two reduced-order models is shown in Fig. 4.29. They consist of models based on dominant fixed-interface normal modes and interface modes, and models based on dominant fixed-interface normal modes and approximate interface modes. When approximate interface modes are considered, two approaches are used: linear and quadratic interpolation schemes (see Sect. 3.1). In the case of linear interpolation, 81 support points are used in the adaptive scheme proposed in Sect. 4.6.7, while 162 are employed in the quadratic case. An average of five independent runs is considered in the figure. First, it is observed that the results of the models based on exact and approximate interface modes are coincident. Thus, the approximation schemes for approximating the interface modes are adequate. Based on the results of the previous section, regarding the accuracy of the reduced-order models, it is expected that the reduced-order model based on dominant fixed-interface normal modes and exact interface modes will produce reliability estimates with sufficient accuracy. Therefore, this case can be considered as the exact one for comparison purposes. From Fig. 4.29, it is also seen

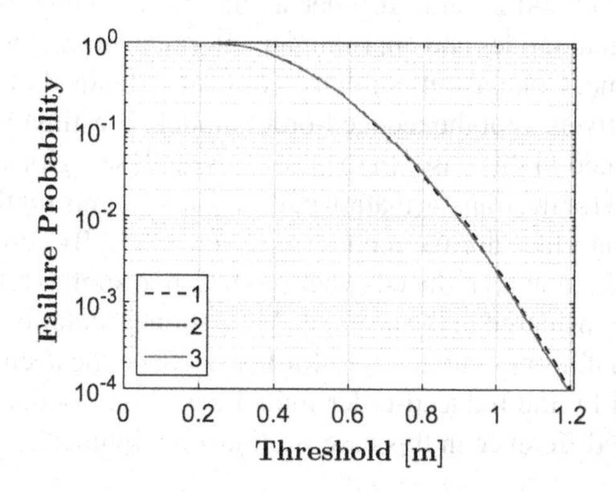

Fig. 4.29 Probability of failure in terms of the threshold level. 1: reduced-order model based on dominant fixed-interface normal modes and exact interface modes. 2: reduced-order model based on dominant fixed-interface normal modes and approximate interface modes (linear interpolation scheme). 3: reduced-order model based on dominant fixed-interface normal modes and approximate interface modes (quadratic interpolation scheme)

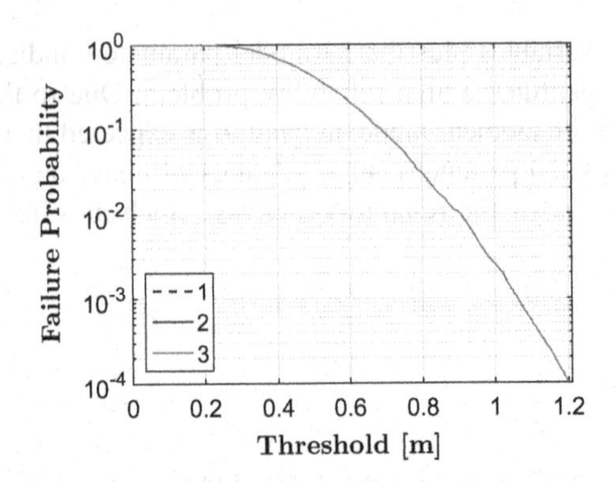

Fig. 4.30 Probability of failure in terms of the threshold level. 1: reduced-order model based on dominant and residual fixed-interface normal modes and exact interface modes. 2: reduced-order model based on dominant and residual fixed-interface normal modes and approximate interface modes (linear interpolation scheme). 3: reduced-order model based on dominant and residual fixed-interface normal modes and approximate interface modes (quadratic interpolation scheme)

that the reliability estimates of both models that use approximate interface modes agree very well. Then, the use of a linear interpolation scheme is sufficient in the context of this application.

The effect of considering the contribution of the residual normal modes on the reliability estimates is shown in Fig. 4.30. This figure presents the probability of failure in terms of the threshold by using the following reduced-order models: a reduced-order model based on dominant and residual fixed-interface normal modes and exact interface modes; a reduced-order model based on dominant and residual fixed-interface normal modes and approximate interface modes by using a linear interpolation scheme; and a reduced-order model based on dominant and residual fixed-interface normal modes and approximate interface modes by using a quadratic interpolation scheme. Conclusions similar to the ones obtained in the previous case regarding the effectiveness of the reduced-order models in estimating the probability of failure are obtained in this case. In the previous analyses, global invariant conditions were assumed for the transformation matrix that accounts for the contribution of the residual fixed-interface normal modes (see Sect. 3.3.5). By comparing Figs. 4.29 and 4.30, it is noticed that all reduced-order models give similar reliability estimates for the thresholds considered in the analysis. Validation calculations indicate that the effect of the residual normal modes is to further enhance the accuracy of reliability estimates obtained by the reduced-order model based on dominant normal modes only. However, the difference in this case is almost negligible.

4.7.5 Computational Effort

Table 4.8 shows the speedup (round to the nearest integer) achieved by the different implementations considered in the previous figures. The corresponding characterization of the reduced-order models is indicated in Table 4.9. Recall that the speedup is the ratio of the execution time by using the full finite element model and the execution time by using a reduced-order model. In this regard, the execution time in performing the reliability analysis by using the full finite element model is approximated as follows. The total number of dynamic analyses involved in the results shown in Figs. 4.29 and 4.30 is approximately 3,700 (four stages of subset simulation). The time for performing one dynamic analysis of the full model is about 4.3 min. Multiplying this time by the total number of dynamic analyses required by the simulation process, the computational effort is expected to be of the order of 265 h (more than 11 days). As in the previous example, the procedure is carried out by using a homemade code based on a Matlab C++ platform.

It is seen that a speedup of six is obtained by the reduced-order model based on dominant fixed-interface normal modes and exact interface modes. This value increases to a speedup of more than 20 when approximate interface modes are considered. Thus, the effect of using approximate interface modes is significant in terms of the computational effort. This reduction in computational time does not compromise the accuracy of the reliability estimates. Furthermore, a speedup value of the order of 10 is achieved when the residual normal modes are explicitly considered in the analysis. For the same number of fixed-interface normal modes per substructure, the computational burden for using residual normal modes is increased by a factor of two for this example. This increase is compensated by the significantly higher accuracy provided by the reduced-order model with residual normal modes (see Figs. 4.24 and 4.26). Based on the previous results, it is noted that for practical purposes, the results obtained from the reduced-order model based on dominant fixed-interface normal modes and approximate interface modes can be used to compute the reliability estimates. Thus, an important reduction in computational efforts is obtained by using the reduced-order model instead of the full finite element model. The gain in computational savings for this structural model is significant considering the complexity associated with the distributed nonlinearities along the height of the building arising from the installation of the vibration control devices.

Table 4.8 Speedup attained for different models. Second application problem

Model	Speedup
Reduced-order-model-1	6
Reduced-order-model-2	23
Reduced-order-model-3	21
Reduced-order-model-4	12
Reduced-order-model-5	9

Table 4.9 Description of reduced-order models. Second application problem

Model	Description
Reduced-order-model-1	Reduced-order model based on dominant normal modes and exact interface modes
Reduced-order-model-2	Reduced-order model based on dominant normal modes and approximate interface modes (linear interpolation)
Reduced-order-model-3	Reduced-order model based on dominant normal modes and approximate interface modes (quadratic interpolation)
Reduced-order-model-4	Reduced-order model based on dominant and residual normal modes and approximate interface modes (linear interpolation)
Reduced-order-model-5	Reduced-order model based on dominant and residual normal modes and approximate interface modes (quadratic interpolation)

Finally, it is noted that once a reduced-order model has been defined, several scenarios in terms of different failure events and system responses can be explored and considered for reliability purposes in an efficient manner. Therefore, even higher speedup values can be obtained for the reliability analysis process as a whole.

References

1. J.G. Anderson, S.E. Hough, A model for the shape of the Fourier amplitude spectrum of acceleration at high frequencies. Bull. Seism. Soc. Am. **74**(5), 1969–1993 (1984)
2. P. Angelikopoulos, C. Papadimitriou, P. Koumoutsakos, Bayesian uncertainty quantification and propagation in molecular dynamics simulations: a high performance computing framework. J. Chem. Phys. **137**, 1441103-1–144103-19 (2012)
3. P. Angelikopoulus, C. Papadimitriou, P. Koumoutsakos, X-TMCMC: adaptive kriging for Bayesian inverse modeling. Comput. Methods Appl. Mech. Eng. **289**, 409–428 (2015)
4. G.M. Atkinson, W. Silva, Stochastic modeling of California ground motions. Bull. Seism. Soc. Am. **90**(2), 255–274 (2000)
5. S.-K. Au, J.L. Beck, Estimation of small failure probabilities in high dimensions by subset simulation. Probabilistic Eng. Mech. **16**(4), 263–277 (2001)
6. S.-K. Au, Y. Wang, *Engineering Risk Assessment with Subset Simulation* (Wiley, New York, 2014)
7. S.-K. Au, E. Patelli, Rare event simulation in finite-infinite dimensional space. Reliab. Eng. Syst. Saf. **148**, 67–77 (2016)
8. B.M. Ayyub, M.M. Gupta, L.N. Kanal, *Analysis and Management of Uncertainty: Theory and Applications* (Elsevier Scientific Publisher, North-Holland, 1992)
9. T.T. Baber, Y. Wen, Random vibration hysteretic, degrading systems. J. Eng. Mech. Div. **107**(6), 1069–1087 (1981)
10. T.T. Baber, M.N. Noori, Modeling general hysteresis behavior and random vibration applications. J. Vib. Acoust. Stress. Reliab. Des. (ASCE) **108**, 411–420 (1986)

11. C.B. Barber, D.P. Dobkin, H. Huhdanpaa, The quickhull algorithm for convex hulls. ACM Trans. Math. Softw. **22**, 469–483 (1996)
12. M. Beer, M. Liebscher, Designing robust structures - a nonlinear simulation based approach. Comput. Struct. **86**, 1102–1122 (2008)
13. S. Bitzarakis, M. Papadrakakis, A. Kotsopulos, Parallel solutions techniques in computational structural mechanics. Comput. Methods Appl. Mech. Eng. **148**, 75–104 (1997)
14. D.M. Boore, Simulation of ground motion using the stochastic method. Pure Appl. Geophys. **160**(3–4), 635–676 (2003)
15. D.M. Boore, W.B. Joyner, T.E. Fumal, Equations for estimating horizontal response spectra and peak acceleration from western north american earthquakes: a summary of recent work. Seism. Res. Lett. **68**(1), 128–153 (1997)
16. J.D. Bray, A. Rodrigues-Marek, Characterization of forward-directivity ground motions in the near-fault region. Soil Dyn. Earthq. Eng. **24**, 815–828 (2004)
17. F. Cérou, P. Del Moral, T. Furon, A. Guyader, Sequential Monte Carlo for rare event estimation. Stat. Comput. **22**(3), 795–808 (2012)
18. J. Ching, S.-K. Au, J.L. Beck, Reliability estimation of dynamical systems subject to stochastic excitation using subset simulation with splitting. Comput. Methods Appl. Mech. Eng. **194**(12–16), 1557–1579 (2005)
19. J. Ching, J.L. Beck, S.K. Au, Hybrid subset simulation method for reliability estimation of dynamical systems subject to stochastic excitation. Probabilistic Eng. Mech. **20**(3), 199–214 (2005)
20. J.P. Conte, B.F. Peng, Full nonstationary analytical earthquake ground-motion model. J. Enginering Mech. **12**, 15–34 (1997)
21. P. Del Moral, A. Doucet, A. Jasra, Sequential Monte Carlo samplers. J. R. Stat. Soc. Ser. B (Statistical Methodology) **68**(3), 411–436 (2006)
22. G. Deodatis, Non-stationary stochastic vector processes: seismic ground motion applications. Probabilistic Eng. Mech. **11**, 149–167 (1996)
23. A. Der Kiureghian, *Enginering Design Reliability Handbook* (CRC Press, New York, 2004)
24. O. Ditlevsen, O.H. Madsen, *Structural Reliability Methods* (Wiley, New York, 1996)
25. B. Ellingwood, T.V. Galambos, Probability-based criteria for structural design. Struct. Saf. **1**(1), 15–26 (1982)
26. M. Ester, H.P. Kriegel, J. Sanders, X. Xu, *A Density-Based Algorithm for Discovering Clusters in Large Spatial Databases with Noise* (AAAI Press, Cambridge, 1996), pp. 226–231
27. W. Fellin, H. Lessmann, M. Oberguggenberger, R. Vieider (eds.), in *Analyzing Uncertainty in Civil Engineering* (Springer, Berlin, 2005)
28. G.S. Fishman, *Monte Carlo: Concepts, Algorithms and Applications* (Springer, New York, 1996)
29. D.M. Frangopol, K. Maute, Life-cycle reliability-based optimization of civil and aerospace structures. Comput. Struct. **81**(7), 397–410 (2003)
30. R. Ghanem, The nonlinear Gaussian spectrum of lognormal stochastic processes and variables. ASME J. Appl. Mech. **66**(4), 964–973 (1999)
31. D.G. Giovanis, I. Papaioannou, D. Straub, V. Papadopoulos, Bayesian updating with subset simulation using artificial neural netwoks. Comput. Methods Appl. Mech. Eng. **319**, 124–145 (2017)
32. D.V. Griffiths, J. Paiboon, J. Huang, G.A. Fenton, Reliability analysis of beams on random elastic foundations. Gotechnique **63**(2), 180–188 (2013)
33. J.E. Hurtado, A.H. Barbat, Monte Carlo techniques in computational stochastic mechanics. Arch. Comput. Methods Eng. **5**(1), 3–29 (1998)
34. H.A. Jensen, M.A. Catalan, On the effects of non-linear elements in the reliability-based optimal design of stochastic dynamical systems. Int. J. Non Linear Mech. **42**(5), 802–816 (2007)
35. H.A. Jensen, J.G. Sepulveda, On the reliability-based design of structures including passive energy dissipation systems. Struct. Saf. **34**, 390–400 (2011)
36. H.A. Jensen, J. Sepulveda, L. Becerra, Robust stochastic design of base-isolated structural systems. Int. J. Uncertain. Quantif. **2**(2), 95–110 (2012)

37. H.A. Jensen, F. Mayorga, M.A. Valdebenito, Reliability sensitivity estimation of nonlinear structural systems under stochastic excitation: a simulation-based approach. Comput. Methods Appl. Mech. Eng. **289**, 1–23 (2015)
38. H.A. Jensen, D.S. Kusanovic, M.A. Valdebenito, G.I. Schuëller, Reliability-based design optimization of uncertain stochastic systems: a gradient-based scheme. J. Eng. Mech. **138**(1), 60–70 (2012)
39. L.S. Katafygiotis, T. Moand, S.H. Cheung, Auxiliary domain method for solving multi-objective dynamic reliability problems for nonlinear structures. Struct. Eng. Mech. **25**(3), 347–363 (2007)
40. P.S. Koutsourelakis, H.J. Pradlwarter, G.I. Schuëller, Reliability of structures in high dimensions, part I: Algorithms and applications. Probabilistic Eng. Mech. **19**(4), 409–417 (2004)
41. S.L. Kramer, *Geotechnical Earthquake Engineering* (Prentice Hall, Englewood Cliffs, 2003)
42. N. Kuschel, R. Rackwitz, Two basic problems in reliability-based structural optimization. Math. Methods Oper. Res. **46**(3), 309–333 (1997)
43. P.C. Mahalanobis, On the generalised distance in statistics, in *Proceedings of the National Institute of Sciences of India* (1936), pp. 49–55
44. G.P. Mavroeidis, A.S. Papageorgiou, A mathematical representation of near-field ground motions. Bull. Seism. Soc. Am. **93**(3), 1099–1131 (2003)
45. N. Metropolis, A.W. Rosenbluth, M.N. Rosenbluth, A.H. Teller, E. Teller, Equation of state calculations by fast computing machines. J. Chem. Phys. **21**(6), 1087–1092 (1953)
46. H. Miyake, T. Iwata, K. Irikura, Source characterization for broadband ground-motion simulation: kinematic heterogeneous model and strong motion generation area. Bull. Seism. Soc. Am. **93**, 2531–2545 (2003)
47. D. Moens, D. Vandepitte, A survey of non-probabilistic uncertainty treatment in finite element analysis. Comput. Methods Appl. Mech. Eng. **194**, 1527–1555 (2005)
48. B. Möller, M. Beer, Engineering computation under uncertainty - capabilities of non-traditional models. Comput. Struct. **86**, 1024–1041 (2008)
49. O. Möller, R.O. Foschi, L.M. Quiroz, M. Rubinstein, Structural optimization for performance-based design in earthquake engineering: applications of neural networks. Struct. Saf. **31**(6), 490–499 (2009)
50. R.M. Neal, Annealed importance sampling. Stat. Comput. **11**(2), 125–139 (2001)
51. A. Nozu, A super asperity model for the 2011 off Pacific coast of Tohoku earthquake. J. Jpn. Assoc. Earthq. Eng. **14**(6), 36–55 (2014)
52. A. Nozu, M. Yamada, T. Nagao, K. Irikura, Generation of strong motion pulses during huge subduction earthquakes and scaling of their generation areas. J. Jpn. Assoc. Earthq. Eng. **14**(6), 96–117 (2014)
53. C. Papadimitriou, *Stochastic Characterization of Strong Ground Motions and Application to Structure Response*. Rep. No. EERL 90-03 (California Institute of Technology, Pasadena, California, USA, 1990)
54. Y.J. Park, Y.K. Wen, A.H. Ang, Random vibration of hysteretic systems under bi-directional ground motions. Earthq. Eng. Struct. Dyn. **14**(4), 543–557 (1986)
55. M.F. Pellissetti, Parallel processing in structural reliability. J. Struct. Eng. Mech. **32**(1), 95–126 (2009)
56. H.J. Pradlwarter, G.I. Schuëller, P.S. Koutsourelakis, D.C. Champris, Application of line sampling simulation method to reliability benchmark problems. Struct. Saf. **29**(3), 208–221 (1998)
57. S. Rezaeian, A. Der Kiureghian, A stochastic ground motion model with separable temporal and spectral nonstationarities. Earthq. Eng. Struct. Dyn. **37**, 1565–1584 (2008)
58. J.O. Royset, A. Der Kiureghian, E. Polak, Reliability-based optimal structural design by the decoupling approach. Reliab. Eng. Syst. Saf. **73**(3), 213–221 (2001)
59. R.Y. Rubinstein, D.P. Kroese, *Simulation and Monte Carlo Method* (Wiley, New York, 2007)
60. F. Sabetta, A. Plugliese, Estimation of response spectra and simulation of nonstationary earthquake ground motions. Bull. Seism. Soc. Am. **86**, 337–352 (1996)
61. G.R. Saragoni, G.C. Hart, Simulation of artificial earthquakes. Earthq. Eng. Struct. Dyn. **2**(3), 249–267 (1974)

62. G.I. Schuëller, H.J. Pradlwarter, Benchmark study on reliability estimation in higher dimensions of structural systems - an overview. Struct. Saf. **29**, 167–182 (2007)
63. T. Soong, M. Grigoriu, *Random Vibration of Mechanical and Structural Systems* (Prentice Hall, Englewood Cliffs, 1993)
64. B. Sudret, *Stochastic Finite Element Methods and Reliability: A State-of-the-Art Report* (University of California, Berkeley, 2000)
65. A. Taflanidis, Robust stochastic design of viscous dampers for base isolation applications, in *ECCOMAS Thematic Conference on Computational Methods in Structural Dynamics and Earthquake Engineering*, Rhodes, Greece, 22–24 June 2009
66. P.K. Umesha, M.T. Venuraju, D. Hartmann, K.R. Leimbach, Optimal design of truss structures using parallel computing. Struct. Multidiscip. Optim. **29**, 285–297 (2005)
67. A. Zerva, *Spatial Variation of Seismic Ground Motions Modeling and Engineering Applications* (CRC Press, Boca Raton, 2003)
68. Y. Zhang, A. Der Kiureghian, First-excursion probability of uncertain structures. Probabilistic Eng. Mech. **9**(1–2), 135–143 (1994)
69. K.M. Zuev, L.S. Katafygiotis, The Horseracing simulation algorithm for evaluation of small failure probabilities. Probabilistic Eng. Mech. **26**(2), 157–164 (2011)
70. K.M. Zuev, J.L. Beck, S.-K. Au, L.S. Katafygiotis, Bayesian post-processor and other enhancements of subset simulation for estimating failure probabilities in high dimensions. Comput. Struct. **92**, 283–296 (2012)

Chapter 5
Reliability Sensitivity Analysis of Dynamical Systems

Abstract The reliability sensitivity analysis of systems subjected to stochastic loading is considered in this chapter. In particular, the change that the probability of failure undergoes due to changes in the distribution parameters of the uncertain model parameters is utilized as a sensitivity measure. A simulation-based approach that corresponds to a simple post-processing step of an advanced sampling-based reliability analysis is used to perform the sensitivity analysis. In particular, subset simulation, introduced in the previous chapter, is applied in the present formulation. The analysis does not require any additional system response evaluations. The feasibility and effectiveness of the approach is demonstrated on a finite element model of a bridge under stochastic ground excitation. The sensitivity analysis is carried out in a reduced space of generalized coordinates. The computational effort involved in the reliability sensitivity analysis of the reduced-order model is significantly decreased with respect to the corresponding analysis of the full finite element model. The reduction is accomplished without compromising the accuracy of the reliability sensitivity estimates.

5.1 Motivation

The level of safety of a structure can be measured in terms of its reliability. Even though this information is essential, it is also important to analyze the sensitivity of the reliability estimates with respect to variations in model parameters [3, 8, 13, 18, 30]. In particular, the determination of the variation in the reliability (or equivalently in the failure probability) due to changes in model parameters can provide useful information. For example, it can be used to identify the most influential model parameters and provide an important insight on system failure for risk-based decision making, such as reliability-based characterization of system responses, robust control, reliability-based design optimization, etc. [2, 6, 15, 24, 26, 34].

The subject of reliability sensitivity has been addressed in a large number of contributions. In fact, many works based on standard approximate methods such as first- and second-order reliability methods and simulation-based methods have been studied in the literature. These methods are quite general, and they have proved to

H. Jensen and C. Papadimitriou, *Sub-structure Coupling for Dynamic Analysis*, Lecture Notes in Applied and Computational Mechanics 89,
https://doi.org/10.1007/978-3-030-12819-7_5

be very effective in a large number of problems, but their range of application is somewhat limited in the context of complex dynamical systems. A representative list of these works is included in the Refs. [1, 3–5, 13, 16, 18, 20, 22, 27, 29].

5.2 Reliability Sensitivity Analysis Formulation

As indicated in Sect. 4.1, the vector of uncertain parameters $\boldsymbol{\theta}$ is characterized in a probabilistic manner by means of a joint probability density function $q(\boldsymbol{\theta})$. For reliability sensitivity purposes, this function depends on a certain number of parameters $\boldsymbol{\tau}$, that is, $q(\boldsymbol{\theta}|\boldsymbol{\tau})$. In practice, the distribution parameters $\boldsymbol{\tau}$ can be considered, for example, as the mean value or standard deviation of $\boldsymbol{\theta}$. In this context, the mean value represents the nominal value, whereas the standard deviation models the uncertainty associated with manufacturing and construction processes. Then, it is clear that the probability of failure depends on several factors, among them, the distribution parameters $\boldsymbol{\tau}$ of the probability density function of the uncertain model parameters. Thus, the probability of failure explicitly depends on the distribution parameter vector, i.e. $P_F(\boldsymbol{\tau})$. In this manner, changes in the distribution parameters will certainly alter the response of the structure and, consequently, its probability of failure. The rate of change that the probability of failure undergoes due to these changes is denoted as reliability sensitivity analysis in the context of this chapter.

A simulation-based approach that is a simple post-processing of subset simulation is considered for performing the corresponding reliability sensitivity analysis [11]. This approach has been validated and illustrated in a series of reliability problems, including complex structural systems such as nonlinear dynamical systems under stochastic excitation and problems involving relatively large finite element models [7, 11, 20, 31]. As in the previous chapter, first excursion probabilities are used to characterize the level of safety of a structure.

5.3 Sensitivity Measure

A classical measure for sensitivity is calculating the gradient of the quantity of interest. In this context, reliability sensitivity is defined as the partial derivative of the failure probability with respect to the distribution parameters of the basic uncertain model parameters. From the definition of the probability of failure in Eq. (4.5), the sensitivity of the failure probability with respect to a distribution parameter τ_j can be written in the form

$$\left.\frac{\partial P_F(\boldsymbol{\tau})}{\partial \tau_j}\right|_{\boldsymbol{\tau}^0} = \int_{\mathbf{z}\in\Omega_z,\boldsymbol{\theta}\in\Omega_\theta} I_F(\mathbf{z},\boldsymbol{\theta})\; p(\mathbf{z})\; \frac{\partial q(\boldsymbol{\theta}|\boldsymbol{\tau})}{\partial \tau_j}\, d\mathbf{z}\, d\boldsymbol{\theta} \tag{5.1}$$

where $\boldsymbol{\tau}^0$ is the value of the distribution parameter vector where the partial derivative is evaluated, and all other terms have been previously defined. In Eq. (5.1), it has been assumed that $q(\boldsymbol{\theta}|\boldsymbol{\tau})$ is differentiable with respect to τ_j and that the integration range does not depend on τ_j. Recall that the previous probability integral represents a high-dimensional problem in the context of dynamical systems under stochastic loading.

The sensitivity can also be defined in terms of the so-called elasticity, which is another measure usually used in the context of sensitivity analysis [7, 19]. Within this context, the elasticity e_{τ_j} of the failure probability with respect to a parameter τ_j, evaluated at $\boldsymbol{\tau}^0$, is defined as

$$e_{\tau_j} = \left.\frac{\partial P_F}{\partial \tau_j}\right|_{\boldsymbol{\tau}^0} \frac{\tau_j^0}{P_F} \tag{5.2}$$

where τ_j^0 is assumed to be non-zero. This dimensionless quantity represents a more objective sensitivity measure when the uncertain model parameters are diverse in dimension. Thus, this sensitivity measure can be used to rank the importance of the model parameters on the system reliability. This measure is also less sensitive to potential bias in the failure probability estimates [7, 19].

5.4 Failure Probability Function Representation

To compute the sensitivity measure, the probability of failure $P_F(\boldsymbol{\tau})$, referred to as failure probability function, is first expressed as a function of the distribution parameter vector $\boldsymbol{\tau}$. The idea is to estimate the failure probability function by using samples and associated intermediate failure events generated by subset simulation under $q(\boldsymbol{\theta}|\boldsymbol{\tau}^0)$, that is, the probability density function of $\boldsymbol{\theta}$ with distribution parameter vector $\boldsymbol{\tau}^0$. Specifically, following the basic ideas of subset simulation (see Sect. 4.3), the probability of the first failure event F_1 can be computed as

$$\begin{aligned} P_{F_1}(\boldsymbol{\tau}) &= \int_{\mathbf{z}\in\Omega_\mathbf{z},\boldsymbol{\theta}\in\Omega_{\boldsymbol{\theta}}} I_{F_1}(\mathbf{z},\boldsymbol{\theta})\; p(\mathbf{z})\; q(\boldsymbol{\theta}|\boldsymbol{\tau})\; d\mathbf{z}\, d\boldsymbol{\theta} \\ &= \int_{\mathbf{z}\in\Omega_\mathbf{z},\boldsymbol{\theta}\in\Omega_{\boldsymbol{\theta}}} I_{F_1}(\mathbf{z},\boldsymbol{\theta})\; \frac{q(\boldsymbol{\theta}|\boldsymbol{\tau})}{q(\boldsymbol{\theta}|\boldsymbol{\tau}^0)}\; p(\mathbf{z}) q(\boldsymbol{\theta}|\boldsymbol{\tau}^0)\; d\mathbf{z}\, d\boldsymbol{\theta} \end{aligned} \tag{5.3}$$

where $q(\boldsymbol{\theta}|\boldsymbol{\tau}^0)$ and $q(\boldsymbol{\theta}|\boldsymbol{\tau})$ are the probability density functions of $\boldsymbol{\theta}$ with distribution parameter vector $\boldsymbol{\tau}^0$ and $\boldsymbol{\tau}$, respectively. Similarly, the probability of the conditional failure event $F_k/F_{k-1}, k = 2, \ldots, m$, can be written as

$$\begin{aligned} P_{F_k/F_{k-1}}(\boldsymbol{\tau}) &= \int_{\mathbf{z}\in\Omega_\mathbf{z},\boldsymbol{\theta}\in\Omega_{\boldsymbol{\theta}}} I_{F_k}(\mathbf{z},\boldsymbol{\theta})\; p(\mathbf{z}|F_{k-1})\; q(\boldsymbol{\theta}|F_{k-1},\boldsymbol{\tau})\; d\mathbf{z}\, d\boldsymbol{\theta} \\ &= \int_{\mathbf{z}\in\Omega_\mathbf{z},\boldsymbol{\theta}\in\Omega_{\boldsymbol{\theta}}} I_{F_k}(\mathbf{z},\boldsymbol{\theta})\; \frac{q(\boldsymbol{\theta}|F_{k-1},\boldsymbol{\tau})}{q(\boldsymbol{\theta}|F_{k-1},\boldsymbol{\tau}^0)}\; p(\mathbf{z}|F_{k-1}) q(\boldsymbol{\theta}|F_{k-1},\boldsymbol{\tau}^0)\; d\mathbf{z}\, d\boldsymbol{\theta} \end{aligned} \tag{5.4}$$

where $p(\mathbf{z}|F_{k-1})$ is the distribution of $\mathbf{z}$ conditional to the failure event F_{k-1}, and $q(\boldsymbol{\theta}|F_{k-1},\boldsymbol{\tau})$ and $q(\boldsymbol{\theta}|F_{k-1},\boldsymbol{\tau}^0)$ are the conditional distributions of $\boldsymbol{\theta}$ given that they lie in F_{k-1} under distribution parameter vectors $\boldsymbol{\tau}$ and $\boldsymbol{\tau}^0$, respectively. By definition, these conditional distributions are equal to

$$q(\boldsymbol{\theta}|F_{k-1},\boldsymbol{\tau}) = \frac{I_{F_{k-1}}(\boldsymbol{\theta})\, q(\boldsymbol{\theta}|\boldsymbol{\tau})}{P_{F_{k-1}}(\boldsymbol{\tau})} \quad , \quad q(\boldsymbol{\theta}|F_{k-1},\boldsymbol{\tau}^0) = \frac{I_{F_{k-1}}(\boldsymbol{\theta})\, q(\boldsymbol{\theta}|\boldsymbol{\tau}^0)}{P_{F_{k-1}}(\boldsymbol{\tau}^0)} \tag{5.5}$$

where $P_{F_{k-1}}(\boldsymbol{\tau})$ and $P_{F_{k-1}}(\boldsymbol{\tau}^0)$ are the probabilities of the failure event F_{k-1} under distribution parameter vectors $\boldsymbol{\tau}$ and $\boldsymbol{\tau}^0$ of the probability density function $q(\cdot)$, respectively. Then, the probability of the conditional failure event F_k/F_{k-1} can be given in the form

$$P_{F_k/F_{k-1}}(\boldsymbol{\tau}) = \frac{P_{F_{k-1}}(\boldsymbol{\tau}^0)}{P_{F_{k-1}}(\boldsymbol{\tau})} \int_{\mathbf{z}\in\Omega_\mathbf{z},\boldsymbol{\theta}\in\Omega_{\boldsymbol{\theta}}} I_{F_k}(\mathbf{z},\boldsymbol{\theta})\, \frac{q(\boldsymbol{\theta}|\boldsymbol{\tau})}{q(\boldsymbol{\theta}|\boldsymbol{\tau}^0)}\, p(\mathbf{z}|F_{k-1}) q(\boldsymbol{\theta}|F_{k-1},\boldsymbol{\tau}^0)\, d\mathbf{z}\, d\boldsymbol{\theta} \tag{5.6}$$

Moreover, by definition, the probability of failure can be expressed as

$$P_F(\boldsymbol{\tau}) = P_{F_m}(\boldsymbol{\tau}) = P_{F_m/F_{m-1}}(\boldsymbol{\tau})\; P_{F_{m-1}}(\boldsymbol{\tau}) \tag{5.7}$$

Thus, from Eqs. (5.6) and (5.7) with $k = m$, it follows that

$$P_F(\boldsymbol{\tau}) = P_{F_{m-1}}(\boldsymbol{\tau}^0) \int_{\mathbf{z}\in\Omega_\mathbf{z},\boldsymbol{\theta}\in\Omega_{\boldsymbol{\theta}}} I_{F_m}(\mathbf{z},\boldsymbol{\theta})\, \frac{q(\boldsymbol{\theta}|\boldsymbol{\tau})}{q(\boldsymbol{\theta}|\boldsymbol{\tau}^0)}\, p(\mathbf{z}|F_{m-1}) q(\boldsymbol{\theta}|F_{m-1},\boldsymbol{\tau}^0)\, d\mathbf{z}\, d\boldsymbol{\theta} \tag{5.8}$$

where from construction $P_{F_{m-1}}(\boldsymbol{\tau}^0) = p_0^{m-1}$. Then,

$$P_F(\boldsymbol{\tau}) = p_0^{m-1} \int_{\mathbf{z}\in\Omega_\mathbf{z},\boldsymbol{\theta}\in\Omega_{\boldsymbol{\theta}}} I_{F_m}(\mathbf{z},\boldsymbol{\theta})\, \frac{q(\boldsymbol{\theta}|\boldsymbol{\tau})}{q(\boldsymbol{\theta}|\boldsymbol{\tau}^0)}\, p(\mathbf{z}|F_{m-1}) q(\boldsymbol{\theta}|F_{m-1},\boldsymbol{\tau}^0)\, d\mathbf{z}\, d\boldsymbol{\theta} \tag{5.9}$$

The last equation represents an analytical characterization, in the framework of subset simulation, of the failure probability function in terms of the distribution parameter vector $\boldsymbol{\tau}$.

5.5 Sensitivity Estimation

Using the previous characterization of the failure probability function, the partial derivative of $P_F(\boldsymbol{\tau})$ with respect to τ_j, evaluated at $\boldsymbol{\tau}^0$, can be written as

$$\left.\frac{\partial P_F}{\partial \tau_j}\right|_{\boldsymbol{\tau}^0} = p_0^{m-1} \int_{\mathbf{z}\in\Omega_\mathbf{z},\boldsymbol{\theta}\in\Omega_{\boldsymbol{\theta}}} I_{F_m}(\mathbf{z},\boldsymbol{\theta})\, \frac{\frac{\partial q}{\partial \tau_j}(\boldsymbol{\theta}|\boldsymbol{\tau}^0)}{q(\boldsymbol{\theta}|\boldsymbol{\tau}^0)}\, p(\mathbf{z}|F_{m-1}) q(\boldsymbol{\theta}|F_{m-1},\boldsymbol{\tau}^0)\, d\mathbf{z}\, d\boldsymbol{\theta} \tag{5.10}$$

or in terms of the expectation operator

$$\left.\frac{\partial P_F}{\partial \tau_j}\right|_{\tau^0} = p_0^{m-1} E_{p(\mathbf{z}|F_{m-1}),q(\boldsymbol{\theta}|F_{m-1},\boldsymbol{\tau}^0)} \left[I_{F_m}(\mathbf{z},\boldsymbol{\theta}) \frac{\frac{\partial q}{\partial \tau_j}(\boldsymbol{\theta}|\boldsymbol{\tau}^0)}{q(\boldsymbol{\theta}|\boldsymbol{\tau}^0)} \right] \tag{5.11}$$

where $E_{p(\mathbf{z}|F_{m-1}),q(\boldsymbol{\theta}|F_{m-1},\boldsymbol{\tau}^0)}[\,\cdot\,]$ is the expectation operator with respect to the distributions $p(\mathbf{z}|F_{m-1})$ and $q(\boldsymbol{\theta}|F_{m-1},\boldsymbol{\tau}^0)$. From the last expression, the sensitivity can be estimated as

$$\left.\frac{\partial P_F}{\partial \tau_j}\right|_{\tau^0} \approx p_0^{m-1} \frac{1}{N_m} \sum_{i=1}^{N_m} I_{F_m}(\mathbf{z}_{m-1,i},\boldsymbol{\theta}^0_{m-1,i}) \frac{\frac{\partial q}{\partial \tau_j}(\boldsymbol{\theta}^0_{m-1,i}|\boldsymbol{\tau}^0)}{q(\boldsymbol{\theta}^0_{m-1,i}|\boldsymbol{\tau}^0)} \tag{5.12}$$

where $\{(\mathbf{z}_{m-1,i},\boldsymbol{\theta}^0_{m-1,i}), i = 1,\ldots,N_m\}$ is the set of samples generated at the last stage of subset simulation under distribution parameter vector $\boldsymbol{\tau}^0$ of the probability density function $q(\cdot)$, and all other terms have been previously defined.

It is observed that a single subset simulation analysis is required for estimating the sensitivity of the probability of failure with respect to the distribution parameters. Therefore, the reliability sensitivity analysis is a simple post-processing of a sampling-based reliability analysis. In other words, the approach does not require any additional sampling of the indicator function (dynamic analysis). In summary, it is seen that the estimation of the failure probability and its gradient can be done with the same samples generated at the last stage of subset simulation. The previous approach can be extended to higher-order derivatives provided that the distribution $q(\boldsymbol{\theta}|\boldsymbol{\tau})$ is sufficiently differentiable. It is noted that the characterization of the partial derivative of the failure probability function with respect to the jth component of $\boldsymbol{\tau}$ (see Eq. (5.10)) can be also expressed in terms of the so-called score function, which is the partial derivative of the logarithm of the distribution $q(\boldsymbol{\theta}|\boldsymbol{\tau}^0)$ [25].

5.6 Sensitivity Versus Threshold

From the formulation of subset simulation (see Sect. 4.3.2), it is clear that the demand function values $\delta_1,\ldots,\delta_m$ at specified probability levels are the ones that are estimated, rather than the conditional failure probabilities. Consequently, subset simulation serves as a method to generate random samples whose response values correspond to specified probability levels, rather than a technique to estimate failure probabilities for specified failure events. As a result, it produces information about the probability of failure versus the threshold and not only for a single value. Since the proposed reliability sensitivity analysis is based on subset simulation, a similar information can be obtained for the sensitivity measures.

For a demand function value $\bar{\delta}$ such that $\delta_{k-1} < \bar{\delta} \leq \delta_k$, $k = 1,\ldots,m$, with $\delta_0 = 0$, the partial derivative of $P_F(\boldsymbol{\tau})$ with respect to τ_j, evaluated at $\boldsymbol{\tau}^0$, can be estimated as [11, 12]

$$\left.\frac{\partial P_F}{\partial \tau_j}\right|_{\boldsymbol{\tau}^0} \approx p_0^{k-1} \frac{1}{N_k} \sum_{i=1}^{N_k} I_{\bar{F}}(\mathbf{z}_{k-1,i}, \boldsymbol{\theta}^0_{k-1,i}) \frac{\frac{\partial q}{\partial \tau_j}(\boldsymbol{\theta}^0_{k-1,i}|\boldsymbol{\tau}^0)}{q(\boldsymbol{\theta}^0_{k-1,i}|\boldsymbol{\tau}^0)} \tag{5.13}$$

where $\bar{F}$ is the failure event defined as

$$\bar{F} = \left\{d(\mathbf{z}, \boldsymbol{\theta}) > \bar{\delta}\right\} \tag{5.14}$$

and $\{(\mathbf{z}_{k-1,i}, \boldsymbol{\theta}^0_{k-1,i}), i = 1, \ldots, N_k\}$ is the set of samples generated at level $k-1$ of subset simulation under distribution parameter vector $\boldsymbol{\tau}^0$ of the probability density function $q(\cdot)$. In this manner, a single simulation run yields reliability sensitivity estimates for all thresholds up to the largest one considered in the analysis. In other words, the whole trend of the sensitivity measure versus the thresholds can be obtained in a direct manner. This feature of the approach is quite desirable because it provides much more information than a point estimate.

5.7 Particular Cases

The previous general formulation can be specialized for different probability distributions of the uncertain model parameters and different distribution parameters as long as Eq. (5.1) holds. Of practical importance is the case when the distribution parameters are represented by the mean values and standard deviations of the system parameters. These distribution parameters can be considered as a control or design variables in a number of important applications such as reliability sensitivity analysis, reliability-based characterization of structural responses, reliability-based design optimization, robust solutions and predictions, robust design optimization, etc. For illustration purposes, some sensitivity measures corresponding to the case of normal and log-normal random variables are given in the following equations.

The partial derivatives of the failure probability with respect to the mean value μ_{θ_j} and standard deviation σ_{θ_j} of the model parameter θ_j, evaluated at μ^0_{θ} and σ^0_{θ} for the case of a normal random variable are estimated as [11, 12]

$$\left.\frac{\partial P_F}{\partial \mu_{\theta_j}}\right|_{\mu^0_{\theta}} \approx p_0^{m-1} \frac{1}{N_m} \sum_{i=1}^{N_m} I_F(\mathbf{z}_{m-1,i}, \boldsymbol{\theta}^0_{m-1,i}) \times \left\{\frac{(\theta^0_{m-1,ij} - \mu^0_{\theta_j})}{{\sigma^0_{\theta_j}}^2}\right\} \tag{5.15}$$

and

$$\left.\frac{\partial P_F}{\partial \sigma_{\theta_j}}\right|_{\sigma_\theta^0} \approx p_0^{m-1} \frac{1}{N_m} \sum_{i=1}^{N_m} I_F\left(\mathbf{z}_{m-1,i}, \boldsymbol{\theta}_{m-1,i}^0\right) \times \left\{ \frac{(\theta_{m-1,ij}^0 - \mu_{\theta_j}^0)^2}{\sigma_{\theta_j}^{0\,2}} - 1 \right\} \frac{1}{\sigma_{\theta_j}^0} \tag{5.16}$$

where $\{(\mathbf{z}_{m-1,i}, \boldsymbol{\theta}_{m-1,i}^0), i = 1, \ldots, N_m\}$ is the set of samples generated at the last stage of subset simulation under distributions $p(\mathbf{z}|F_{m-1})$ and $q(\boldsymbol{\theta}|F_{m-1}, \boldsymbol{\tau}^0)$, respectively, and $\boldsymbol{\theta}_{m-1,ij}^0$ is the jth component of the sample vector $\boldsymbol{\theta}_{m-1,i}^0$.

For the case of a log-normal random variable, the estimators are written as [11, 12]

$$\left.\frac{\partial P_F}{\partial \mu_{\theta_j}}\right|_{\mu_\theta^0} \approx p_0^{m-1} \frac{1}{N_m} \sum_{i=1}^{N_m} I_F(\mathbf{z}_{m-1,i}, \boldsymbol{\theta}_{m-1,i}^0) \times \left\{ \left[\frac{(\ln(\theta_{m-1,ij}^0) - \mu_j)}{\sigma_j{}^2} \right] \alpha_j + \left[\frac{(\ln(\theta_{m-1,ij}^0) - \mu_j)^2}{\sigma_j{}^2} - 1 \right] \frac{1}{\sigma_j^2} \beta_j \right\} \tag{5.17}$$

and

$$\left.\frac{\partial P_F}{\partial \sigma_{\theta_j}}\right|_{\sigma_\theta^0} \approx p_0^{m-1} \frac{1}{N_m} \sum_{i=1}^{N_m} I_F\left(\mathbf{z}_{m-1,i}, \boldsymbol{\theta}_{m-1,i}^0\right) \times \left\{ \left[\frac{(\ln(\theta_{m-1,ij}^0) - \mu_j)}{\sigma_j^2} \right] \lambda_j - \left[\frac{(\ln(\theta_{m-1,ij}^0) - \mu_j)^2}{\sigma_j^2} - 1 \right] \frac{1}{\sigma_j^2} \lambda_j \right\} \tag{5.18}$$

where

$$\alpha_j = \left[2 - \exp\left(-\sigma_j{}^2\right)\right] \exp\left(-\left(\mu_j + \sigma_j{}^2/2\right)\right) \tag{5.19}$$

$$\beta_j = \left[\exp\left(-\sigma_j{}^2\right) - 1\right] \exp\left(-\left(\mu_j + \sigma_j{}^2/2\right)\right) \tag{5.20}$$

$$\lambda_j = -[\exp(\sigma_j^2) - 1]^{1/2} \exp(-(\mu_j + 3\sigma_j^2/2)) \tag{5.21}$$

with

$$\mu_j = \ln\left(\left(\mu_{\theta_j}^0\right)^2 / \sqrt{\left(\mu_{\theta_j}^0\right)^2 + \left(\sigma_{\theta_j}^0\right)^2}\right) \tag{5.22}$$

$$\sigma_j = \sqrt{\ln\left(1 + \left(\sigma_{\theta_j}^0 / \mu_{\theta_j}^0\right)^2\right)} \tag{5.23}$$

and all other terms have been previously defined. Similar expressions can be derived for higher order derivatives and for other types of probability distributions of the uncertain model parameters provided that the distribution $q(\boldsymbol{\theta}|\boldsymbol{\tau})$ is sufficiently differentiable.

5.8 Application Problem

The objective of the application problem is to determine the feasibility and effectiveness of the proposed reliability sensitivity analysis approach in an involved model. Even though the proposed sensitivity analysis is a simple post-processing of the subset simulation, it can be computationally very demanding due to the large number of dynamic analyses required during the reliability sensitivity estimation (evaluation of the indicator function). Therefore, the total computational demand may become excessive when the computational time for performing a dynamic finite element analysis is significant. To deal with this difficulty, the reliability sensitivity analysis is carried out in a reduced-order model.

5.8.1 *Model Description*

A three-dimensional bridge finite element model of 10,068 degrees of freedom is considered in the application problem. The bridge model, shown in Fig. 5.1, is curved in plan and has a total length of 119 m. It has five spans of lengths equal to 24 m, 20 m, 23 m, 25 m, and 27 m, respectively, and four piers of 8 m height that monolithically support the girder. Each pier is founded on an array of four piles

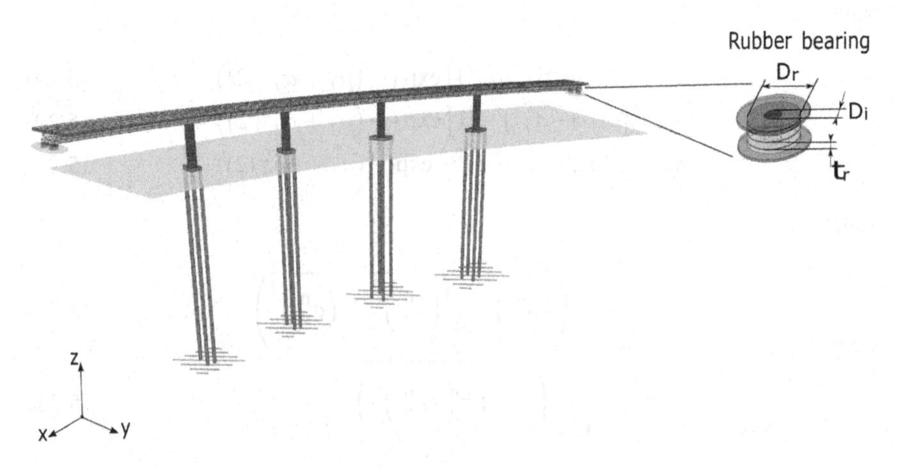

Fig. 5.1 Finite element model of bridge structure

of 35 m height. The piers and piles are modeled as column elements of circular cross-section with $D_c = 1.6$ m and $D_p = 0.6$ m diameter, respectively, while the deck cross-section is a box girder. The deck girder is modeled by beam and shell elements and rests on each abutment through two rubber bearings, which are used as an isolation system.

The rubber bearings consist of layers of rubber and steel, with the rubber being vulcanized to the steel plates. A schematic representation of a rubber bearing is also shown in the figure, where D_r represents the external diameter, D_i is the internal diameter, and $H_r = t_r n_r$ is the total height of rubber in the bearing where t_r is the layer thickness and n_r is the number of rubber layers. The nominal values of the rubber bearings parameters are set equal to $D_r = 0.80$ m, $D_i = 0.10$ m, and $H_r = 0.17$ m. The interaction between the piles and the soil is modeled by a series of translational springs in the x and y direction along the height of the piles, with stiffnesses varying linearly from $K_s = 11{,}200$ T/m at the base to 0 at the surface. The net effect of these elements is to increase the translational stiffness of the column elements that model the piles. Material properties of the structural model have been assumed as follows: Young's modulus $E = 2.0 \times 10^{10}$ N/m^2; Poisson's ratio $\nu = 0.2$, and mass density $\rho = 2{,}500$ kg/m^3. In addition, 3% of critical damping is added to the model. It is assumed that the structural components, such as the piers, piles, and the deck girder, remain linear during the analysis, while the nonlinearities are localized in the rubber bearings response. In addition, the axial deformation of the piers and piles is neglected with respect to their bending deformation.

The bridge structure is subjected to a ground acceleration applied in a direction defined at 25° with respect to the x-axis. It is modeled as the non-stationary stochastic process described in Sect. 4.5. The values of the various parameters involved in the model are taken as the ones considered in the previous chapter. The duration of the excitation is taken equal to $T = 30$ s with a sampling interval equal to $\Delta T = 0.01$ s. Thus, the vector of uncertain parameters $\mathbf{z}$ involves more than 3,000 uncertain parameters, as $\mathbf{z}^T = < z_1, z_2, \ldots, z_{3001} >$. Consequently, the corresponding reliability problem and, therefore, the reliability sensitivity analysis problem is high-dimensional.

5.8.2 Rubber Bearings

5.8.2.1 Description

Rubber bearings have been used over many years in a number of seismically isolated structures worldwide [14, 21, 28]. They require minimal initial cost and maintenance compared to other passive, semi-active, and active energy absorption devices. Rubber bearing systems, in principle, are able to provide horizontal flexibility together with the restoring force and supply the required hysteretic damping. An analytical model that simulates measured restoring forces under bidirectional loadings is considered. The model is based on a series of experimental tests conducted for real-size rubber

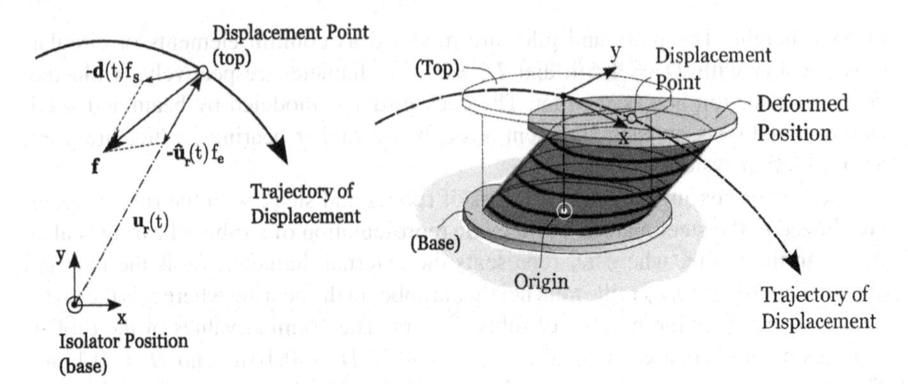

Fig. 5.2 Decomposition of the restoring force

bearings [23, 32]. The loading tests of seven full-scale isolators were carried out using the Caltrans Seismic Response Modification Device Test Facility at the University of California, San Diego. The specimens used in the tests were made with high damping rubber compounds. In particular, horizontal bidirectional loading tests for isolators with a diameter of 0.7 m and 1.3 m were conducted. On the basis of the test results, the model assumes that the restoring force on the rubber bearing is composed of a force directed to the origin of the isolator and another force approximately opposite to the direction of the movement of the isolator. This decomposition of the restoring force is schematically shown in Fig. 5.2.

According to the model, a vector approximately in the direction of the motion $\mathbf{d}(t)$, can be defined in terms of the isolator displacement vector $\mathbf{u}_r(t)$ in the x and y direction by means of the nonlinear differential equation [9, 10, 33]

$$\dot{\mathbf{d}}(t) = \frac{1}{\alpha} \parallel \dot{\mathbf{u}}_r(t) \parallel \left[\hat{\mathbf{u}}_r(t) - \parallel \mathbf{d}(t) \parallel^{\beta} \hat{\mathbf{d}}(t) \right] \ , \ \mathbf{u}_r(0) = \mathbf{0} \ , \ \mathbf{d}(0) = \mathbf{0} \tag{5.24}$$

where $\dot{\mathbf{u}}_r(t)$ is the velocity vector, $\hat{\mathbf{u}}_r(t)$ and $\hat{\mathbf{d}}(t)$ are the unit directional vectors of $\dot{\mathbf{u}}_r(t)$ and $\mathbf{d}(t)$, respectively, and $\parallel \cdot \parallel$ indicates the Euclidean norm. The parameters α and β are positive constants that relate to the yield displacement and smoothness of yielding, respectively. Once the vector $\mathbf{d}(t)$ has been derived, the restoring force $\mathbf{f}(t)$ on the isolator (in the x and y direction) is expressed in terms of the unit directional vector $\hat{\mathbf{u}}_r(t)$ and the vector $\mathbf{d}(t)$ as

$$\mathbf{f}(t) = -\hat{\mathbf{u}}_r(t) f_e(t) - \mathbf{d}(t) f_s(t) \tag{5.25}$$

where $f_e(t)$ is the nonlinear elastic component and $f_s(t)$ is the elastoplastic component. Based on the results reported in [33], it was concluded that the model is able to accurately simulate the test results for both bidirectional and unidirectional loading.

5.8.2.2 Model Parameter Identification and Validation

The mathematical model for the description of the isolator behavior can be used to calibrate the model parameters by using a specific set of loading tests carried out for real-size bearings. First, the parameters α and β, which define the transition curve from the elastic to inelastic regime, are calibrated. They are estimated as $0.2H_r$ and 0.7, respectively, where H_r is the total height of rubber, as indicated before [33]. Next, the stress-strain relationships for $\tau_e(t) = f_e(t)/A$, and $\tau_s(t) = f_s(t)/A$ are calibrated by means of quadratic and cubic curves as [17]

$$\tau_e(t) = \begin{cases} 0.35\gamma(t) & \text{if } 0 \leq \gamma(t) \leq 1.8 \\ 0.35\gamma(t) + 0.2(\gamma(t) - 1.8)^2 & \text{if } \gamma(t) \geq 1.8 \end{cases} \quad (5.26)$$

and

$$\tau_s(t) = 0.125 + 0.015\gamma(t) + 0.012\gamma(t)^3 \quad (5.27)$$

where A is the cross-sectional area of the rubber, and $\gamma(t) = \| \mathbf{u}_r(t) \| / H_r$ is the average shear-strain. The test results show that the scatter of the experimental data around these calibrated curves is relatively small for average shear strains of less than 200%. Test-restoring forces and those calculated by the model under unidirectional loading are compared in Fig 5.3 for two specimens. They correspond to a medium- and large-sized rubber bearing, respectively. The test results were conducted for a maximum average shear strain of 150%. It is seen that the analytical model simulates the test results very well. The extra loop shown in the figures is generated by the analytical model to illustrate the predicted behavior of the rubber bearings for large average shear strains (250%). Additional validation calculations have shown that the

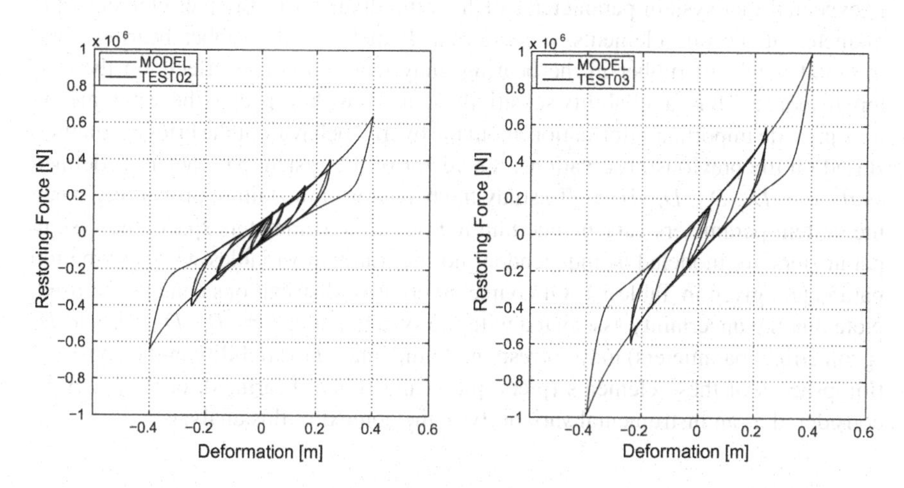

Fig. 5.3 Comparison of analytical and experimental hysteresis loops. Left figure: medium size rubber bearing. $D_r = 0.8\,\text{m}$, $H_r = 0.16\,\text{m}$, $D_i = 0.15\,\text{m}$. Right figure: large size rubber bearing. $D_r = 1.0\,\text{m}$, $H_r = 0.16\,\text{m}$, $D_i = 0.15\,\text{m}$

analytical model is also able to accurately simulate the test results for bidirectional loadings [9, 10, 33].

5.8.3 Reliability Sensitivity Analysis Formulation

The performance of the bridge structure is characterized in terms of the probability of occurrence of three failure events. The events are related to the maximum absolute acceleration at the middle of the deck girder, the maximum relative displacement between the top of the piers and their connections with the pile foundation, and the maximum relative displacement between the deck girder and the base of the rubber bearings at each abutment. Mathematically, the failure events are defined as

$$\begin{aligned} F_1(\mathbf{z}, \boldsymbol{\theta}) &= \left\{ \max_{t\in[0,T]} \left(\left| \frac{\ddot{x}_{\text{absolute}}(t, \mathbf{z}, \boldsymbol{\theta})}{2.00\,\text{m/s}^2} \right| \right) > 1 \right\}, \\ F_2(\mathbf{z}, \boldsymbol{\theta}) &= \left\{ \max_{t\in[0,T]} \left(\left| \frac{\delta x(t, \mathbf{z}, \boldsymbol{\theta})}{0.07\,\text{m}} \right| \right) > 1 \right\}, \\ F_3(\mathbf{z}, \boldsymbol{\theta}) &= \left\{ \max_{t\in[0,T]} \left(\left| \frac{\delta r(t, \mathbf{z}, \boldsymbol{\theta})}{0.10\,\text{m}} \right| \right) > 1 \right\} \end{aligned} \tag{5.28}$$

where $\ddot{x}_{\text{absolute}}(t, \mathbf{z}, \boldsymbol{\theta})$ represents the absolute acceleration at the middle of the deck girder (in the x or y direction), $\delta x(t, \mathbf{z}, \boldsymbol{\theta})$ denotes the relative displacement between the top of the piers and their connections with the pile foundation (in the x or y direction), and $\delta r(t, \mathbf{z}, \boldsymbol{\theta})$ describes the relative displacement between the deck girder and the base of the rubber bearings at each abutment (in the x or y direction). It is expected that system parameters, such as the diameter of the pier elements, the diameter of the pile elements, the external diameter of the rubber bearings, and the total height of rubber in the bearing, may have important effects on the system response. Thus, a reliability sensitivity analysis with respect to these parameters may provide important information about the overall behavior of the bridge structure. Based on the previous observations, the vector of the system parameters is defined as $\boldsymbol{\theta}^T =< D_c, D_p, D_r, H_r >$. To study the behavior of the failure probability when the system parameters vary in a certain region of the parameters space, the system parameters are modeled as independent normal random variables with distribution parameters given in Table 5.1. Of course, alternative distributions can also be used. Note that the uncertainty associated with the system parameters D_c, D_p, D_r and H_r (geometrical parameters) may correspond to the inherent variability in the construction process of these elements (piers, piles, and rubber bearings), or they may be considered as an instrumental variability in the context of this analysis.

Table 5.1 Distribution of system parameters

System parameter	Mean value	C.O.V.
D_c (diameter of pier elements)	$\mu_{D_c} = 1.6\,\text{m}$	0.10
D_p (diameter of pile elements)	$\mu_{D_p} = 0.6\,\text{m}$	0.10
D_r (external diameter of rubber bearings)	$\mu_{D_r} = 0.8\,\text{m}$	0.10
H_r (total height of rubber)	$\mu_{H_r} = 0.16\,\text{m}$	0.10

5.8.4 Reduced-Order Model

To carry out the reliability sensitivity analysis in a reduced-order model, the bridge model is divided into a number of substructures. In particular, the structural model is subdivided into nine linear substructures and two nonlinear substructures, as shown in Fig. 5.4.

Substructures S_1, S_2, S_3, and S_4 are composed of the different pile elements, substructures S_5, S_6, S_7 and S_8 include the different pier elements, and substructure S_9 corresponds to the deck girder. Finally, substructures S_{10} and S_{11} are the nonlinear substructures composed of the rubber bearings located at the left and right abutment, respectively. Based on the previous definition of substructures, it is clear that substructures S_1, S_2, S_3, and S_4 depend on the system parameter D_p, substructures S_5, S_6, S_7 and S_8 depend on the system parameter D_c, while substructure S_9 is independent of the system parameters. Furthermore, the nonlinear substructures depend on the system parameters D_r and H_r. In connection with Chap. 2, the corresponding parametrization functions are given by $h^j(\theta_j) = \theta_j^4$ (parameter related to the inertia

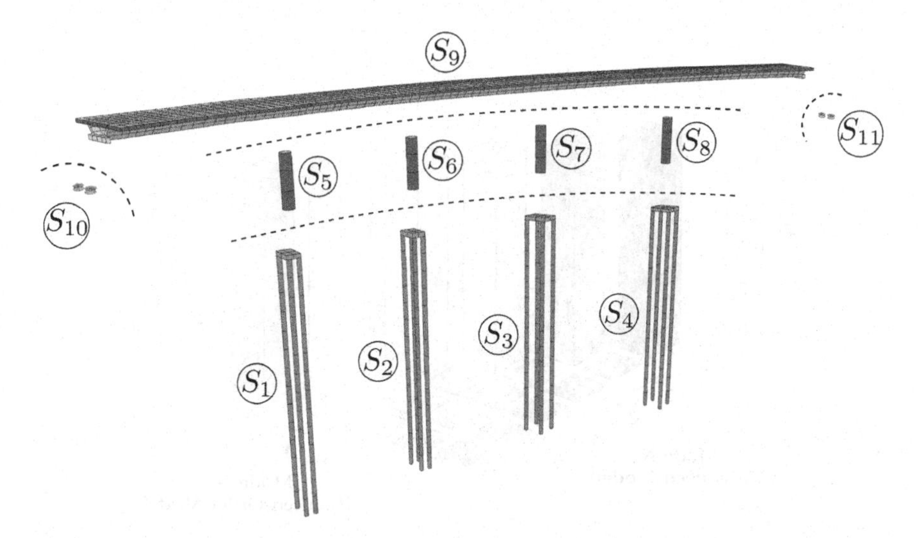

Fig. 5.4 Linear and nonlinear substructures of the finite element model

Table 5.2 Modal frequency difference between the modal frequencies of the full model and the reduced-order model based on dominant fixed-interface modes

Frequency number	Unreduced model	Reduced-order model	Error
	ω (rad/s)	ω (rad/s)	| Difference |
1	4.214	4.216	2.0×10^{-3}
2	4.282	4.284	2.0×10^{-3}
3	4.569	4.572	3.0×10^{-3}
4	12.197	12.249	5.2×10^{-2}
5	15.424	15.462	3.8×10^{-2}
6	23.419	23.421	2.0×10^{-3}

term in the stiffness matrices) and $g^j(\theta_j) = \theta_j^2$ (parameter related to the area term in the mass matrices), where the model parameter θ_j is either D_c or D_p, normalized by its mean value.

Validation calculations indicate that retaining three generalized coordinates (dominant fixed-interface normal modes) for each one of substructures S_1, S_2, S_3, and S_4, two for each one of substructures S_5, S_6, S_7 and S_8, and 10 for substructure S_9 are adequate in the context of this application. The absolute value of the difference between the modal frequencies using the full nominal reference finite element model and the modal frequencies computed using the reduced-order model based on dominant fixed-interface normal modes is shown in Table 5.2. The modal frequencies for both models are computed by considering only the linear components of the structural system. A small difference is observed with this number of generalized coordinates. The corresponding matrix of MAC-values between the first six modal vectors com-

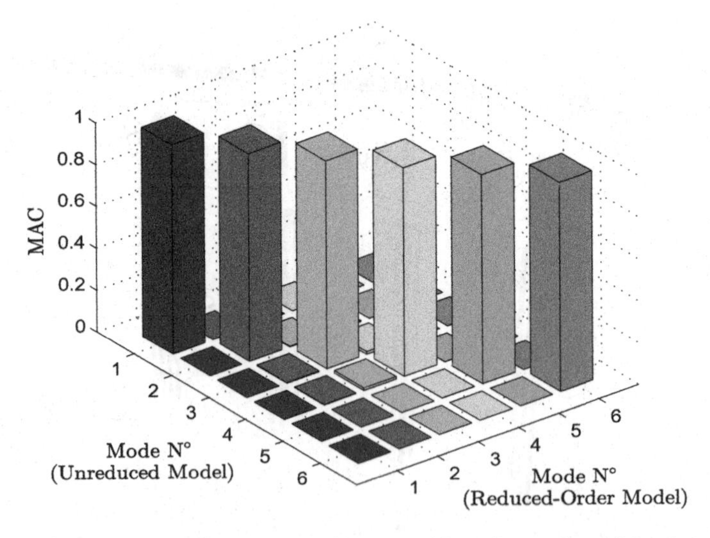

Fig. 5.5 MAC-values between the mode shapes computed from the unreduced finite element model and from the reduced-order model based on dominant fixed-interface modes

puted from the unreduced finite element model and from the reduced-order model is shown in terms of a 3-D representation in Fig. 5.5.

It is observed that the values at the diagonal terms are close to one and almost zero at the off-diagonal terms. Thus, the modal vectors of both models are consistent. The comparison with the lowest six modes is based on the fact that the contribution of the higher order modes (higher than the 6th mode) in the dynamic response of the model is negligible. In fact, the dynamic response of the magnitudes associated with the failure events, that is, $\ddot{x}_{\text{absolute}}(t, \mathbf{z}, \boldsymbol{\theta})$, $\delta x(t, \mathbf{z}, \boldsymbol{\theta})$, and $\delta r(t, \mathbf{z}, \boldsymbol{\theta})$, obtained from the reduced-order model, coincides with the response obtained from the unreduced finite element model. Note that residual normal modes and interface modes are not involved in the construction of the reduced-order model. In summary, a total of 30 generalized coordinates, corresponding to the fixed-interface normal modes of the linear substructures, out of 10,008 internal degrees of freedom of the original model, are retained for the nine linear substructures. Therefore, the number of interface degrees of freedom is equal to 60 in this case. With this reduction, the total number of generalized coordinates of the reduced-order model represents a 99% reduction with respect to the unreduced model. Thus, a drastic reduction in the number of generalized coordinates is obtained with respect to the number of the degrees of freedom of the original unreduced finite element model. Based on the previous analysis, it is concluded that the reduced-order model and the full finite element model are equivalent in the context of this application problem. Therefore, the reliability sensitivity analysis of the bridge structural model is carried out by using the reduced-order model. From the practical point of view, it is important to note that the selection of the fixed-interface modes per substructure, necessary to achieve a prescribed accuracy, can be done offline, before the reliability sensitivity analysis takes place.

In the following, the reliability sensitivity analysis corresponding to the three failure events is presented. The sensitivity measures with respect to a given parameter are estimated by considering the other parameters fixed at their mean values. This is done to isolate the effect of the parameter variation on the system reliability.

Table 5.3 Sensitivity measures in terms of partial derivatives. (·%) sample coefficient of variation. Failure event F_1

	Average value (c.o.v.)
$\frac{\partial P_{F_1}}{\partial \mu_{D_p}}$	3.85×10^{-2} (27%)
$\frac{\partial P_{F_1}}{\partial \mu_{D_c}}$	5.37×10^{-3} (47%)

Table 5.4 Sensitivity measures in terms of elasticity coefficients. (·%) sample coefficient of variation. Failure event F_1

	Average value (c.o.v.)
$e^{F_1}_{\mu_{D_p}}$	7.26 (21%)
$e^{F_1}_{\mu_{D_c}}$	3.28 (47%)

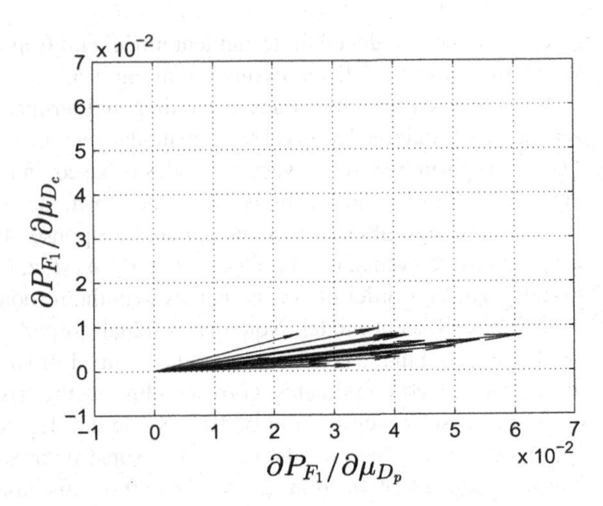

Fig. 5.6 Sensitivity of the failure probability P_{F_1} with respect to the mean value of the diameter of the pier and pile elements: 20 independent estimations

5.8.5 Results: Failure Event F_1

The results of the reliability sensitivity analysis corresponding to the failure event associated with the maximum absolute acceleration at the middle of the deck girder are given in Tables 5.3 and 5.4.

Table 5.3 shows the sensitivity analysis in term of the partial derivatives of the failure probability with respect to the mean value of the system parameters D_p and D_c, while Table 5.4 gives the corresponding sensitivity measures in terms of the elasticity coefficients. The proposed approach is implemented by using 1,000 samples at each conditional level of subset simulation with conditional failure probabilities equal to $p_0 = 0.1$. The estimates shown in the tables correspond to an average of 20 independent runs. The sensitivity information provided in Tables 5.3 and 5.4 is also showed in Fig. 5.6 in form of arrows indicating the magnitude and sign of the sensitivity. Twenty representative estimations are considered in the figure.

It is observed that the sensitivity measures with respect to the mean value of the parameters D_p and D_c are positive. Thus, an increase in the value of these parameters increases the probability of failure. In fact, an increase in the diameter of the pier and pile elements tends to increase the maximum absolute acceleration at the middle of the deck girder, which is reasonable from a structural point of view. It is also seen that failure appears to be most sensitive to the mean value of the diameter of the piles, as expected. The estimates generated by the proposed simulation-based approach present some level of dispersion, which can be observed from Fig. 5.6. However, on average, the estimates converge to the reference value. This result is shown in Fig. 5.7 in terms of the elasticity coefficient estimates, where the reference result is obtained directly by Monte Carlo simulation with a large number of samples (100,000 in this case). The corresponding direct simulation is carried out by using the reduced-order model. The actual variability of the sensitivity estimates is given in parentheses in Tables 5.3 and 5.4. That number corresponds to the sample coefficient of variation of the estimates over 20 independent runs.

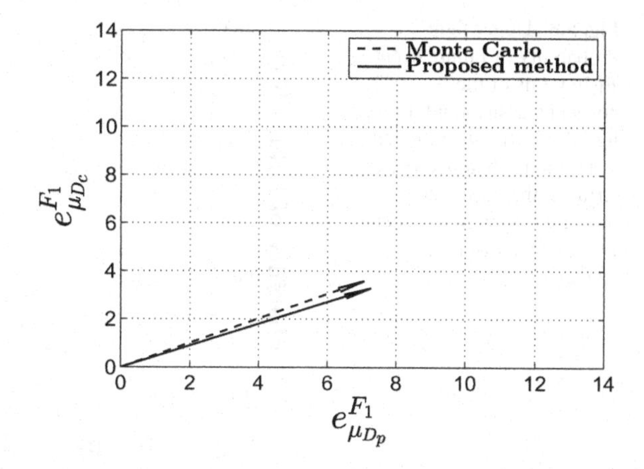

Fig. 5.7 Average of the elasticity coefficient estimates generated by the proposed approach from 20 simulation runs compared to the reference estimate (Monte Carlo). Failure event F_1. System parameters D_p and D_c

A similar reliability sensitivity analysis can be performed with respect to the standard deviation of the system parameters D_p and D_c. It turns out that the elasticity of the failure probability with respect to the standard deviation of these parameters is positive. Therefore, the probability of failure increases, i.e., the structural reliability reduces, with an increase in the standard deviation (variability) of the diameters of the pier and pile elements. The information provided by the sensitivity analysis with respect to the standard deviation of the system parameters can be used to identify the parameters whose uncertainty plays a major role in affecting the failure probability.

As stated in Sect. 5.6, the proposed method yields with a single subset simulation run reliability sensitivity estimates for all thresholds up to the largest one considered in the analysis. In this context, Figs. 5.8 and 5.9 show the probability of failure and the corresponding elasticity coefficients in terms of the threshold. The results with respect to the mean value of the diameter of the pile elements are shown in Fig. 5.8, while Fig. 5.9 shows the results related to the mean value of the diameter of the pier elements. In these figures, an average of 20 independent runs is considered, where the threshold is normalized by the acceptable acceleration response level equal to 2.0 m/s^2 (see Sect. 5.8.3). These figures illustrate the whole trend of the probability of failure and sensitivity measure in terms of the threshold, not only for the normalized target value equal to one. This feature of the proposed method is quite useful, since the whole trend of the sensitivity measures versus the threshold is obtained. It is observed from the figures that the elasticity coefficients increase as the threshold level increases and, therefore, the failure probability becomes more sensitive as the failure probability becomes smaller.

5.8.6 Results: Failure Event F_2

The results of the reliability sensitivity analysis associated with the second failure event are given in Tables 5.5 and 5.6. Table 5.5 shows the sensitivity analysis in terms

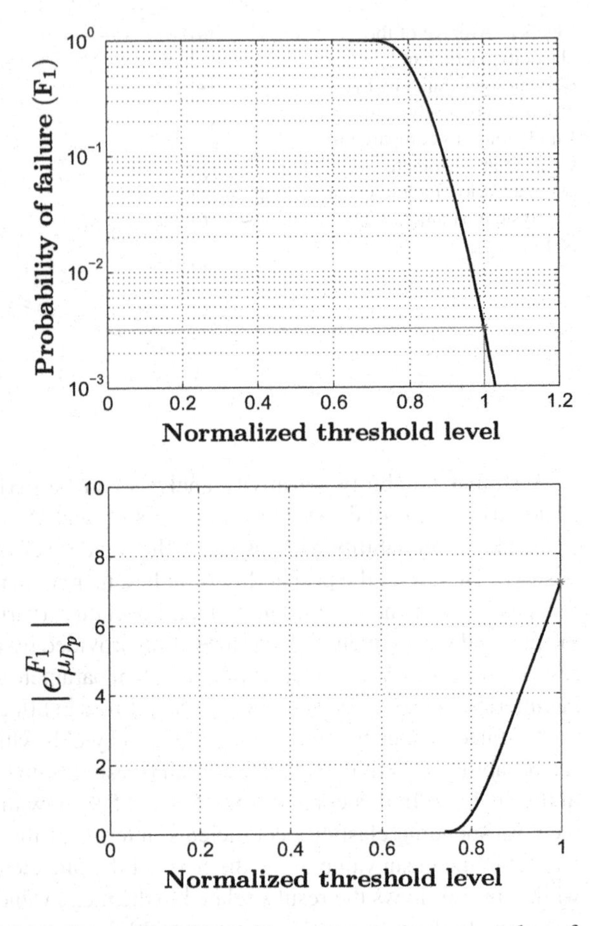

Fig. 5.8 Upper figure: Probability of failure event F_1 in terms of the normalized threshold. Lower figure: Elasticity coefficient of failure probability P_{F_1} in terms of the normalized threshold. System parameter D_p (diameter of pile elements)

of the partial derivatives of the failure probability with respect to the mean value of the system parameters D_p and D_c, while Table 5.6 gives the corresponding sensitivity measures in terms of the elasticity coefficients.

This information is also shown in Fig. 5.10 in form of arrows indicating the magnitude and sign of the sensitivity. As in the previous case, the results are based on 20 independent runs. For this failure event, the sensitivity measures with respect to the mean value of the parameters D_p and D_c are negative. Thus, an increase in the value of these parameters decreases the probability of failure. In this case, an increase in the diameter of the pier and pile elements tends to decrease the maximum relative displacement between the piers and the piles, as expected. The results indicate that both parameters, that is, the mean value of the diameter of the pile and pier elements, have an important effect on P_{F_2}.

A close examination of the results reveals that the probability of failure event F_2 is more sensitive to the diameter of the pier elements than to the diameter of the pile elements, which makes sense from a physical point of view. The numerical results also show that, on average, the estimates obtained from the proposed approach

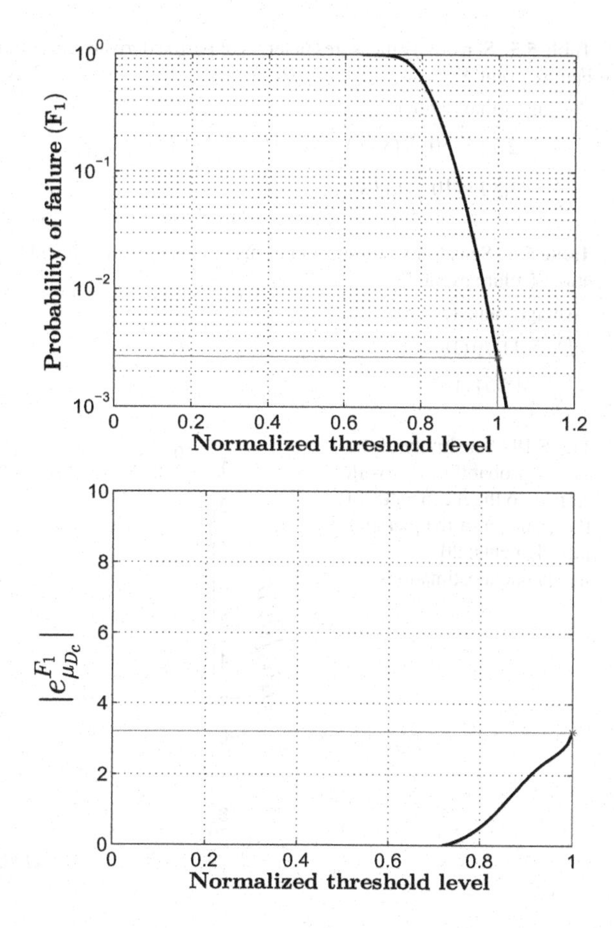

Fig. 5.9 Upper figure: Probability of failure event F_1 in terms of the normalized threshold. Lower figure: Elasticity coefficient of failure probability P_{F_1} in terms of the normalized threshold. System parameter D_c (diameter of pier elements)

converge to the reference result, as demonstrated in Fig. 5.11. In fact, the average estimate coincides with the reference value, which was directly obtained by Monte Carlo simulation with the same number of samples as used in the previous case.

The sample average of the failure probability elasticities in terms of the number of independent simulation runs is shown in Fig. 5.12. For comparison, the results obtained by Monte Carlo simulation are also shown in the figure (with a square symbol). It is seen that the sample average of the elasticity coefficients stabilizes very fast to the reference result. Thus, the estimate obtained by the proposed sensitivity measure is practically unbiased. The corresponding sample coefficient of variation of the estimates, based on 20 independent runs, is given in parentheses in Tables 5.5 and 5.6. The trend of the probability of failure and the sensitivity measure in terms of the threshold is shown in Figs. 5.13 and 5.14, respectively. The results corresponding to the mean value of the diameter of the pile elements are shown in Fig. 5.13, while Fig. 5.14 shows the results related to the mean value of the diameter of the pier elements. An average of 20 independent runs is considered in the figures, where the threshold level is normalized by the acceptable relative displacement response level

Table 5.5 Sensitivity measures in terms of partial derivatives. (·%) sample coefficient of variation. Failure event F_2

Average value (c.o.v.)
$\frac{\partial P_{F_2}}{\partial \mu_{D_p}}$ -2.73×10^{-2} (33%)
$\frac{\partial P_{F_2}}{\partial \mu_{D_c}}$ -3.30×10^{-3} (37%)

Table 5.6 Sensitivity measures in terms of elasticity coefficients. (·%) sample coefficient of variation. Failure event F_2

Average value (c.o.v.)
$e^{F_2}_{\mu_{D_p}}$ -11.79 (18%)
$e^{F_2}_{\mu_{D_c}}$ -18.64 (14%)

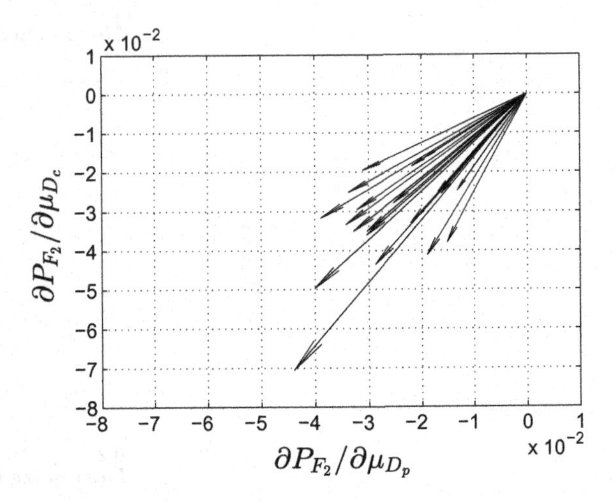

Fig. 5.10 Sensitivity of the failure probability P_{F_2} with respect to the mean value of the diameter of the pier and pile elements: 20 independent estimations

equal to 0.07 m (see Sect. 5.8.3). Once again, it is clear from the figures that the elasticity coefficients increase in magnitude as the threshold increases and, therefore, the failure probability becomes more sensitive as the failure probability becomes smaller.

5.8.7 Results: Failure Event F_3

The results of the reliability sensitivity analysis associated with the failure event related to the maximum relative displacement between the deck girder and the base of the rubber bearings at each abutment are given in Tables 5.7 and 5.8.

Table 5.7 shows the sensitivity analysis in terms of the partial derivatives of the failure probability with respect to the mean value of the system parameters D_r and H_r, while Table 5.8 gives the corresponding sensitivity measures in terms of the

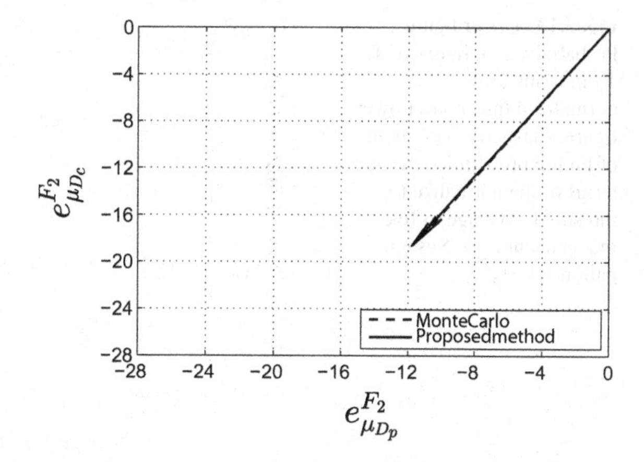

Fig. 5.11 Average of the elasticity coefficient estimates generated by the proposed approach from 20 simulation runs compared to the reference estimate (Monte Carlo). Failure event F_2. System parameters D_p and D_c

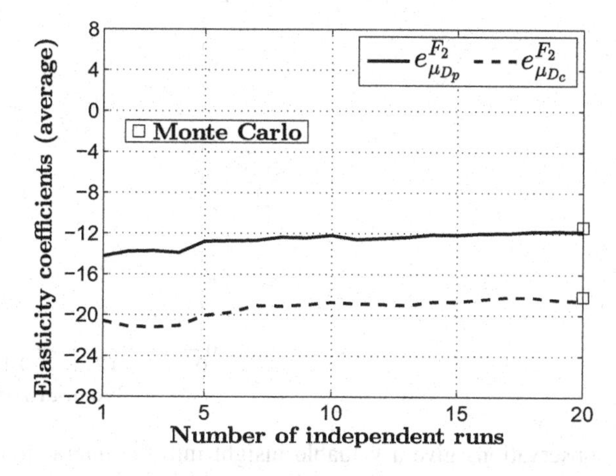

Fig. 5.12 Sample average of elasticity coefficients corresponding to the probability of failure event F_2 in terms of the number of independent simulation runs

elasticity coefficients. The information is also shown in Fig. 5.15 in form of arrows indicating the magnitude and sign of the sensitivity. As in the previous cases, the results are based on 20 independent runs.

From the tables, it is seen that the sensitivity of the failure probability with respect to the mean value of the external diameter D_r is negative. Thus, an increase in the external diameter of the isolators decreases the probability of failure. This is reasonable since the base isolation system becomes stiffer and, therefore, the relative displacement between the deck girder and the base of the rubber bearings tends to decrease. On the other hand, the sensitivity of the failure probability with respect to the mean value of the total height of rubber H_r is positive. In this case, an increase in the total height of rubber in the isolator increases the probability of failure. This is consistent with the fact that the isolation system becomes more flexible increasing in this manner the relative displacement between the deck girder and the rubber bearings. The corresponding elasticity coefficients indicate that the external diameter of the isolators plays a significant role in affecting the probability of failure. These

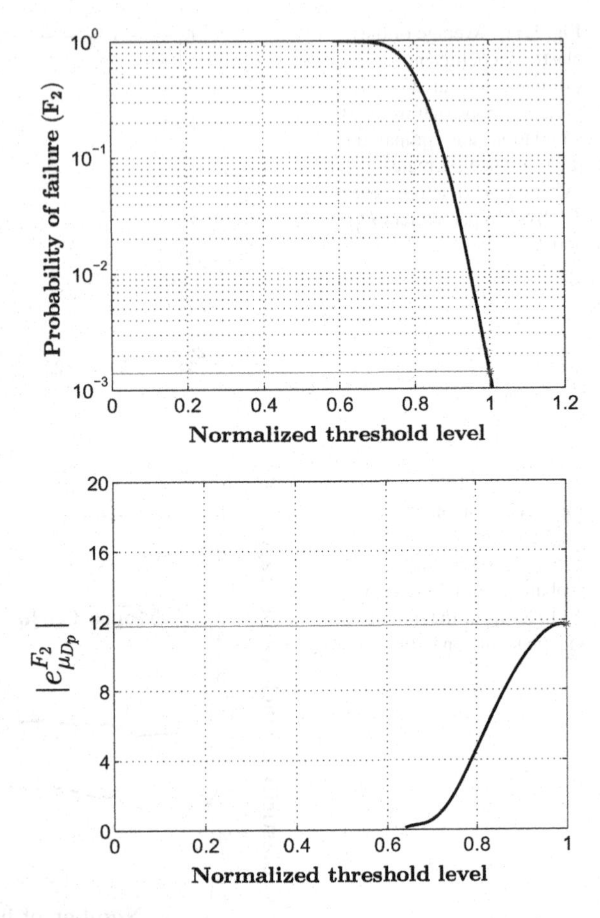

Fig. 5.13 Upper figure: Probability of failure event F_2 in terms of the normalized threshold. Lower figure: Elasticity coefficient of failure probability P_{F_2} in terms of the normalized threshold. Average of five independent runs. System parameter D_p

observations give a valuable insight into the interaction and effect of the isolator parameters on the failure event associated with the maximum relative displacement between the deck girder and the base of the rubber bearings.

As in the previous cases, the average estimates obtained by the proposed approach coincide with the reference values as shown in Fig. 5.16. Information about the sample behavior of the failure probability elasticities in terms of the number of independent simulation runs is shown in Fig. 5.17. This figure shows the sample average of the elasticity with respect to the mean value of the system parameters D_r and H_r.

It is seen that the average stabilizes extremely fast. For comparison, the results obtained by Monte Carlo simulation are also shown in the figure (with a square symbol). The average of the elasticity coefficients coincides with the Monte Carlo results. This result indicates that the sensitivity estimation in terms of the elasticity coefficients is practically unbiased, as in the previous cases. The corresponding sample coefficient of variation is shown in Fig. 5.18. The coefficient of variation of the elasticity of the failure probability with respect to the external diameter D_r and the total height of rubber H_r tends to 5% and 17%, respectively (number in parentheses

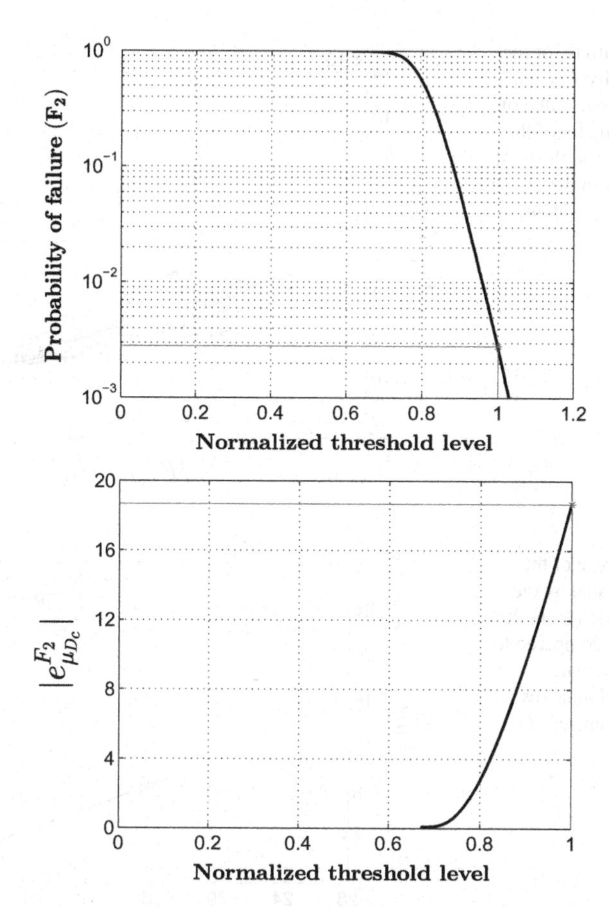

Fig. 5.14 Upper figure: Probability corresponding to failure event F_2 in terms of the normalized threshold. Lower figure: Elasticity coefficient of failure probability P_{F_2} in terms of the normalized threshold. Average of five independent runs. System parameter D_c

Table 5.7 Sensitivity measures in terms of partial derivatives. (·%) sample coefficient of variation. Failure event F_3

	Average value (c.o.v.)
$\frac{\partial P_{F_3}}{\partial \mu_{D_r}}$	-9.48×10^{-1} (16%)
$\frac{\partial P_{F_3}}{\partial \mu_{H_r}}$	3.67×10^{-1} (29%)

Table 5.8 Sensitivity measures in terms of elasticity coefficients. (·%) sample coefficient of variation. Failure event F_3

	Average value (c.o.v.)
$e^{F_3}_{\mu_{D_r}}$	−16.75 (5%)
$e^{F_3}_{\mu_{H_r}}$	9.06 (17%)

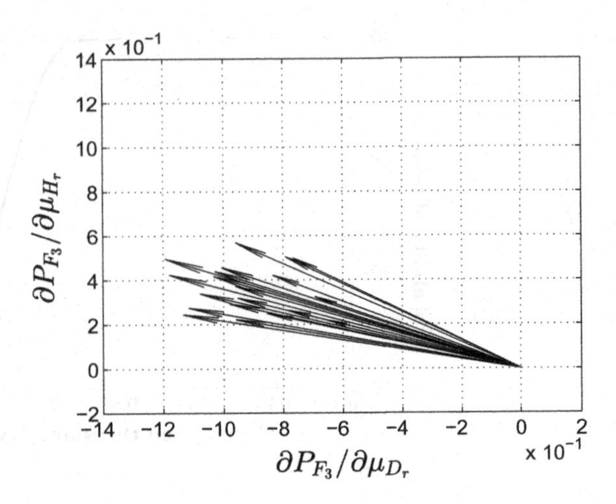

Fig. 5.15 Sensitivity of the failure probability P_{F_3} with respect to the mean value of the external diameter of the rubber bearings and the total height of rubber in the bearings: 20 independent estimations

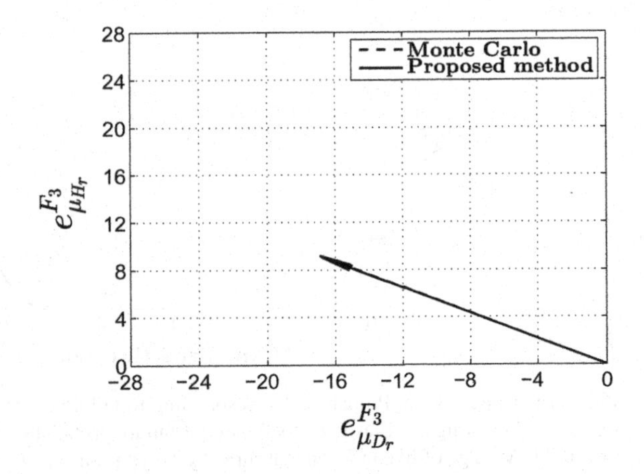

Fig. 5.16 Average of the estimates generated by the proposed approach from 20 simulation runs compared to the reference estimate (Monte Carlo). Failure event F_3. System parameters D_r and H_r

in Table 5.8). These small values correspond to the actual variability of the failure probability elasticity estimates.

The trend of the probability of failure and the sensitivity measure in terms of the threshold is shown in Figs. 5.19 and 5.20, respectively. The results corresponding to the mean value of the external diameter of the rubber bearings are shown in Fig. 5.19, while Fig. 5.20 shows the results related to the mean value of the total height of rubber in the bearings. An average of 20 independent runs is considered in the figures, where the threshold is normalized by the acceptable relative displacement response level equal to 0.10 m (see Sect. 5.8.3).

As in the previous failure events, the elasticity coefficients increase in magnitude as the threshold level increases and, therefore, the failure probability becomes more sensitive as the failure probability becomes smaller. The importance of the external diameter of the rubber bearings on failure event F_3, compared to the total height of rubber, can also be seen from the probability curves of the previous figures. In

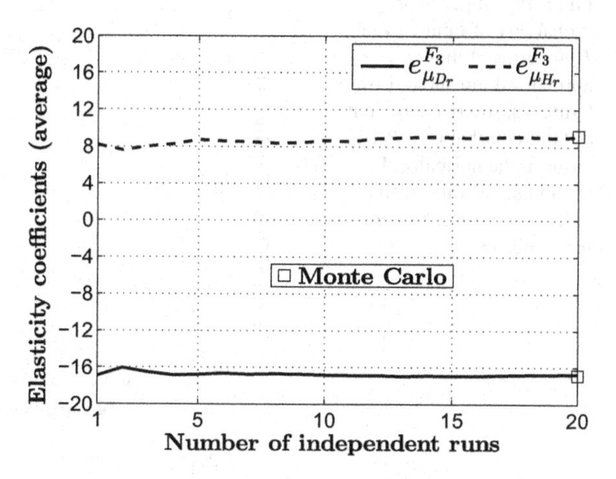

Fig. 5.17 Sample average of elasticity coefficients corresponding to the probability of failure event F_3 in terms of the number of independent simulation runs

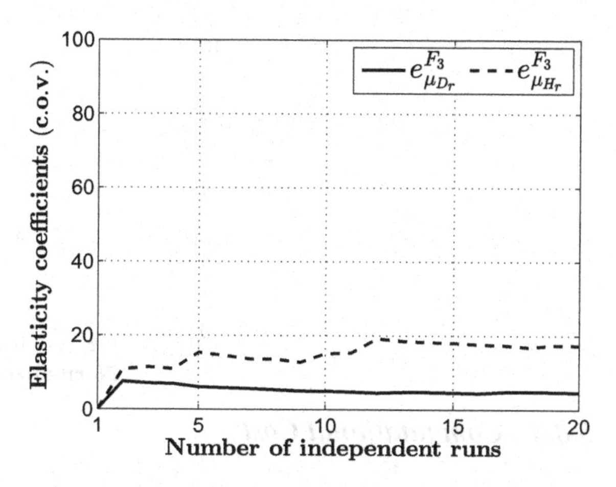

Fig. 5.18 Sample coefficient of variation of elasticity coefficients corresponding to the probability of failure event F_3 in terms of the number of independent simulation runs

fact, the probability of failure considering the external diameter of the bearings as uncertain is estimated as $P_{F_3} = 4.5 \times 10^{-2}$, for a normalized threshold equal to one, while a probability of failure $P_{F_3} = 6.5 \times 10^{-3}$ is obtained when the total height of rubber in the bearings is considered as uncertain. Note that the difference is almost one order of magnitude.

In summary, the previous results corresponding to the different failure events represent valuable and practical information about the global performance of the bridge model.

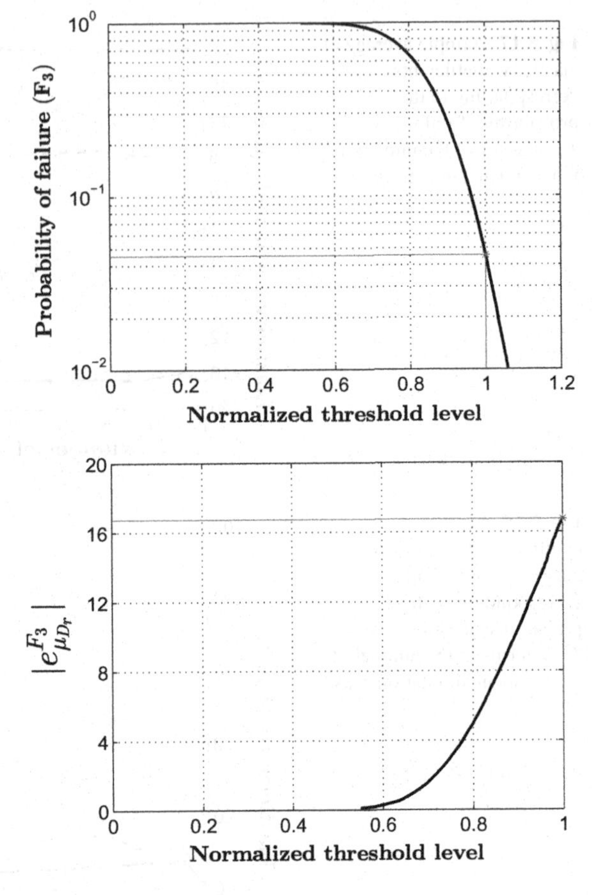

Fig. 5.19 Upper figure: Probability of failure event F_3 in terms of the normalized threshold. Lower figure: Elasticity coefficient of failure probability P_{F_3} in terms of the normalized threshold. Average of five independent runs. System parameter D_r

5.8.8 Computational Cost

The number of finite element runs required during the reliability sensitivity analysis mainly depends on the number of simulations needed by the proposed approach. Such a number is related to the number of levels or stages carried out by subset simulation. Thus, the computational effort for assembling the finite element model and obtaining its nonlinear dynamic response for a given set of system parameters is the fundamental factor for comparison purposes. In this regard, the proposed model reduction technique is quite effective. In fact, the execution time for assembling the reduced-order model represents 0.03% of the time required for the original unreduced finite element model. Overall, the use of the reduced-order model for estimating the reliability sensitivity measures results in a drastic reduction of the computational effort of almost two orders of magnitude. In other words, the ratio of the execution time for obtaining the reliability sensitivity measures by using the full finite element model and the execution time for obtaining the reliability sensitivity measures by using the reduced-order model is about 90 in this case. Thus, a significant reduction in

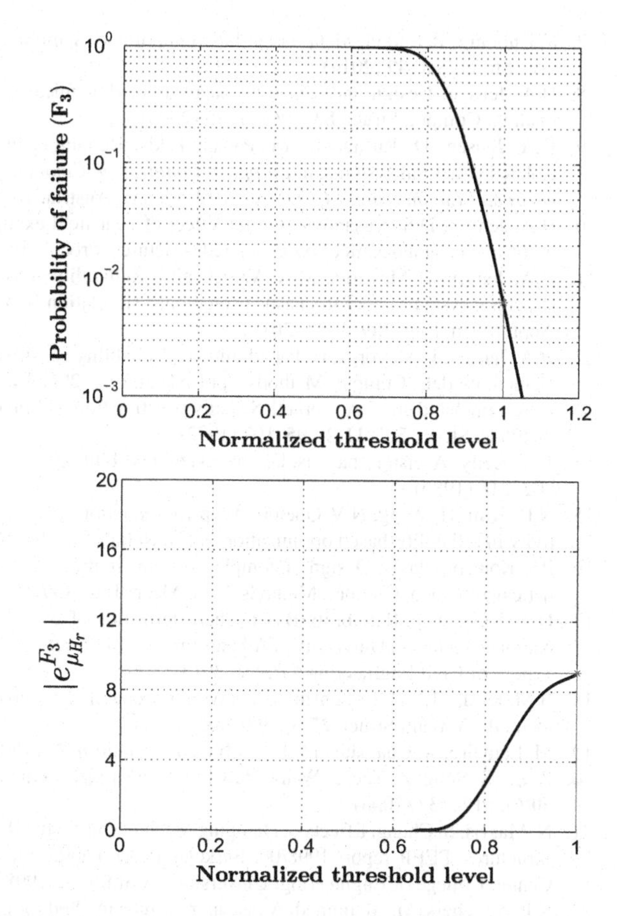

Fig. 5.20 Upper figure: Probability of failure event F_3 in terms of the normalized threshold. Lower figure: Elasticity coefficient of failure probability P_{F_3} in terms of the normalized threshold. Average of five independent runs. System parameter H_r

computational effort is achieved without compromising the accuracy of the reliability sensitivity estimates.

References

1. S.K. Au, Reliability-based design sensitivity by efficient simulation. Comput. Struct. **83**(14), 1048–1061 (2005)
2. H.-G. Beyer, B. Sendhoff, Robust optimization - a comprehensive survey. Comput. Methods Appl. Mech. Eng. **196**(33–34), 3190–3218 (2007)
3. P. Bjerager, S. Krenk, Parametric sensitivity in first order reliability theory. J. Eng. Mech. **115**(7), 1577–1582 (1989)
4. J. Ching, Y.H. Hsieh, Local estimation of failure probability function and its confidence interval with maximum entropy principle. Probab. Eng. Mech. **22**(1), 39–49 (2007)
5. O. Ditlevsen, H.O. Madsen, *Structural Reliability Methods* (Wiley, Chichester, 1996)
6. I. Doltsinis, Z. Kang, Robust design of structures using optimization methods. Comput. Methods Appl. Mech. Eng. **193**(23–26), 2221–2237 (2004)

7. V. Dubourg, B. Sudret, Meta-model-based importance sampling for reliability sensitivity analysis. Struct. Saf. **49** (2014)
8. H.A. Jensen, Design and sensitivity analysis of dynamical systems subjected to stochastic loading. Comput. Struct. **83**, 1062–1075 (2005)
9. H.A. Jensen, D. Kusanovic, M. Papadrakakis, Reliability-based characterization of base-isolated structural systems, in *European Congress on Computational Methods in Applied Sciences and Engineering. ECCOMAS 2012*, Vienna, Austria, 10–14 September 2012
10. H.A. Jensen, D.S. Kusanovic, On the effect of near-field excitations on the reliability-based performance and design of base-isolated structures. Probab. Eng. Mech. **36**, 28–44 (2014)
11. H.A. Jensen, F. Mayorga, M.A. Valdebenito, Reliability sensitivity estimation of nonlinear structural systems under stochastic excitation: a simulation-based approach. Comput. Methods Appl. Mech. Eng. **289**, 1–23 (2015)
12. H.A. Jensen, F. Mayorga, C. Papadimitriou, Reliability sensitivity analysis of stochastic finite element models. Comput. Methods Appl. Mech. Eng. **296**, 327–351 (2015)
13. A. Karamchandani, C.A. Cornell, Sensitivity estimation within first and second order reliability methods. Struct. Saf. **11**(2), 95–107 (1992)
14. J.M. Kelly, Aseismic base isolation: review and bibliography. Soil Dyn. Earthq. Eng. **5**(4), 202–216 (1986)
15. N.H. Kim, H. Wang, N.V. Queipo, Adaptive reduction of design variables using global sensitivity in reliability-based optimization. Int. J. Reliab. Saf. **1**(1–2), 102–119 (2006)
16. P.S. Koutsourelakis, Design of complex systems in the presence of large uncertainties: a statistical approach. Comput. Methods Appl. Mech. Eng. **197**(49–50), 4092–4103 (2008)
17. D.A. Kusanovic, Reliability-Based Characterization of Base-Isolated Buildings. MSc thesis, National Technical University of Athens, Institute of Structural Analysis and Seismic Research, School of Civil Engineering, Greece, 2013
18. B.M. Kwak, T.W. Lee, Sensitivity analysis for reliability-based optimization using an AFOSM method. Comput. Struct. **27**(3), 399–406 (1987)
19. M. Lemaire, A. Chateauneuf, J.-C. Mitteau, *Structural Reliability* (Wiley, New York, 2009)
20. Z. Lu, S. Song, Z. Yue, J. Wang, Reliability sensitivity method by line sampling. Struct. Saf. **30**(6), 517–532 (2008)
21. N. Makris, S. Chang, Effects of Damping Mechanisms on the Response of Seismically Isolated Structures. PEER report 1998/06. Berkeley (CA): Pacific Earthquake Engineering Research Center. College of Engineering, University of California, 1998
22. R.E. Melchers, M. Ahammed, A fast approximate method for parameter sensitivity estimation in Monte Carlo structural reliability. Comput. Struct. **82**(1), 55–61 (2004)
23. S. Minewaki, M. Yamamoto, M. Higashino, H. Hamaguchi, H. Kyuke, T. Sone, H. Yoneda, Performance tests of full size isolators for super high-rise isolated buildings. J. Struct. Eng. AIJ **55**(B), 469–477 (2009)
24. S. Rahman, D. Wei, Design sensitivity and reliability-based structural optimization by univariate decomposition. Struct. Multidiscip. Optim. **35**(3), 245–261 (2008)
25. R.Y. Rubinstein, D.P. Kroese, *Simulation and Monte Carlo Method* (Wiley, New York, 2007)
26. G.I. Schuëller, H.A. Jensen, Computational methods in optimization considering uncertainties – an overview. Comput. Methods Appl. Mech. Eng. **198**(1), 2–13 (2008)
27. S. Song, Z. Lu, H. Qiao, Subset simulation for structural reliability sensitivity analysis. Reliab. Eng. Syst. Saf. **94**(2), 658–665 (2009)
28. L. Su, G. Ahmadi, J.G. Tadjbakhsh, A comparative study of performances of various base isolation systems, part II: sensitivity analysis. Earthq. Eng. Struct. Dyn. **19**, 21–33 (1990)
29. A.A. Taflanidis, G. Jia, A simulation-based framework for risk assessment and probabilistic sensitivity analysis of base-isolated structures. Earthq. Eng. Struct. Dyn. **40**(14), 1629–1651 (2011)
30. M.A. Valdebenito, H.A. Jensen, G.I. Schuëller, F.E. Caro, Reliability sensitivity estimation of linear systems under stochastic excitation. Comput. Struct. **92–93**, 257–268 (2012)
31. Y.T. Wu, Computational methods for efficient structural reliability and reliability sensitivity analysis. AIAA J. **32**(8), 1717–1723 (1994)

32. M. Yamamoto, S. Minewaki, M. Higashino, H. Hamaguchi, H. Kyuke, T. Sone, H. Yoneda, Performance tests of full size rubber bearings for isolated superhigh-rise buildings, in *International Symposiumon Seismic Response Controlled Buildings for Sustainable Society*, Tokyo, Japan, 2009
33. M. Yamamoto, S. Minewaki, H. Yoneda, M. Higashino, Nonlinear behavior of high-damping rubber bearings under horizontal bidirectional loading: full-scale test and analitycal modeling. Earthq. Eng. Struct. Dyn. **41**(13), 1845–1860 (2012)
34. E. Zio, N. Pedroni, Monte Carlo simulation-based sensitivity analysis of the model of a thermal-hydraulic passive system. Reliab. Eng. Syst. Saf. **107**, 90–106 (2012)

Chapter 6
Reliability-Based Design Optimization

Abstract The solution of reliability-based design optimization problems by using reduced-order models is considered in this chapter. Specifically, problems involving high-dimensional stochastic dynamical systems are analyzed. The design process is formulated in terms of a constrained nonlinear optimization problem, which is solved by a class of interior point algorithms based on feasible directions. Search directions are estimated in an efficient manner as a by-product of reliability analyses. The design process generates a sequence of steadily-improved feasible designs. Three numerical examples are presented to evaluate the performance of the interior point algorithm and the effectiveness of reduced-order models in the context of complex reliability-based optimization problems. High speedup values can be obtained for the design process without changing the accuracy of the final designs.

6.1 Motivation

Structural optimization by means of deterministic mathematical programming techniques has been widely accepted as a viable tool for engineering design [2, 17]. However, in many structural engineering applications response predictions are based on models whose parameters are uncertain. This is due to a lack of information about the value of system parameters either external to the structure, such as environmental loads, or internal, such as system behavior. Although traditional approaches have been used successfully in many practical applications, a proper design procedure must explicitly consider the effects of uncertainties as they may cause significant changes in the global performance of final designs [12, 15, 31]. Under uncertain conditions, probabilistic approaches such as reliability-based formulations provide a realistic and rational framework for structural optimization, which explicitly accounts for the uncertainties [13, 28, 32, 41]. In this chapter, a reliability-based formulation characterized in terms of the minimization of an objective function subject to multiple design requirements, including standard deterministic constraints and reliability constraints, is considered.

H. Jensen and C. Papadimitriou, *Sub-structure Coupling for Dynamic Analysis*,
Lecture Notes in Applied and Computational Mechanics 89,
https://doi.org/10.1007/978-3-030-12819-7_6

6.2 Optimization Problem Formulation

A reliability-based design optimization problem can be characterized as a constrained nonlinear optimization problem of the form

$$\begin{aligned} \text{Min}_{\mathbf{x}} \;& c(\mathbf{x}) \\ \text{s.t.} \;& g_i(\mathbf{x}) \leq 0 \;\; i = 1, \ldots, n_c \\ & s_i(\mathbf{x}) \leq 0 \;\; i = 1, \ldots, n_f \end{aligned} \tag{6.1}$$

where $\mathbf{x}(x_i, i = 1, \ldots, n_d)$ is the vector of deterministic design variables, $c(\mathbf{x})$ is the objective function, $g_i(\mathbf{x}) \leq 0,\; i = 1, \ldots, n_c$, are standard constraints, and $s_i(\mathbf{x}) \leq 0\;,\; i = 1, \ldots, n_f$, are reliability constraints. The side constraints are defined as

$$\mathbf{x} \in X \;\;,\;\; x_i \in X_i = \{x_i | x_i^l \leq x_i \leq x_i^u\} \;\;,\;\; i = 1, \ldots, n_d \tag{6.2}$$

where x_i^l and x_i^u are the lower and upper limits of the design variable x_i, respectively. It is assumed that the objective and constraint functions are smooth and differentiable functions of the design variables. The objective function c can be defined in terms of initial, construction, repair or downtime costs, structural weight, structural performance, or general cost functions. Standard constraints are related to general design requirements such as geometric conditions, material cost components, availability of materials, etc. Finally, reliability constraints are associated with design specifications characterized by means of reliability measures. Reliability measures given in terms of failure probabilities with respect to specific failure criteria, such as serviceability and partial or total collapse failure, are considered in the present formulation. Then, the reliability constraint functions are written in terms of failure probability functions as

$$s_i(\mathbf{x}) = P_{F_i}(\mathbf{x}) - P^*_{F_i} \;, i = 1, \ldots, n_f \tag{6.3}$$

where $P_{F_i}(\mathbf{x})$ is the probability function for the failure event F_i evaluated at the design $\mathbf{x}$, and $P^*_{F_i}$ is the target failure probability for the ith failure event. As indicated in previous chapters, the probability that design conditions are satisfied within a particular reference period T provides a useful reliability measure for structural systems under stochastic excitation. Within this context, the probability of failure evaluated at the design $\mathbf{x}$ is formally defined as

$$P_{F_i}(\mathbf{x}) = P\left[\max_{j=1,\ldots,n_r} \max_{t \in [0,T]} \frac{| r^i_j(t, \mathbf{z}, \boldsymbol{\theta}, \mathbf{x}) |}{r^{i*}_j} > 1\right] \tag{6.4}$$

where $r_j^i(t, \mathbf{z}, \boldsymbol{\theta}, \mathbf{x})$, $j = 1, \ldots, n_r$, are the response functions associated with the failure event F_i, r_j^{i*} is the acceptable response level for the response r_j^i, $P[\cdot]$ is the probability that the expression in parentheses is true, and all other terms have been previously defined (see Chap. 4). Equivalently, the failure probability function evaluated at the design $\mathbf{x}$ can be written in terms of the multidimensional probability integral

$$P_{F_i}(\mathbf{x}) = \int_{d_i(\mathbf{z},\boldsymbol{\theta},\mathbf{x})>1} p(\mathbf{z})q(\boldsymbol{\theta})d\mathbf{z}d\boldsymbol{\theta} \tag{6.5}$$

where d_i is the normalized demand function corresponding to failure event F_i, that is,

$$d_i(\mathbf{z}, \boldsymbol{\theta}, \mathbf{x}) = \max_{j=1,\ldots,n_r} \max_{t\in[0,T]} \frac{| r_j^i(t, \mathbf{z}, \boldsymbol{\theta}, \mathbf{x}) |}{r_j^{i*}} \tag{6.6}$$

The above formulation can be extended in a direct manner if the cost of partial or total failure consequences is also included in the definition of the objective function. In the same manner, constraints related to statistics of structural responses (i.e., mean value and/or higher-order statistical moments) can be included in the formulation as well. Finally, design variables defined as distribution parameters of uncertain model parameters can also be considered. Thus, the formulation is quite general in the sense that different reliability-based optimization formulations can be considered, such as life-cycle cost optimal design, robust design optimization, risk-based design optimization, multi-objective or compromise optimization, etc. Finally, it is noted that the multidimensional probability integral given in Eq. (6.5) involves a large number of uncertain parameters (hundreds or thousands) in the context of dynamical systems under stochastic excitation, as previously pointed out [3, 23, 27, 39]. Therefore, the reliability estimation for a given design constitutes a high-dimensional problem, which is extremely demanding from a numerical point of view [7, 11, 33].

6.3 Method of Solution

The solution of the reliability-based optimization problem defined in Eq. (6.1) can be obtained in principle by a number of techniques such as standard deterministic optimization schemes based on zero- or first-order algorithms [1, 2, 8, 9, 15–17, 21, 22, 26, 27, 35, 40, 41, 44, 49] or stochastic search algorithms [29, 42, 45–47, 51, 54]. The algorithms based on standard optimization techniques are usually combined with approximation concepts in order to construct approximate representations of the different reliability quantities involved in the problem as explicit functions of the design variables. This strategy has been adopted since the numerical efforts associated with the solution of reliability-based optimization problems is dominated by the reliability assessment step. On the other hand, stochastic search algorithms are usually based on direct search schemes, where only the values of the functions involved in the optimization problem are used directly as inputs to the optimization algorithm.

The use of the above optimization approaches has been found useful in a number of structural optimization applications. However, the application of reliability-based design optimization to stochastic dynamical systems involving medium/large finite element models remains to some extent limited [43, 44]. In fact, the solution of reliability-based design problems of stochastic finite element models requires a large number of analyses to be performed during the design process. These analyses correspond to finite element re-analyses over the design space (required by the optimizer) and system responses over the uncertain parameter space (required by the simulation technique for reliability estimation). Consequently, the computational demands depend highly on the number of finite element analyses and the time taken for performing an individual analysis. Thus, the computational demands in solving reliability-based design problems may be large or even excessive. In addition, most of the proposed methodologies for solving this class of problems do not possess proven convergence properties.

To deal with the previous difficulties, an interior point scheme based on feasible directions combined with reduced-order models is selected for solving the aforementioned optimization problem.

6.4 Interior Point Algorithm

A class of feasible direction algorithms based on the solution of the Karush–Kuhn–Tucker (KKT) first-order optimality conditions is considered for the solution of the optimization problem given in Eq. (6.1) [18, 30, 34]. At each iteration, the search direction is a descent feasible direction of the objective function. A one-dimensional line search is then carried out in order to obtain a new feasible design better than the previous one. The process continues until convergence is achieved. By construction, the method generates a sequence of steadily-improved feasible designs. This class of algorithms has proved to be quite effective in deterministic optimization problems [4, 18]. In fact, a large number of test problems [20] have been solved very efficiently where the number of iterations remains comparable when the size of the problem is increased. Additionally, the above scheme has also been useful for solving a certain class of stochastic optimization problems [24].

6.4.1 Search Direction

The KKT first-order optimality conditions corresponding to the inequality constrained optimization problem (6.1) can be expressed as [34]

$$\begin{aligned}
&\nabla c(\mathbf{x}) + \nabla \mathbf{g}(\mathbf{x})\boldsymbol{\lambda}_g + \nabla \mathbf{s}(\mathbf{x})\boldsymbol{\lambda}_s = \mathbf{0} \\
&\mathbf{G}(\mathbf{x})\boldsymbol{\lambda}_g = \mathbf{0} \ , \ \mathbf{S}(\mathbf{x})\boldsymbol{\lambda}_s = \mathbf{0} \\
&g_i(\mathbf{x}) \le 0 \ , \ i = 1, \ldots, n_c \ , \ s_i(\mathbf{x}) \le 0 \ , \ i = 1, \ldots, n_f \\
&\boldsymbol{\lambda}_g \ge \mathbf{0} \ , \ \boldsymbol{\lambda}_s \ge \mathbf{0}
\end{aligned} \tag{6.7}$$

where $\boldsymbol{\lambda}_g \in R^{n_c}$ and $\boldsymbol{\lambda}_s \in R^{n_f}$ are the vectors of dual variables, $\nabla \mathbf{g}(\mathbf{x}) \in R^{n_d \times n_c}$ and $\nabla \mathbf{s}(\mathbf{x}) \in R^{n_d \times n_f}$ are the matrices of derivatives of the standard and reliability constraint functions, respectively, given by

$$\begin{aligned}
\nabla \mathbf{g}(\mathbf{x}) &= [\nabla g_1(\mathbf{x}), \nabla g_2(\mathbf{x}), \ldots, \nabla g_{n_c}(\mathbf{x})] \\
\nabla \mathbf{s}(\mathbf{x}) &= [\nabla s_1(\mathbf{x}), \nabla s_2(\mathbf{x}), \ldots, \nabla s_{n_f}(\mathbf{x}))]
\end{aligned} \tag{6.8}$$

and $\mathbf{G}(\mathbf{x})$ and $\mathbf{S}(\mathbf{x})$ are diagonal matrices such that $G_{ii}(\mathbf{x}) = g_i(\mathbf{x}), i = 1, \ldots, n_c$, and $S_{ii}(\mathbf{x}) = s_i(\mathbf{x}), i = 1, \ldots, n_f$. Under certain regularity conditions, the KKT conditions are necessary for optimality [5, 30]. In order to solve the nonlinear system of equations (6.7) for $(\mathbf{x}, \boldsymbol{\lambda}_g, \boldsymbol{\lambda}_s)$, a Newton-like iteration is considered [6, 36]. The iteration is written as

$$\begin{bmatrix} \mathbf{B}^k & \nabla \mathbf{g}(\mathbf{x}^k) & \nabla \mathbf{s}(\mathbf{x}^k) \\ \boldsymbol{\Lambda}_g^k \nabla \mathbf{g}(\mathbf{x}^k)^T & \mathbf{G}(\mathbf{x}^k) & \mathbf{0} \\ \boldsymbol{\Lambda}_s^k \nabla \mathbf{s}(\mathbf{x}^k)^T & \mathbf{0} & \mathbf{S}(\mathbf{x}^k) \end{bmatrix} \begin{Bmatrix} \mathbf{v}_1^k \\ \boldsymbol{\lambda}_g^{k+1} \\ \boldsymbol{\lambda}_s^{k+1} \end{Bmatrix} = - \begin{Bmatrix} \nabla c(\mathbf{x}^k) \\ \mathbf{0} \\ \mathbf{0} \end{Bmatrix} \quad k = 0, 1, 2, \ldots \tag{6.9}$$

where $(\mathbf{x}^k, \boldsymbol{\lambda}_g^k, \boldsymbol{\lambda}_s^k)$ is the starting point at the kth iteration, $\mathbf{v}_1^k$ is a direction in the design space, $\boldsymbol{\Lambda}_g^k$ and $\boldsymbol{\Lambda}_s^k$ are diagonal matrices with $\Lambda_{gii}^k = \lambda_{gi}^k, i = 1, \ldots, n_c$, and $\Lambda_{sii}^k = \lambda_{si}^k, i = 1, \ldots, n_f$, and $\mathbf{B}^k$ is a symmetric matrix that represents an approximation of the Hessian matrix of the Lagrangian function associated with the constrained nonlinear optimization problem. Depending on the choice of $\mathbf{B}^k$, the system of equations (6.9) may represent a second-order, a quasi-Newton, or a first-order iteration. This matrix is updated during the iterations by employing a BFGS-type updating rule [30, 37]. From Eq. (6.9), it can be proved that the vector $\mathbf{v}_1^k$ is a descent direction of the objective function $c(\mathbf{x})$ [18, 24]. However, this direction is not necessarily feasible since it is tangent to the active constraints and, therefore, is not useful as a search direction in the context of interior point algorithms. In order to obtain a feasible direction, an auxiliary linear system is considered, namely

$$\begin{bmatrix} \mathbf{B}^k & \nabla \mathbf{g}(\mathbf{x}^k) & \nabla \mathbf{s}(\mathbf{x}^k) \\ \boldsymbol{\Lambda}_g^k \nabla \mathbf{g}(\mathbf{x}^k)^T & \mathbf{G}(\mathbf{x}^k) & \mathbf{0} \\ \boldsymbol{\Lambda}_s^k \nabla \mathbf{s}(\mathbf{x}^k)^T & \mathbf{0} & \mathbf{S}(\mathbf{x}^k) \end{bmatrix} \begin{Bmatrix} \mathbf{v}_2^k \\ \bar{\boldsymbol{\lambda}}_g^k \\ \bar{\boldsymbol{\lambda}}_s^k \end{Bmatrix} = - \begin{Bmatrix} \mathbf{0} \\ \boldsymbol{\lambda}_g^k \\ \boldsymbol{\lambda}_s^k \end{Bmatrix} \quad k = 0, 1, 2, \ldots \tag{6.10}$$

where $\mathbf{v}_2^k$ is an auxiliary direction that points to the interior of the feasible domain [18], and $\bar{\boldsymbol{\lambda}}_g^k$ and $\bar{\boldsymbol{\lambda}}_s^k$ are auxiliary dual variables. Based on the solution of equations (6.9) and (6.10), a descent feasible direction $\mathbf{v}^k$ can be defined as $\mathbf{v}^k = \mathbf{v}_1^k + \rho^k \mathbf{v}_2^k$, where ρ^k

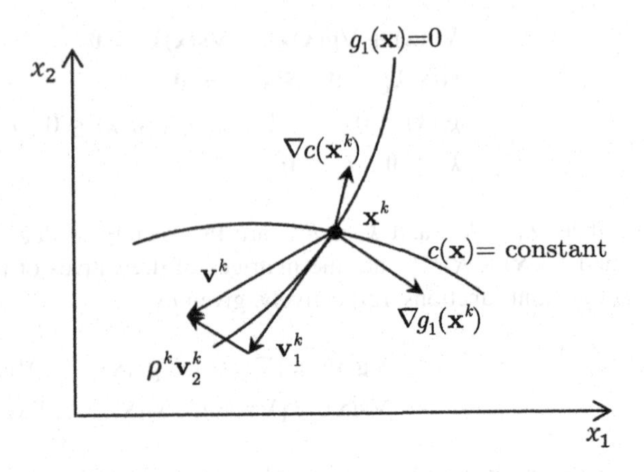

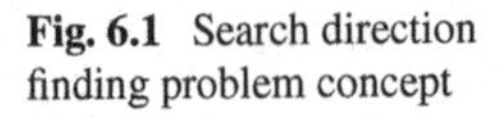

Fig. 6.1 Search direction finding problem concept

is a positive number given by $\rho^k = (\alpha - 1){\mathbf{v}_1^k}^T \nabla c(\mathbf{x}^k)/({\mathbf{v}_2^k}^T \nabla c(\mathbf{x}^k))$, with $\alpha \in (0, 1)$ [18, 24]. The search direction $\mathbf{v}^k$ is then obtained by solving two linear systems with the same coefficient matrix (Eqs. (6.9) and (6.10)). Thus, the factorization phase for solving the primal-dual systems is done once per iteration only. The reader is referred to [18, 24, 30, 37] for a detailed numerical implementation of the search direction-finding problem.

6.4.2 Descent Feasible Direction Concept

For illustration purposes, Fig. 6.1 shows the concept of the search direction-finding problem for an optimization problem with two design variables and one constraint. At $\mathbf{x}^k$, the descent direction $\mathbf{v}_1^k$ is tangent to the active constraint. As previously pointed out, it can be shown that the direction $\mathbf{v}_2^k$ points to the interior of the feasible domain, improving the feasibility of the search direction $\mathbf{v}^k = \mathbf{v}_1^k + \rho^k \mathbf{v}_2^k$. It can be seen that the rate of descent of the objective function c along $\mathbf{v}^k$ is smaller than along $\mathbf{v}_1^k$. This is the cost that the algorithm imposes for obtaining a feasible descent direction.

Once the search direction has been established, a new point is determined by a line search strategy along $\mathbf{v}^k$ with a satisfactory decrease of the objective function. Such a strategy is discussed in the following section.

6.4.3 Line Search

A search scheme based on Armijo's and Wolfe's criteria for unconstrained optimization is considered in the present formulation [19]. Such criteria require the evaluation of standard and reliability constraint functions at a number of points along the search direction. As indicated before, the evaluation of the reliability constraint functions is in general quite involved. Then, an inexact line-search procedure, where approximate failure probability functions are used, is implemented. The approximation of the failure probability function P_{F_i} in terms of the design variables along the search direction $\mathbf{v}$ is taken as [15, 25]

$$P_{F_i}(\mathbf{x}+\tau\mathbf{v}) \approx \bar{P}_{F_i}(\tau) = \exp\left[\alpha_{i0} + \alpha_{i1}\tau + \alpha_{i2}\tau^2\right], \quad \tau \geq 0 \tag{6.11}$$

where the coefficients α_{ij}, $j = 0, 1, 2$ are computed by solving a least squares problem that involves the estimates of the failure probability $P_{F_i}(\mathbf{x})$ and its directional derivative along $\mathbf{v}$, that is, $\nabla P_{F_i}^T(\mathbf{x})\mathbf{v}$, at n_p different points defined in terms of the coefficients τ_l, $l = 1, \ldots, n_p$, where the first point corresponds to the current design $\mathbf{x}$, i.e. $\tau_1 = 0$ [24]. With the information on probabilities and directional sensitivities, the coefficients α_{ij}, $l = 0, 1, 2$, are determined by minimizing the residual

$$\begin{aligned} J(\boldsymbol{\alpha}_i) = &\frac{\sum_{l=1}^{n_p}\left[p_{F_i}(\tau_l) - \left(\alpha_{i0} + \alpha_{i1}\tau_l + \alpha_{i2}\tau_l^2\right)\right]^2}{\left[\max_{l=1,\ldots,n_p}\left(p_{F_i}(\tau_l)\right) - \min_{l=1,\ldots,n_p}\left(p_{F_i}(\tau_l)\right)\right]^2} \\ &+ \frac{4\sum_{l=1}^{n_p}\left[p'_{F_i}(\tau_l) - (\alpha_{i1} + 2\alpha_{i2}\tau_l)\right]^2}{\left[|\max_{l=1,\ldots,n_p}\left(p'_{F_i}(\tau_l)\right)| + |\min_{l=1,\ldots,n_p}\left(p'_{F_i}(\tau_l)\right)|\right]^2} \end{aligned} \tag{6.12}$$

where $\boldsymbol{\alpha}_i =< \alpha_{i0}, \alpha_{i1}, \alpha_{i2} >^T$, $p_{F_i}(\cdot)$ is equal to $\text{Ln}(P_{F_i}(\cdot))$, and $p'_{F_i}(\cdot)$ denotes the derivative of $p_{F_i}(\cdot)$. This approximation strategy corresponds to a particular case of a general approach proposed in [50] for constructing multidimensional response surfaces using information on both function values and sensitivities. Once the failure probability functions have been approximated, the implementation of Armijo's and Wolfe's criteria to determine whether a step length is adequate or not can be carried out in an efficient manner [24].

The above strategy for approximating the failure probability functions along the search directions has proved to be adequate for coping with the variability of the reliability estimates computed during the optimization process. As information on failure probabilities and directional sensitivities, collected at several points, is used simultaneously, the resulting approximation has been found of sufficient accuracy for a number of applications, including the numerical examples presented in this chapter. A detailed implementation of the proposed inexact line-search procedure, including the implementation of Armijo's and Wolfe's criteria and the selection of the different points along the search direction $\mathbf{v}$, can be found in [18, 24].

6.5 Gradient Estimation

The determination of the search direction (see Eqs. (6.9) and (6.10)) requires the gradient of the functions associated with the optimization problem with respect to the design variables. In the framework of this formulation, it is assumed that the gradient of the functions related to the objective and standard constraint functions is readily available, either analytically or numerically. Contrarily, the estimation of the gradient of failure probability functions is not direct. In what follows, an approach to estimate the required gradients is presented.

6.5.1 Approximate Gradient of Failure Probability Function

By definition, the gradient of the ith failure probability function at the design $\mathbf{x}^k$ can be estimated by means of the limit

$$\left.\frac{\partial P_{F_i}(\mathbf{x})}{\partial x_l}\right|_{\mathbf{x}=\mathbf{x}^k} = \lim_{\Delta x_l \to 0} \frac{P_{F_i}(\mathbf{x}^k + \mathbf{i}(l)\Delta x_l) - P_{F_i}(\mathbf{x}^k)}{\Delta x_l}, \quad l = 1, \ldots, n_d \tag{6.13}$$

where $\mathbf{i}(l)$ is a vector of length n_d with all entries equal to zero except for the lth entry, which is equal to one. The calculation of this limit is a challenging task, as failure probabilities must be evaluated using simulation techniques. In addition, failure probability functions are in general non-smooth in terms of design variables due to the variability of the reliability estimates. Even though some approaches have been proposed for evaluating the gradient of failure probability functions for some particular cases [41], a more general approach is considered here [49]. Note that the reliability sensitivity analysis approach introduced in Chap. 5 is not applicable within the scope of this formulation, as the focus is on sensitivity with respect to deterministic design variables and not on probability distribution parameters.

The idea of the approach is to generate approximations of the failure probability functions with respect to the design variables. In other words, non-smooth failure probability functions are approximated by differentiable representations, which are the quantities used for computing the gradients [22, 48, 49]. Specifically, for estimating the limit in Eq. (6.13), two approximate representations of different quantities are introduced.

The first of these approximations involves the performance function κ_i, which is given in terms of the normalized demand function as $\kappa_i(\mathbf{z}, \boldsymbol{\theta}, \mathbf{x}) = 1 - d_i(\mathbf{z}, \boldsymbol{\theta}, \mathbf{x})$. If $\mathbf{x}^k$ is the current design, the performance function κ_i is approximated in the vicinity of $\mathbf{x}^k$ as

$$\bar{\kappa}_i(\mathbf{z}, \boldsymbol{\theta}, \mathbf{x}) = \kappa_i(\mathbf{z}, \boldsymbol{\theta}, \mathbf{x}^k) + \boldsymbol{\delta}_i{}^T \Delta\mathbf{x} \tag{6.14}$$

where $\mathbf{x} = \mathbf{x}^k + \Delta\mathbf{x}$, and $\boldsymbol{\delta}_i$ is a set of coefficients. The determination of such coefficients is discussed in the next section.

Next, the failure probability function is expressed in terms of a threshold d_i^* of the normalized demand function in the vicinity of one as

$$P\left[d_i(\mathbf{z},\boldsymbol{\theta},\mathbf{x}) > d_i^*\right] \approx e^{\psi_0+\psi_1(d_i^*-1)}, \tag{6.15}$$

where ψ_0 and ψ_1 are real constants and $d_i^* \in [1-\varepsilon, 1+\varepsilon]$, being ε a small number. By replacing the approximations introduced in Eqs. (6.14) and (6.15) in (6.13), it can be shown that the limit associated with the gradient of the probability function can be approximated by [48]

$$\left.\frac{\partial P_{F_i}(\mathbf{x})}{\partial x_l}\right|_{\mathbf{x}=\mathbf{x}^k} \approx \psi_1 \delta_{il} P_{F_i}(\mathbf{x}^k), \quad l = 1,\ldots,n_d \tag{6.16}$$

where δ_{il} is the lth element of the vector $\boldsymbol{\delta}_i$. It is noted that the information generated by subset simulation allows to estimate the failure probability $P_{F_i}(\mathbf{x}^k)$ and the coefficient ψ_1 [22]. It should be noted that the previous approach can also be used for estimating the directional derivative of failure probability functions. Actually, the derivative along a direction $\mathbf{v}$ of the failure probability function P_{F_i}, at the design $\mathbf{x}^k$, can be estimated as

$$\nabla_{\mathbf{v}} P_{F_i}(\mathbf{x}^k) = \nabla P_{F_i}(\mathbf{x})^T \mathbf{v}\,\Big|_{\mathbf{x}=\mathbf{x}^k} \approx \psi_1 \boldsymbol{\delta}_i^T \mathbf{v}\; P_{F_i}(\mathbf{x}^k) \tag{6.17}$$

6.5.2 Coefficient Estimation

The estimation of the coefficients $\boldsymbol{\delta}_i$, in Eq. (6.14), is carried out by the following two steps.

First, for samples $\{(\mathbf{z}_j,\boldsymbol{\theta}_j),\, j=1,\ldots,M\}$ near the limit state surface, that is, $\kappa_i(\mathbf{z}_j,\boldsymbol{\theta}_j,\mathbf{x}) \approx 0$, or in terms of the normalized demand function $d_i(\mathbf{z}_j,\boldsymbol{\theta}_j,\mathbf{x}) \approx 1$, the performance function is evaluated at a number of points in the neighborhood of $\mathbf{x}^k$. These points are generated as

$$\mathbf{x}^{kp} - \mathbf{x}^k = \frac{\boldsymbol{\xi}_p}{\parallel \boldsymbol{\xi}_p \parallel} R\;,\; p = 1,\ldots,N = Q \times M \tag{6.18}$$

where the components of the vector $\boldsymbol{\xi}_p$ are independent, identically distributed standard Gaussian random variables, N and Q positive integers and R is a user-defined small positive number. This number defines the radius of the hypersphere $\boldsymbol{\xi}_p/\parallel \boldsymbol{\xi}_p \parallel R$, centered at the current design $\mathbf{x}^k$.

Second, the coefficients $\boldsymbol{\delta}_i$ of the approximation (6.14) are computed by least squares. To this end, the following set of equations is generated

$$\kappa_i(\mathbf{z}_j,\boldsymbol{\theta}_j,\mathbf{x}^{kp}) = \kappa_i(\mathbf{z}_j,\boldsymbol{\theta}_j,\mathbf{x}^k) + \boldsymbol{\delta}_i{}^T \frac{\boldsymbol{\xi}_p}{\parallel \boldsymbol{\xi}_p \parallel} R$$
$$p = j + (q-1)\times M\;, q = 1,\ldots,Q\;, j = 1,\ldots,M \tag{6.19}$$

Since the samples $\{(\mathbf{z}_j, \boldsymbol{\theta}_j), j = 1, \ldots, M\}$ are chosen near the limit state surface, the approximate performance function $\bar{\kappa}_i$ is representative of the behavior of the actual limit state surface in the vicinity of the design $\mathbf{x}^k$. Numerical experience has shown that the approximation introduced in Eq. (6.16) is adequate in the context of the proposed optimization scheme [22, 49].

In summary, a single reliability analysis plus the evaluation of the demand function in the vicinity of a given design suffices for estimating the gradient of failure probability functions. The reader is referred to [48] for issues such as the number of points required for performing least squares (Q and M), and the generation of design points in the vicinity of the current design (calibration of the radius R).

6.6 Final Remarks

The current optimization strategy is based on a local optimization algorithm. Therefore, the optimization process can converge to a local optimum. This situation may occur in structural optimization problems with, for example, non-convex objective functions, or non-convex or disjoint design spaces. Thus, the solution of the optimization process does not ensure the identification of the global optimum. However, as previously pointed out, all iterations given by the algorithm strictly verify the constraints and, therefore, the iterations can be stopped at any time still leading to better feasible designs than the initial design. This property is particularly important and useful when dealing with involved problems such as reliability-based optimization of high-dimensional stochastic dynamical systems. In addition, the information obtained during the design process gives a valuable insight into the complex interaction of the design variables on the performance and reliability of complex systems. Therefore, the usefulness of the proposed optimization strategy is evident. Attempts to find the global optimum can be based on physical considerations of the problem or the use of global optimization algorithms based on, for example, state space search, evolutionary algorithms, genetic algorithms, simulated annealing, random optimization, etc. [14, 52]. In this last case, a large number of function evaluations (i.e., reliability estimations) is expected to be required. In this regard, the exploitation of all the parallelization features of the optimization scheme should be considered as a remedy for the large computational effort associated with finding the global optimum.

As indicated before, the solution of the reliability-based optimization problem (6.1) is computationally very demanding due to the large number of dynamic analyses required during the design process. This is due to the reliability estimation and the iterative nature of the optimization strategy. Consequently, the computational cost may become excessive when the computational time for performing a dynamic analysis is significant. To cope with this difficulty, the design process is carried out on reduced-order models as previously pointed out.

6.7 Numerical Examples

Three numerical examples are presented in this section. The objective of the first two examples is to demonstrate that although the proposed optimization scheme is quite effective (in terms of the number of reliability evaluations to be performed), the computational cost for solving this class of problems may be significant, even for relatively simple structural models. On the other hand, the goal of the third example problem is to evaluate the effectiveness and efficiency of using reduced-order models in the reliability-based design optimization of an involved structural model.

6.7.1 Example 1: Model Description

One of the moment-resisting frames of a six-story building under ground motion is considered as the first example problem. The isometric view of the six-story building is shown in Fig. 6.2, while the details of the moment-resisting frame are shown in Fig. 6.3. The frame has a total length of 6.0 m and floor thicknesses $e_i, i = 1, \ldots, 6$. The active masses are 5.0×10^4 kg for all floors. The floor height of 2.5 m is constant for all floors, leading to a total height of 15 m. Young's modulus E and the damping ratios are treated as uncertain system parameters.

Young's modulus is modeled by a log-normal random variable (θ_E) with most probable value $\bar{\theta}_E = 2.45 \times 10^{10}\,\text{N/m}^2$ and coefficient of variation of 10%. For simplicity, it is assumed that the modulus of elasticity is a homogeneous and fully correlated random field. On the other hand, the damping ratios, which are assumed to be identical, are modeled by independent log-normal random variables ($\theta_{\zeta_i}, i = 1, \ldots, 6$) with mean value $\bar{\theta}_\zeta = 0.05$ and coefficient of variation of 40%. Due to the simplicity of the structural model, it is used directly during the design process. In other words, a reduced-order model is not considered in this example.

The building is excited horizontally by a ground acceleration, which is modeled as a nonstationary filtered white noise process. In particular, the Clough–Penzien model, which is based on a filtered white noise process, is considered [10]. According to this model the ground acceleration is defined as

$$a(t) = < \omega_{1g}^2 \,,\, 2\zeta_{1g}\omega_{1g} \,,\, -\omega_{2g}^2 \,,\, -2\zeta_{2g}\omega_{2g} > \mathbf{a}(t) \tag{6.20}$$

where $\omega_{1g}, \zeta_{1g}, \omega_{2g}$ and ζ_{2g} are model parameters related to soil conditions [53], and $\mathbf{a}(t)$ represents the state-space variables of the filter, which satisfies the first-order differential equation

$$\dot{\mathbf{a}}(t) = \mathbf{A}\,\mathbf{a}(t) + \mathbf{a}_g\, e(t)\, w(t) \tag{6.21}$$

where the matrix $\mathbf{A}$ specifies the filter and it is given by

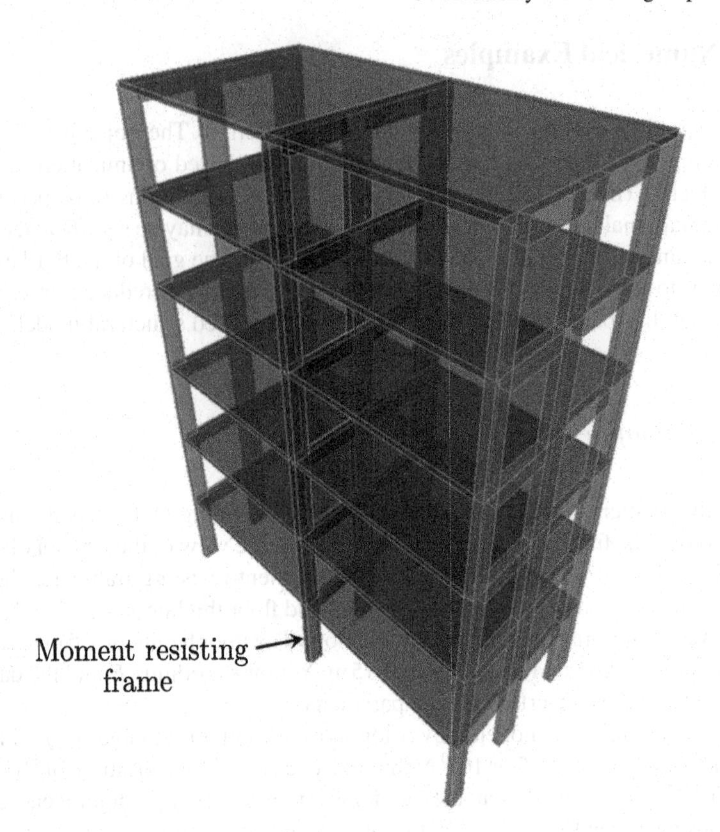

Fig. 6.2 Isometric view of the six-story shear building model

$$\mathbf{A} = \begin{bmatrix} 0 & 1 & 0 & 0 \\ -\omega_{1g}^2 & -2\zeta_{1g}\omega_{1g} & 0 & 0 \\ 0 & 0 & 0 & 1 \\ \omega_{1g}^2 & 2\zeta_{1g}\omega_{1g} & -\omega_{2g}^2 & -2\zeta_{2g}\omega_{2g} \end{bmatrix}, \tag{6.22}$$

the vector $\mathbf{a}_g = < 0, 1, 0, 0 >$ relates the white noise process $w(t)$ with the filter excitation, and $e(t)$ is an envelope function of time. The driven white noise process $w(t)$ is characterized by its autocorrelation function as $E(w(t)w(t+\tau)) = I\delta(\tau)$, where I denotes the white noise intensity and $\delta(\tau)$ the Dirac delta function. In practice, the white noise process is evaluated at discrete time steps and specified by a set of independent, identically distributed Gaussian random variables. The above representation of the random load allows describing a wide range of Gaussian processes, including band-limited white noise processes, colored excitations, and non-stationary excitations. The values $\omega_{1g} = 15.0$ rad/s, $\zeta_{1g} = 0.6$, $\omega_{2g} = 1.0$ rad/s, $\zeta_{2g} = 0.9$, $I = 1.26 \times 10^{-1}$ m^2/s^3 and the envelope function equal to

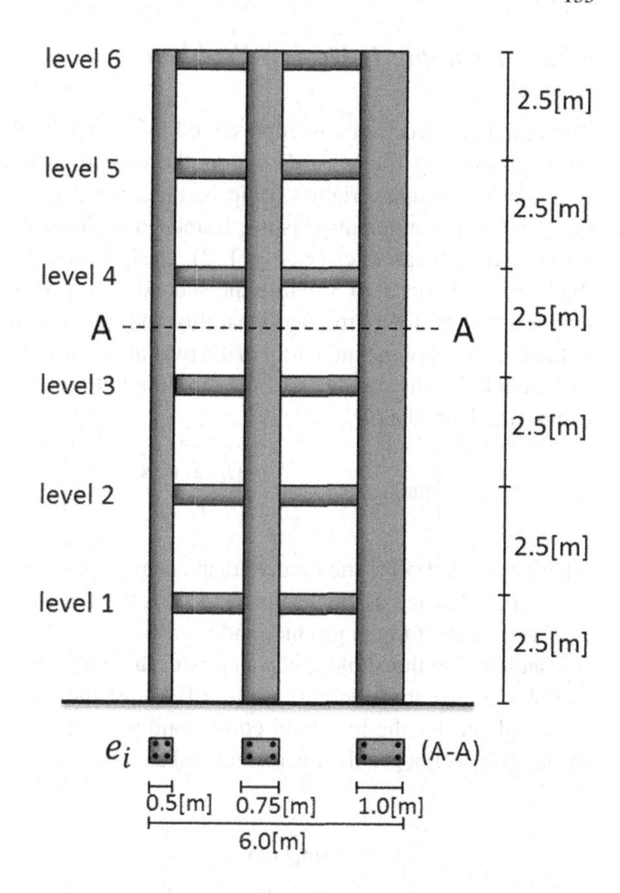

Fig. 6.3 Details of the moment-resisting frame

$$e(t) = \frac{e^{-0.5t} - e^{-t}}{\max_t(e^{-0.5t} - e^{-t})} \quad , \quad 0 < t < 10 \text{ s} \tag{6.23}$$

are used in this example. As previously pointed out, the white noise process is evaluated at discrete time steps. The sampling interval is assumed to be $\Delta t = 0.01$ s, and the duration of the excitation is $T = 10$ s. Then, the discrete-time white noise sequence $\omega(t_j) = \sqrt{I/\Delta t}\, z_j$, where $z_j, j = 1, \ldots, 1001$, are independent, identically distributed standard Gaussian random variables, is considered in this case. Thus, the total number of uncertain parameters involved in the problem is more than one thousand. In fact, the random variable vectors **z** and $\boldsymbol{\theta}$ are given by $\mathbf{z}^T =< z_1, z_2, \ldots, z_{1001} >$ and $\boldsymbol{\theta}^T =< \theta_E, \theta_{\zeta_1}, \ldots, \theta_{\zeta_6} >$, respectively.

6.7.2 *Example 1: Design Problem*

The initial construction cost represented by the total volume of the columns of the moment-resisting frame is chosen as the objective function for the optimization problem. The design variables comprise the column thicknesses of the different floors ($x_i = e_i$) of the moment-resisting frame. First, the column thicknesses are linked into two design variables ($x_i, i = 1, 2$). Design variable number one represents the thickness of floors 1 to 3, while the second design variable controls the thickness of floors 4 to 6. To control serviceability and minor damage, the design criteria are defined in terms of the interstory drifts over all stories of the moment-resisting frame and the relative displacement of the top floor with respect to the ground. The failure events are defined as

$$F_i = \left\{ \max_{t_k, k=1,\ldots,1001} \frac{|r_i(t_k, \mathbf{z}, \boldsymbol{\theta}, \mathbf{x})|}{r_i^*} > 1 \right\} \ , \ i = 1, \ldots, 7 \tag{6.24}$$

where $r_i(t_k, \mathbf{z}, \boldsymbol{\theta}, \mathbf{x})$ is the relative displacement between the $(i-1, i)$th floors ($i = 1, \ldots, 6$) evaluated at the design $\mathbf{x}$, $r_7(t_k, \mathbf{z}, \boldsymbol{\theta}, \mathbf{x})$ is the relative displacement of the roof with respect to the ground, and $r_i^*, i = 1, \ldots, 7$, are the corresponding critical thresholds. The threshold associated with the interstory drift failure events is equal to 0.2% of the story height, while a 0.1% of the moment-resisting frame height is considered for the threshold corresponding to the top floor displacement failure event. The design problem is formulated as

$$\begin{aligned} &\text{Min}_{\mathbf{x}} \ c(\mathbf{x}) \\ &\text{s.t.} \ P_{F_i}(\mathbf{x}) \leq 10^{-3} \quad i = 1, \ldots, 7 \\ &x_2 \leq x_1 \\ &0.2\text{m} \leq x_i \leq 0.8\text{m} \quad i = 1, 2 \end{aligned} \tag{6.25}$$

where $c(x_1, x_2) = 5.625(3.0x_1 + 3.0x_2)$ is the volume of the columns of the moment-resisting frame. There are seven reliability constraints, one geometric constraint, and two side constraints. For the dynamic analysis, it is assumed that the moment-resisting frame remains linear throughout the duration of the excitation. It is noted that the estimation of the probability of failure for a given design represents a high-dimensional reliability problem. In fact, as previously indicated, more than a thousand random variables are involved in the corresponding multidimensional probability integrals.

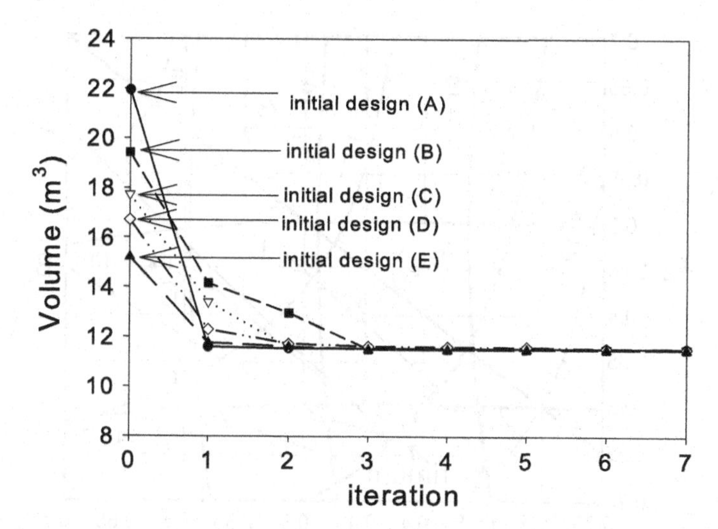

Fig. 6.4 Iteration history of the optimization process in terms of the objective function for different initial designs. Example 1 (linked design variables)

6.7.3 Example 1: Results - Linked Design Variables Case

In terms of the numerical implementation, the optimization problem is solved using an exterior sampling approximation [42]. That is, the same stream of random numbers is used throughout all iterations in the optimization process. In this way, the relative importance of failure probability estimation errors, when comparing similar designs in the design space, can be reduced. The iteration history of the design process in terms of the objective function is shown in Fig. 6.4 for different initial feasible designs. The initial design A is set equal to $\mathbf{x}_A^T =< 0.7, 0.6 >$, design B equal to $\mathbf{x}_B^T =< 0.7, 0.45 >$, design C equal to $\mathbf{x}_C^T =< 0.6, 0.45 >$, design D equal to $\mathbf{x}_D^T =< 0.52, 0.47 >$, and design E equal to $\mathbf{x}_E^T =< 0.6, 0.3 >$, where the units are in meters. It is seen that the process converges in few iterations for all cases. In fact, most of the improvements of the objective function take place in the first optimization cycles. The trajectory of the optimizer as well as some objective contours and iso-probability curves are shown in Fig. 6.5.

It is observed that the iso-probability curves associated with the interstory drift of the first floor (P_{F_1}) are almost independent of the column thicknesses of the upper floors (x_2). On the other hand, the iso-probability curves related to the interstory drift of the fourth floor (P_{F_4}) are almost independent of the column thicknesses of the lower floors (x_1). Finally, the iso-probability curves associated with the roof displacement (P_{F_7}) show a strong interaction between the column thicknesses of all floors, as expected. These results give a valuable insight into the interaction and effect of the design variables on the reliability of the moment-resisting frame. From the optimization point of view, it is observed that the trajectory of the optimizer to the optimum is quite direct for all cases. The geometric and side constraints are inactive

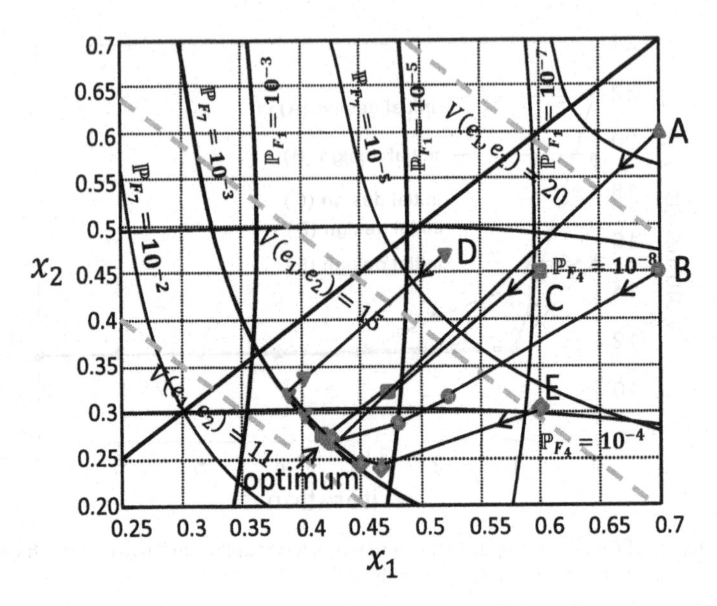

Fig. 6.5 Trajectory of the optimizer, some objective contours ($V(e_1, e_2) = constant$) and some iso-probability curves. Example 1 (linked design variables case)

at the final design. Among the reliability constraints, the one related to the relative displacement of the top floor with respect to the ground is active at the final design.

6.7.4 *Example 1: Results - Independent Design Variables Case*

Next, the column thicknesses of all floors of the moment-resisting frame are assumed to be independent with side constraints $0.20\ \text{m} \leq x_i \leq 0.80\ \text{m}, i = 1, \ldots, 6$, and geometric constraints $x_{i+1} \leq x_i, i = 1, \ldots, 5$. Then, the optimization problem has seven reliability constraints and five standard (geometric) constraints. The iteration history of the design process in terms of the objective function is shown in Fig. 6.6 for five different initial feasible designs. These initial designs are defined as: $P_1 = (0.70, 0.70, 0.70, 0.60, 0.60, 0.60)$, $P_2 = (0.65, 0.65, 0.60, 0.60, 0.55, 0.55)$, $P_3 = (0.70, 0.65, 0.65, 0.50, 0.50, 0.40)$, $P_4 = (0.60, 0.60, 0.60, 0.45, 0.45, 0.45)$, and $P_5 = (0.70, 0.60, 0.50, 0.45, 0.40, 0.35)$, where the units of the column thicknesses are in meters. It is seen that the design process converges in few iterations for all cases. As previously pointed out, all the iterations verified the constraints. In consequence, the iterations can be stopped at any time still leading to feasible designs that are better than the initial estimate.

The corresponding final volume for different initial designs is given in Table 6.1. For comparison, this table also shows the final design of the system with deterministic

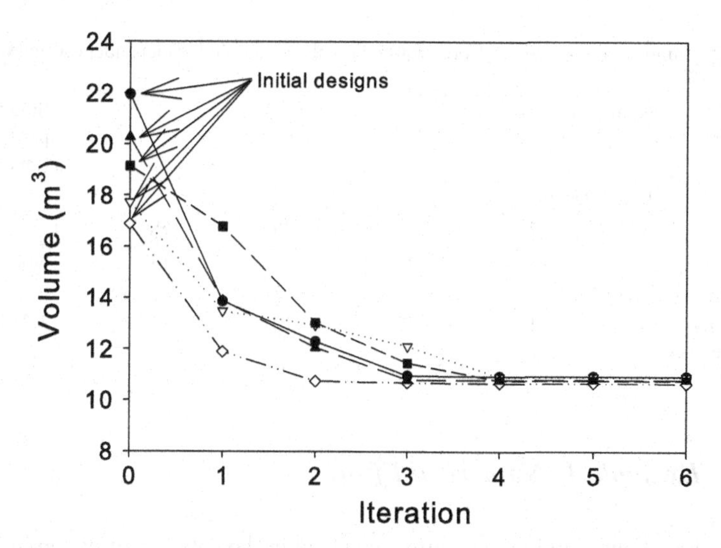

Fig. 6.6 Iteration history of the optimization process in terms of the objective function for different initial designs. Example 1 (independent design variables case)

Table 6.1 Final volume for five different initial designs (P_i, $i = 1, \ldots, 5$). Example 1 (independent design variables case)

Initial design	Final design	
	Uncertain model	Deterministic model
Point	Volume (m ^{3})	
P_1	10.93	8.66
P_2	10.91	8.72
P_3	10.75	8.74
P_4	10.80	8.68
P_5	10.77	8.71

structural parameters (deterministic model). In this case, Young's modulus and the damping ratios are equal to their most probable values. It is observed that the volume of the moment-resisting frame at the final design of the system with uncertain structural parameters is greater than the corresponding volume of the model with deterministic parameters, as expected. The volume of the uncertain model increases by more than 20% with respect to the volume of the deterministic model. This result stresses the fact that uncertainty in the structural parameters may cause important changes in the performance and reliability of final designs. The small differences observed in the final volumes for different initial designs are due to the use of approximate quantities and the variability of the reliability and sensitivity estimates computed during the optimization process. However, the differences are negligible from a practical point of view.

Table 6.2 Summary of numerical efforts required to solve Example 1 (independent design variables case)

Type of analysis	Number of analyses					Simulations per analysis (on average)
	Initial design					
	P_1	P_2	P_3	P_4	P_5	
Reliability estimate	13	15	12	10	13	4,000
Sensitivity estimate	13	15	12	10	13	1,200

6.7.5 Example 1: Numerical Effort

Information about the numerical effort involved in the solution of the optimization problem is provided in Table 6.2. The numerical effort is due to the estimation of the reliability (by means of subset simulation) and its sensitivity during the optimization process.

The table summarizes the numerical effort as follows. The first column indicates the type of analysis performed, while the next five columns show the number of times the aforementioned analysis was repeated throughout the optimization procedure for different initial designs. Finally, the seventh column indicates the average number of simulations required for performing one particular type of analysis. For example, a total of 13 reliability analyses are required for solving the problem corresponding to initial design P_1; each of these analyses involves (on average) 4,000 simulations (dynamic analyses). These results stress the fact that even though the optimization scheme is quite effective (in terms of the number of reliability evaluations), the computational cost for solving the problem may be important.

6.7.6 Example 2: Structural Model

A finite element model consisting of a nonlinear 52-story building under stochastic earthquake excitation is analyzed in the second example. Two isometric views of the structural system are shown in Fig. 6.7. The plan view and the dimensions of each floor are shown in Fig. 6.8. The interstory height is 3.6 m for all floors except the first one, which has a height of 14 m.

The building has a reinforced concrete core of shear walls and a reinforced concrete perimeter moment frame as shown in Fig. 6.8. The columns in the perimeter have a circular cross section. Properties of the reinforced concrete have been assumed as follows: Young's modulus $E = 2.45 \times 10^{10}\,\text{N/m}^2$, Poisson's ratio $\mu = 0.3$, and mass density $\rho = 2{,}500\,\text{kg/m}^3$. For the dynamic analysis, it is assumed that each

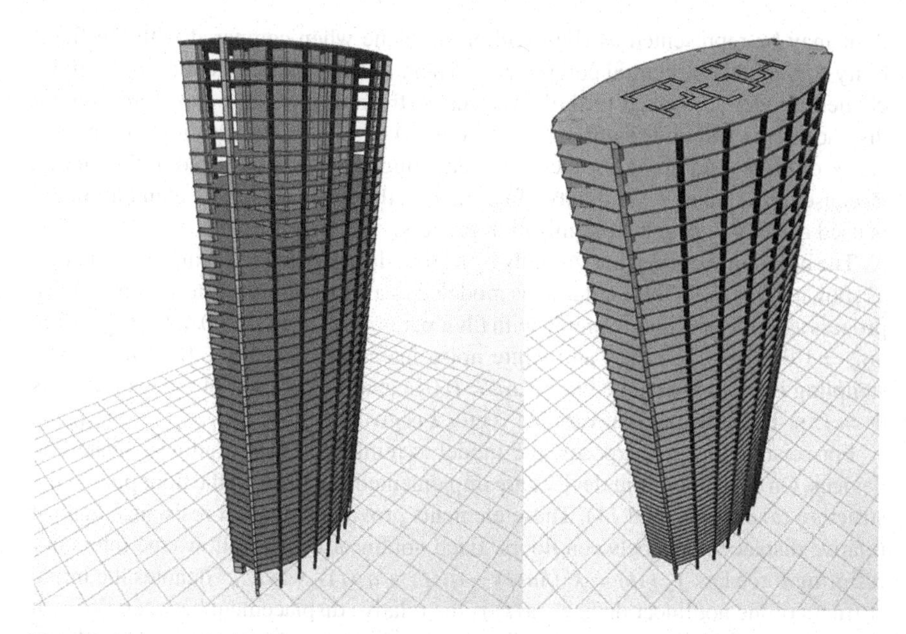

Fig. 6.7 Isometric views of the 52-story building model

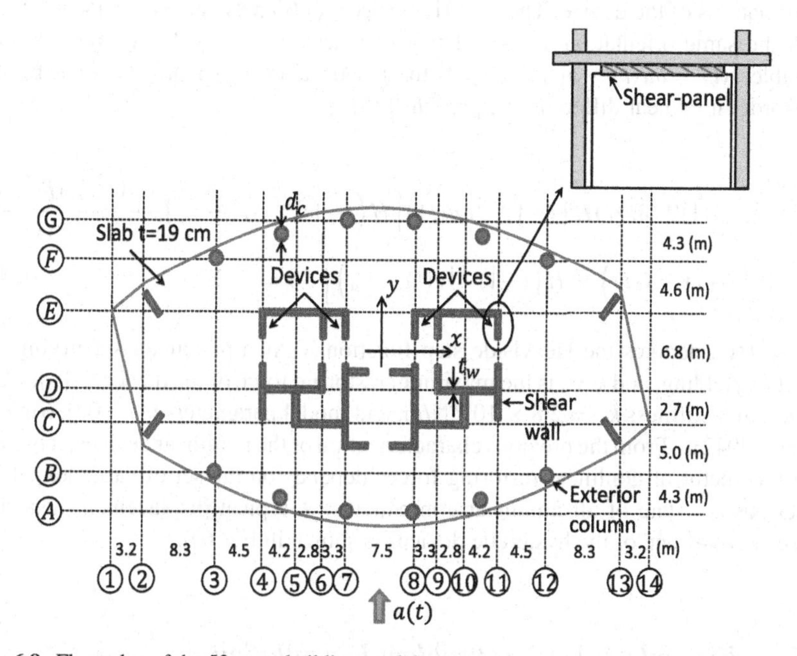

Fig. 6.8 Floor plan of the 52-story building model

floor may be represented as rigid within the plane when compared with the flexibility of the other structural components. Then, the degrees of freedom of the finite element model are linked to three degrees of freedom per floor (two translational displacements and one rotational displacement) by using condensation techniques. A 5% of critical damping for the modal damping ratios is introduced in the model. Because of the relative simplicity of the model, the unreduced finite element model is used directly during the optimization process.

The building is excited horizontally by a ground acceleration $a(t)$ in the y direction as shown in Fig. 6.8. The excitation is modeled as a nonstationary filtered white noise process as in the previous example with filter parameters $\omega_{1g} = 15.0\,\text{rad/s}$, $\xi_{1g} = 0.6$, $\omega_{2g} = 1.0$ rad/s, $\xi_{2g} = 0.9$, and white noise intensity $I = 1.26 \times 10^{-1}\,\text{m}^2/\text{s}^3$. The sampling interval is assumed to be $\Delta t = 0.01\text{s}$ and the duration of the excitation is $T = 15\,\text{s}$. Thus, the random variable vector $\mathbf{z}$ is given by $\mathbf{z}^T = < z_1, z_2, \ldots, z_{1501} >$.

For seismic design purposes, the model is reinforced with nonlinear hysteretic devices. On each floor, four devices are implemented as shown on the floor plan of the structure (axes 4,7,8, and 11). These elements provide additional resistance against relative displacements between floors. Each nonlinear device follows the interstory restoring force law $f_{NL}(t) = k^e(\delta u(t) - q_1(t) + q_2(t))$, where k^e denotes the initial stiffness of the nonlinear device, $\delta u(t)$ is the relative displacement between floors at the position of the device in the y direction, and $q_1(t)$ and $q_2(t)$ denote the plastic deformations of the device. The restoring force $f_{NL}(t)$ acts between adjacent floors with the same orientation as the relative displacement $\delta u(t)$. Using the auxiliary variable $v(t) = \delta u(t) - q_1(t) + q_2(t)$, the plastic elongations are specified by the first-order nonlinear differential equations [38].

$$\dot{q}_i(t) = (-1)^{i+1}\dot{\delta u}(t) H\Big((-1)^{i+1}\dot{\delta u}(t)\Big)\Big[H\Big((-1)^{i+1}v(t) - v_y\Big)\frac{(-1)^{i+1}v(t) - v_y}{v_p - v_y}$$

$$H\Big(v_p - (-1)^{i+1}v(t)\Big) + H\Big((-1)^{i+1}v(t) - v_p\Big)\Big],\ i = 1, 2 \tag{6.26}$$

where $H(\cdot)$ denotes the Heaviside step function, v_y is a parameter specifying the onset of yielding, and $k^e v_p$ is the maximum restoring force of the device. All devices have initial stiffness $k^e = 2.8 \times 10^{10}\,\text{N/m}$, and model parameters $v_p = 0.006\,\text{m}$ and $v_y = 0.0042\,\text{m}$. From the previous characterization of the nonlinear devices, it is clear that the vector of nonlinear restoring forces depends on the set of variables which describes the state of all nonlinear components. A typical displacement-restoring force curve of one of the hysteretic devices is shown in Fig. 6.9.

6.7.7 Example 2: Design Problem Formulation

The variables to be controlled are the thicknesses of the shear walls (t_w) and the diameters of the exterior columns (d_c). The dimensions of these structural components at

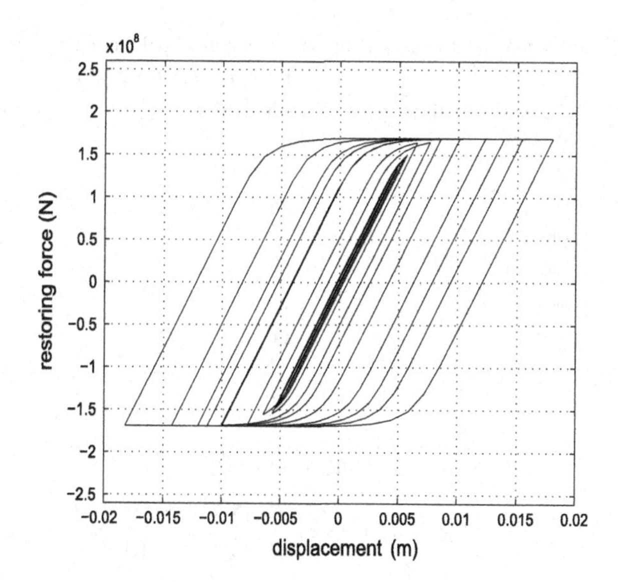

Fig. 6.9 Typical displacement-restoring force curve of the hysteretic devices (shear panels)

each floor are linked to one intermediate optimization variable (x^{inter}) as $t_w = x^{\text{inter}}\hat{t}_w$ and $d_c = x^{\text{inter}}\hat{d}_c$, where $\hat{t}_w$ and $\hat{d}_c$ are the nominal values of the thickness of the shear walls and the diameter of the exterior columns at each floor, respectively. The intermediate parameters are grouped into six optimization variables. The definition of these variables over the height of the building is given in Table 6.3. The weight of the resisting elements (shear walls and exterior columns) is chosen as the objective function for the design problem. The reliability constraints are given in terms of the interstory drifts and the roof displacement. The failure events associated with the interstory displacements are given by

$$F_i = \left\{\max_{t_k, k=1,\ldots,1501} |\delta u_i(t_k, \mathbf{x}^{\text{inter}}, \mathbf{z})| > \delta u_i^*\right\} \quad , \quad i = 1, \ldots, 52 \tag{6.27}$$

where $\delta u_i(t_k, \mathbf{x}^{\text{inter}}, \mathbf{z})$ is the relative displacement between the $(i-1, i)$th floors evaluated at the intermediate design variable $\mathbf{x}^{\text{inter}}$, $\mathbf{z}$ is the vector of random variables that describes the earthquake excitation, and δu_i^* is the critical threshold level equal to 0.1% of the story height. On the other hand, the failure event related to the roof displacement is given by

$$F_r = \left\{\max_{t_k, k=1,\ldots,1501} |u_{52}(t_k, \mathbf{x}^{\text{inter}}, \mathbf{z})| > u_{52}^*\right\} \tag{6.28}$$

where $u_{52}(t_k, \mathbf{x}^{\text{inter}}, \mathbf{z})$ is the displacement of the roof with respect to the base, and u_{52}^* is the critical threshold level equal to 0.1% of the building height. The target failure probabilities are taken equal to 10^{-4} for all events. The reliability-based optimal design problem is written in terms of the objective function $c(\mathbf{x}^{\text{inter}}) = \sum_{i=1}^{6} x_i^{\text{inter}}$ as

Table 6.3 Intermediate optimization variables and linking detail

Intermediate optimization variables	Design elements (floors)	
	Initial floor	Final floor
x_1^{inter}	1	8
x_2^{inter}	9	16
x_3^{inter}	17	24
x_4^{inter}	25	32
x_5^{inter}	33	42
x_6^{inter}	43	52

$$\begin{aligned} &\text{Min}_{\mathbf{x}^{\text{inter}}} \; c(\mathbf{x}^{\text{inter}}) \\ &\text{s.t.} \;\; P_{F_i}(\mathbf{x}^{\text{inter}}) \leq 10^{-4} \quad i = 1, \ldots, 52 \\ &\qquad P_{F_r}(\mathbf{x}^{\text{inter}}) \leq 10^{-4} \\ &\qquad x_i^{\text{inter}} > x_{i+1}^{\text{inter}} \;, i = 1, \ldots, 5 \\ &\qquad x_i^{\text{inter}} \geq 0.20 \;, i = 1, \ldots, 6 \end{aligned} \tag{6.29}$$

where reliability, geometric and side constraints have been considered. Note that there are six optimization variables and 53 reliability constraints. The size of the above optimization problem can be considered quite involved, even in the context of deterministic optimization. As in the first application problem, the estimation of the probability of failure for a given design represents a high-dimensional reliability problem.

6.7.8 *Example 2: Results*

The final design obtained by the proposed scheme, which has been solved using an exterior sampling approximation, is presented in Table 6.4. For comparison, the final design of the model without the hysteretic devices (linear model) is also indicated. It is seen that the dimensions of the structural components (shear walls and exterior columns) at the final design of the linear model are greater than the corresponding components of the nonlinear model. The nonlinear behavior of the hysteretic devices has a positive impact on the overall performance of the system. In fact, the hysteresis of the devices increases the energy dissipation reducing in this manner the response of the structural system. The total weight of the linear model increases more than 15% with respect to the weight of the nonlinear model at the final design.

Figure 6.10 shows the iteration history of the optimization process in terms of the objective function for three different initial points. It is observed that the optimization process converges in about five iterations. The fast convergence implies that the

Table 6.4 Final Design. Example problem 2

Optimization variable	Initial point	Final point	
		Linear model	Nonlinear model
x_1^{inter}	1.10	1.06	0.93
x_2^{inter}	1.00	0.67	0.63
x_3^{inter}	0.80	0.56	0.42
x_4^{inter}	0.70	0.49	0.40
x_5^{inter}	0.60	0.45	0.38
x_6^{inter}	0.50	0.38	0.29
Resisting elements total weight (kg)	3.65×10^7	2.85×10^7	2.45×10^7

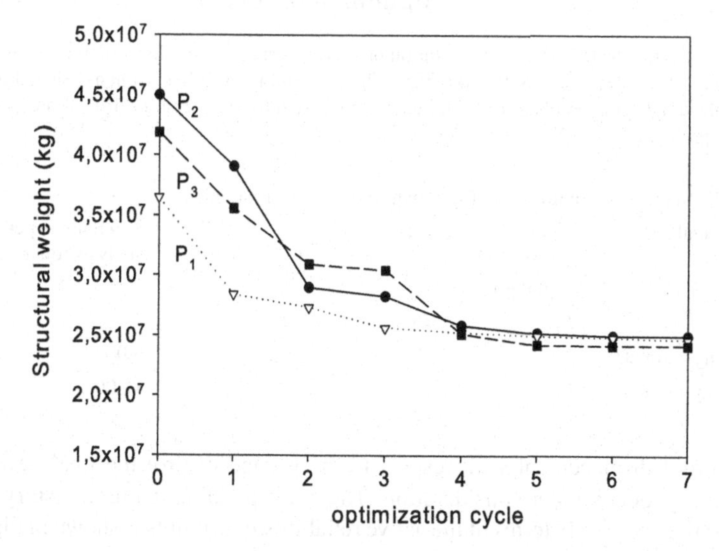

Fig. 6.10 Iteration history of the optimization process in terms of the objective function. Initial point $P_1 = (1.2, 1.0, 1.0, 1.0, 0.9, 0.7)$. Initial point $P_2 = (1.1, 1.0, 0.8, 0.7, 0.6, 0.5)$. Initial point $P_3 = (1.1, 1.0, 1.0, 0.8, 0.8, 0.7)$

optimization process takes few excursion probability and sensitivity estimations for all cases. It is also seen that the different initial points lead to basically the same final design in this case. Of course, the selection of a particular initial point may lead to different final designs in the general case. In such cases, engineering criteria and the background knowledge of the problem at hand should be applied in order to select an appropriate initial design and assess the properties of the final design.

The results of the optimization process indicate that the active constraints at the final design are the reliability constraints related to the events associated with the relative displacement of the roof with respect to the ground (P_{F_r}) and the interstory drift of the first floor (P_{F_1}). Therefore, the relative displacement of the first floor

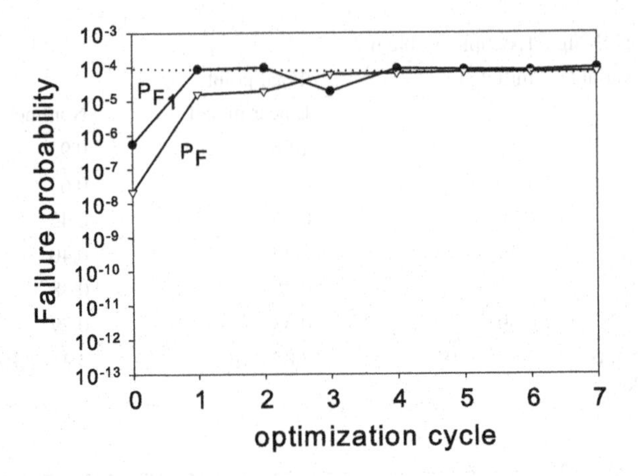

Fig. 6.11 Iteration history of the optimization process in terms of the active reliability constraints. Initial point $P_2 = (1.1, 1.0, 0.8, 0.7, 0.6, 0.5)$. P_F: probability of failure event associated with the roof displacement. P_{F_1}: probability of failure event related to the interstory drift of the first floor. Example problem 2

Table 6.5 Summary of numerical efforts required to solve Example 2

Type of analysis	Number of analyses			Simulations per analysis (on average)
	Initial point			
	P_1	P_2	P_3	
Reliability estimate	16	17	13	5,000
Sensitivity estimate	16	17	13	1,200

and the roof displacement with respect to the ground control the final design for the building model under consideration. The corresponding iteration history of the optimization process in terms of the active reliability constraints is shown in Fig. 6.11 for the initial design corresponding to initial point P_2.

6.7.9 Example 2: Numerical Considerations

The numerical efforts associated with the solution of the optimization problem for the different initial points are summarized in Table 6.5. As in the previous example, the first column of this table indicates the type of analysis performed, while the next three columns show the number of times the aforementioned analysis was repeated throughout the optimization procedure for different initial points. Finally, the fifth column indicates the average number of simulations required for performing one particular type of analysis. Once again, these results emphasize the fact that although the optimization scheme is quite effective, the cost for solving the problem may be substantial, especially for involved finite element models. Such a case is explored in the next example.

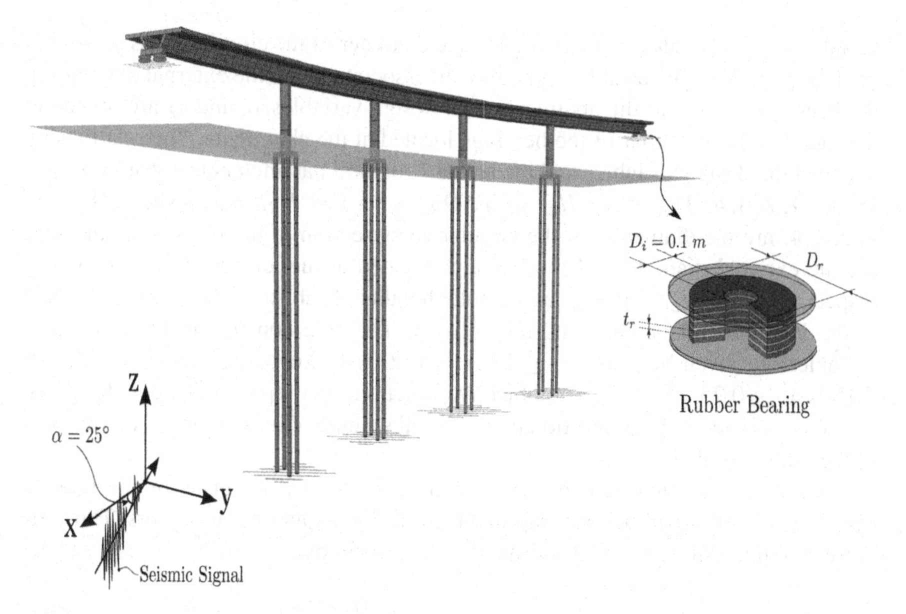

Fig. 6.12 Bridge structural model

6.7.10 Example 3: Reliability-Based Design Formulation

The bridge structural model shown in Fig. 6.12 is considered as the third example. The model has been borrowed from Sect. 5.8 of Chap. 5. The corresponding description of the nonlinear model as well as the characterization of the excitation can be found in the aforementioned section. The ground acceleration is applied at 25° with respect to the x-axis as shown in the figure. Material properties are assumed to be known in the context of this example. Thus, the vector of uncertain parameters involved in the problem is only related to the characterization of the ground acceleration, that is, $\mathbf{z}^T = < z_1, z_2, \ldots, z_{3001} >$.

The reliability-based design problem is defined in terms of the following constrained nonlinear optimization problem as

$$\begin{aligned} &\text{Min}_{\mathbf{x}} \ c(\mathbf{x}) \\ &\text{s.t.} \ P_{F_i}(\mathbf{x}) \leq P^*_{F_i} \ , i = 1, 2 \\ &\mathbf{x} \in X \end{aligned} \tag{6.30}$$

where $\mathbf{x}(x_i, i = 1, \ldots, 8)$, is the vector of design variables, $c(\mathbf{x})$ is the objective function, $P_{F_i}(\mathbf{x}), i = 1, 2$, are the failure probability functions, and $P^*_{F_i}, i = 1, 2$, are the corresponding target failure probabilities, which are taken equal to $P^*_{F_i} = 10^{-4}, i = 1, 2$. The design variables include the diameter of the piers circular cross section, and the external diameter and total height of rubber in the bearings. Design variables x_1, x_2, x_3 and x_4 are related to the diameter of the circular cross section of the four piers, design variables x_5 and x_7 are associated with the external diameter of the bearings located at the abutments, and design variables x_6 and x_8 are related to the total height of rubber of the bearings located at the abutments. The relationship between the design variables and the actual structural parameters is given by $D_{pi} = x_i, i = 1, 2, 3, 4$, $D_{Lr} = x_5$, $H_{Lr} = x_6$, $D_{Rr} = x_7$, and $H_{Rr} = x_8$, where $D_{pi}, i = 1, 2, 3, 4$, are the diameters of the circular cross section of the piers, D_{Lr} and D_{Rr} are the external diameters of the bearings located at the left and right abutments, respectively, and H_{Lr} and H_{Rr} are the total heights of rubber of the bearings located at the left and right abutments, respectively. The side constraints for the design variables are given by $1.2 \leq x_i \leq 2.0, i = 1, 2, 3, 4$; $0.6 \leq x_i \leq 1.0, i = 5, 7$, and $0.15 \leq x_i \leq 0.25, i = 6, 8$. The objective function $c(\mathbf{x})$ represents a cost function, which is assumed to be proportional to the total volume of rubber in the bearings and to the total volume of the piers.

Failure, that is, unacceptable performance, is defined in terms of the relative displacement of the rubber bearings and the relative displacement of piers. Thus, the corresponding failure probability functions are given by

$$P_{F_1}(\mathbf{x}) = P\left[\max_{t \in [0,T]} \frac{|\, u_{b\max}(t, \mathbf{z}, \mathbf{x}) \,|}{0.10\text{m}} > 1\right] \tag{6.31}$$

$$P_{F_2}(\mathbf{x}) = P\left[\max_{t \in [0,T]} \frac{|\, \delta_{\max}(t, \mathbf{z}, \mathbf{x}) \,|}{0.07\text{m}} > 1\right] \tag{6.32}$$

where $u_{b\max}(t, \mathbf{z}, \mathbf{x})$ represents the relative displacement between the deck girder and the base of the four rubber bearings located at the abutments (in the x or y direction), and $\delta_{\max}(t, \mathbf{z}, \mathbf{x})$ denotes the relative displacement between the top of the four piers and their connection with the pile foundation (in the x or y direction). Note that the estimation of the failure probability functions for a given design $\mathbf{x}$ represents a high-dimensional reliability problem, as in the previous example problem.

6.7.11 Example 3: Substructures Characterization

Considering the previous design formulation, the bridge structure is divided into a number of substructures. The division is guided by a parametrization scheme so that the substructure matrices for each one of the introduced substructures depend at most on only one of the design variables. In particular, the structural model is subdivided into six linear substructures and two nonlinear substructures, as shown in Fig. 6.13. Substructure S_1 is composed of the pile elements; substructures S_2, S_3, S_4 and S_5 include the different pier elements; and substructure S_6 corresponds to the deck girder. Finally, substructures S_7 and S_8 are the nonlinear substructures comprising the rubber

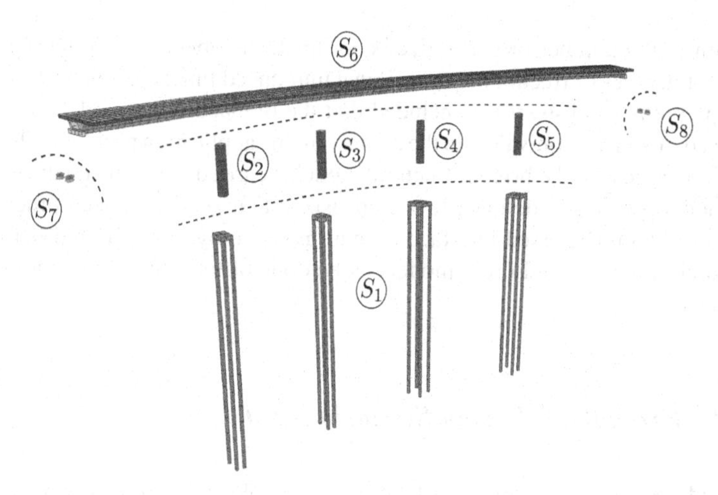

Fig. 6.13 Substructures of the finite element model used for design purposes

bearings located at the left and right abutment, respectively. With this subdivision, substructures S_1 and S_6 do not depend on the design variables, while substructures S_2, S_3, S_4 and S_5 depend on the design variables x_1, x_2, x_3, and x_4, respectively, and design variables x_5, x_6, x_7, and x_8 are associated with the nonlinear substructures S_7 and S_8. The nonlinear parametrization functions of the design variables x_j, $j = 1, 2, 3, 4$, corresponding to the linear substructures, can be defined as $h^j(x_j) = x_j^4$ (variable associated with the inertia term in the stiffness matrices), and $g^j(x_j) = x_j^2$ (variable associated with the area term in the mass matrices), respectively. Note that the parametrization of the reduced-order matrices is considered with respect to the design variables, instead of uncertain model parameters, in this example problem.

Validation calculations similar to the ones performed in Sect. 5.8.4 indicate that retaining ten generalized coordinates for substructure S_1, two for each one of substructures S_2, S_3, S_4 and S_5, and ten for substructure S_6 is adequate in the context of this application. In fact, the absolute value of the difference between the modal frequencies using the full nominal-reference finite element model and the modal frequencies computed using the reduced-order model based on dominant fixed-interface normal modes is very small. In addition, the corresponding matrix of MAC-values between the first six modal vectors computed from the unreduced finite element model and from the reduced-order model shows that the values at the diagonal terms are close to one and almost zero at the off-diagonal terms. Thus, both models are consistent.

In summary, a total of 28 generalized coordinates, corresponding to the dominant fixed-interface normal modes of the linear substructures, out of 10,008 internal DOFs of the original model, are retained for the six linear substructures. On the other hand, the number of interface degrees of freedom is equal to 60 in this case. With this reduction, the total number of generalized coordinates of the reduced-order model represents a 99% reduction with respect to the unreduced model. Thus, a drastic reduction in the number of generalized coordinates is obtained with respect to the number of degrees of freedom of the original unreduced finite element model. Based on the previous analysis, it is concluded that the reduced-order model and the full unreduced model are equivalent in the context of this application problem. Therefore, the design process of the bridge structural model is carried out by using the reduced-order model previously defined. It is emphasized that such a reduced-order model is based on dominant fixed-interface normal modes only. As indicated before, the calibration of the reduced-order model can be done offline, before the optimization takes place.

6.7.12 *Example 3: Design Scenario No. 1*

Taking advantage of the reduced-order model, a couple of design scenarios are investigated in detail to get insight into the reliability and general performance of the bridge structure under consideration. First, the design of the rubber bearings (isolators) located on the abutments is considered. In particular, the effect of the external diameter and the total height of rubber in the bearings on the design of such elements are studied. To this end, the design variables x_1, x_2, x_3, and x_4, which control the diameters of the circular cross sections of the piers, are kept constant and equal to their upper bound values $x_i = 2.0, i = 1, 2, 3, 4$. This is done in order to isolate the effect of the design variables associated with the rubber bearings. Moreover, these variables are linked into two design variables, one related to the external diameter ($D_r = x_5 = x_7$) and the other related to the total height of rubber ($H_r = x_6 = x_8$). In other words, all rubber bearings are assumed to have the same geometrical properties. With the previous setting, Fig. 6.14 shows some iso-probability curves and objective function contours as well as the final design.

The objective contours are normalized by a cost factor and the iso-probability curves are constructed by using a set of failure estimates distributed over the design space. These curves have been smoothed for presentation purposes. It is important to note that these curves were constructed by using the reduced-order model in a reasonable computational time. The construction of these curves from the full finite element model is not practical due to the excessive computational time required to estimate the failure probabilities over the design space. The figure indicates that the probability of failure event F_1 decreases as the external diameter of the isolators increases. This is reasonable since the isolation system becomes stiffer with rubber bearings having larger external diameters and thus the relative displacements between the deck girder and the base of the rubber bearings at each abutment, which control the failure event F_1, are expected to decrease. On the contrary, the failure probability increases as the height of rubber increases. In this case, the isolation system becomes more flexible and, therefore, the relative displacements between the deck girder and the base of the rubber bearings increase. A similar effect is observed with respect to the failure event related to the relative displacement between the top of the piers and their connection with the pile foundation (F_2). That is, an increase of the external

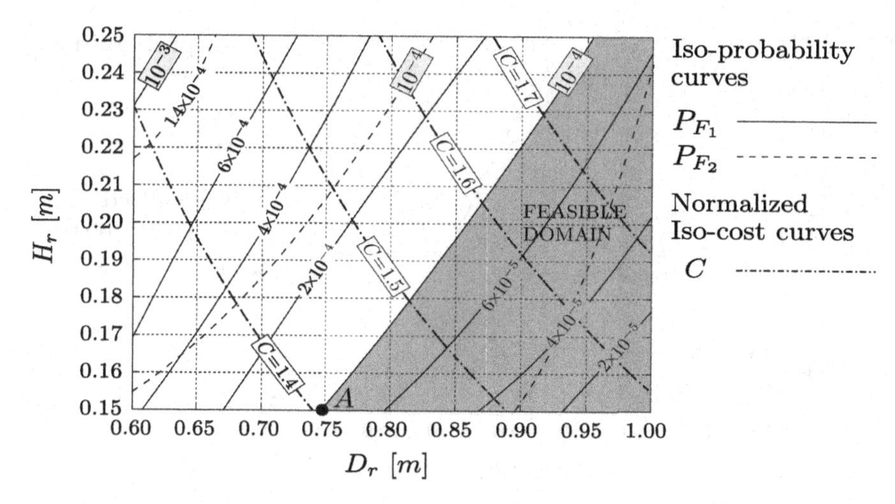

Fig. 6.14 Design space in terms of the design variables associated with the rubber bearings (D_r, H_r). P_{F_1}: iso-probability curves of failure event F_1. P_{F_2}: iso-probability curves of failure event F_2. C: normalized objective function contours

diameter of the isolators decreases the probability of failure event F_2, since the overall system becomes stiffer. On the other hand, an increase of the total height of rubber in the isolators increases the probability of failure event F_2, since in this case the structural system becomes more flexible. The corresponding final design is given by $D_r = 0.75$ m and $H_r = 0.15$ m. The side constraint associated with the height of rubber and the reliability constraint related to the maximum relative displacement between the deck girder and the base of the rubber bearings at each abutment are active in the final design. Contrarily, the reliability constraint associated with the maximum relative displacement between the top of the piers and their connection with the pile foundation is inactive.

6.7.13 *Example 3: Design Scenario No. 2*

The interaction between bridge structural components and rubber bearing parameters is considered in this scenario. Specifically, the design space in terms of the diameter of the circular cross sections of the piers and the external diameter of the rubber bearings is constructed. For illustration purposes, the design variables associated with the diameters of the circular cross sections of the piers are linked to one design variable $D_p = x_1 = x_2 = x_3 = x_4$, while the design variables x_5 and x_7 associated with the external diameters of the rubber bearings are linked to one design variable $D_r = x_5 = x_7$. Design variables related to the total heights of rubber in the bearings are kept constant and equal to their lower bound values, namely, $x_6 = x_8 = 0.15$.

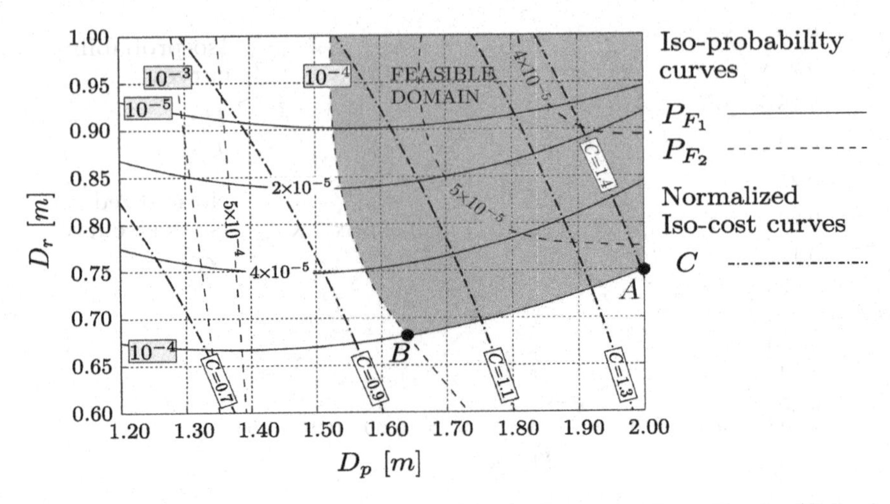

Fig. 6.15 Design space in terms of the diameter of the circular cross sections of the piers (D_p) and the external diameter of the rubber bearings (D_r). P_{F_1}: iso-probability curves of failure event F_1. P_{F_2}: iso-probability curves of failure event F_2. C: normalized objective function contours

Figure 6.15 shows some objective contours and iso-probability curves as well as the final design. The figure demonstrates that the probability of failure event F_2 decreases as the diameter of the circular cross sections of the piers increases. In this case, the piers become stiffer and, therefore, the relative displacements between the top of the piers and their connection with the piles foundations decrease. Moreover, the failure event F_2 is controlled by the diameter of the circular cross sections of the piers for values of this quantity close to its lower bound, that is, $D_p \leq 1.45\,\text{m}$. In this range of values, the iso-probability curves are almost perpendicular. So the effect of the external diameter of the isolator is negligible. In other words, the flexibility of the pier elements controls the relative displacements between the top of the piers and their connection with the piles foundations, as expected. Contrarily, for values of this quantity close to its upper bound, $D_p \geq 1.70\,\text{m}$, a strong interaction between the diameter of the circular cross sections of the piers and the external diameter of the rubber bearings is observed. Thus, for rigid pier elements the relative displacements between the top of the piers and their connection with the pile foundation is controlled by both design variables, that is, D_p and D_r. In fact, the iso-probability curves indicate that, for example, an increase in the diameter of the circular cross sections of the piers is compensated by a decrease in the external diameter of the bearings. In other words, for such combinations of the design variables D_p and D_r the probability of failure remains invariant. On the other hand, the failure event F_1 is mainly controlled by the external diameter of the rubber bearings. Actually, the iso-probability curves associated with failure event F_1 show a relatively weak interaction between the diameter of the circular cross sections of the piers and the external diameter of the rubber bearings. The probability of failure of this event decreases as the external diameter of the isolators increases, which is the same behavior observed in Fig. 6.14.

The final design for this scenario is given by $D_p = 1.64$ m and $D_r = 0.67$ m (point B in the figure) where both reliability constraints are active.

The results shown in Fig. 6.15 can also be used to demonstrate the benefits of designing the isolators and the bridge structure simultaneously. For example, if the design process involves only the isolators and the diameter of the circular cross sections of the piers are kept constant at their upper bound values ($D_p = 2.0$ m), the optimal design is given by $D_r = 0.75$ m (point A in the figure) with a corresponding normalized cost equal to $C = 1.4$. On the other hand, if the diameter of the circular cross sections of the piers is also considered as a design variable, the final design moves from point A to point B, with a decrease of the normalized cost by about 30%. Thus, taking into account the interaction between the design variables associated with the bridge structure and the isolators during the design process is quite beneficial in terms of the cost of the final design.

From the optimization point of view, the corresponding design process converges in less than eight iterations starting from the initial feasible design given by $D_p = 2.0$ m, $D_r = 1.0$ m and $H_r = 0.25$ m. Of course, the design process can be carried out by considering all design variables as independent during the design process as well. In this case, the corresponding iteration history of the optimization process in terms of the objective function indicates that the cost decreases by about 5% with respect to the final design shown in Fig. 6.15. Finally, it is noted that the above observations and remarks give a valuable insight into the complex interaction of the design variables on the performance and reliability of the bridge structural system.

6.7.14 Example 3: Computational Cost

Table 6.6 shows the online computational costs involved in the assemblage of the finite element model and the computation of the dynamic response for a given design considering the full finite element model and the reduced-order model. These operations and procedures are performed at each iteration of the design process. The last column of the table indicates the speedup (rounded to the nearest integer) achieved by the reduced-order model for the different tasks. As previously pointed out, the speedup is the ratio of the execution time by using the unreduced model and the execution time by using the reduced-order model. It is seen that the difference is quite significant. In fact, the overall speedup for the online calculations is more than 90 in this case.

The offline computational cost, that is, the cost of calculations related to the definition of the reduced-order model and the characterization of the transformation matrix that maps the generalized coordinates to the physical coordinates of the unreduced model, which are performed once during the design process, corresponds to approximately two full analyses (finite element model generation and dynamic response) of the unreduced model in this case. Considering this cost an overall speedup value of about 10 is obtained by the proposed methodology in solving this particular design problem. However, once the reduced-order model has been defined, several design

Table 6.6 Online computational cost of different tasks for a given design

	Full finite element model	Reduced-order model	
Tasks	Time (s)	Time (s)	Speedup
Finite element generation and dynamic response	55.34	0.58	95

scenarios, that is, in terms of different objective functions and reliability constraints, can be explored and solved efficiently. So high speedup values can be obtained for the design process as a whole. As previously pointed out, this reduction in computational effort is achieved without compromising the accuracy of the final design.

References

1. N.M. Alexandrov, J.E. Dennis Jr., R.M. Lewis, V. Torczon, A trust-region framework for managing the use of approximation models in optimization. Struct. Multidiscip. Optim. **15**(1), 16–23 (1998)
2. J. Arora, *Optimization of Structural and Mechanical Systems* (World Scientific, Singapore, 2007)
3. S.K. Au, J.L. Beck, Estimation of small failure probabilities in high dimensions by subset simulation. Probabilistic Eng. Mech. **16**(4), 263–277 (2001)
4. F.J. Baronand, O. Pironneau, Multidisciplinary optimal design of a wing profile, in *Proceedings of Structural Optimization*, vol. 93 (COPPE, Rio de Janeiro, Brazil, 1993), pp. 61–68
5. A.D. Belegundu, T.R. Chandrupatla, *Optimization Concepts and Applications in Engineering* (Prentice Hall, Upper Sadle River, 1999)
6. R.H. Byrd, J. Nocedal, R.B. Schanabel, Representation of Quasi-Newton matrices and their use in limited memory methods. Math. Program. **63**(4), 129–156 (1994)
7. F. Cérou, P. Del Moral, T. Furon, A. Guyader, Sequential Monte Carlo for rare event estimation. Stat. Comput. **22**(3), 795–808 (2012)
8. J. Ching, Y.H. Hsieh, Local estimation of failure probability function and its confidence interval with maximum entropy principle. Probabilistic Eng. Mech. **22**(1), 39–49 (2007)
9. J. Ching, W.C. Hsu, Transforming reliability limit-state constraints into deterministic limit-state constraints. Struct. Saf. **30**(1), 11–33 (2008)
10. R.W. Clough, J. Penzien, *Dynamics of Structures* (McGraw-Hill, New York, 1975)
11. P. Del Moral, A. Doucet, A. Jasra, Sequential Monte Carlo samplers. J. R. Stat. Soc. Ser. B (Stat. Methodol.) **68**(3), 411–436 (2006)
12. O. Ditlevsen, O.H. Madsen, *Structural Reliability Methods* (Wiley, Hoboken, 1996)
13. I. Enevoldsen, J.D. Sørensen, Reliability-based optimization in structural engineering. Struct. Saf. **15**(3), 169–196 (1994)
14. C.C. Floudas, P.M. Pardalos, in *A Collection of Test Problems for Constrained Global Optimization Algorithms*, Lecture notes in computer science, ed. by G. Goos, J. Hartmanis (Springer, Berlin, Heidelberg, 1990)
15. M. Gasser, G.I. Schuëller, Reliability-based optimization of structural systems. Math. Methods Oper. Res. **46**(3), 287–307 (1997)
16. D.E. Goldberg, *Genetic Algorithms in Search, Optimization and Machine Learning* (Addison-Wesley, Reading (MA), 1989)

17. R.T. Haftka, Z. Gürdal, in *Elements of Structural Optimization*, 3rd edn., ed. by Kluwer (Springer, Berlin, 1992)
18. J. Herskovits, G. Santos, On the computer implementation of feasible direction interior point algorithms for nonlinear optimization. Struct. Optim. **14**(2–3), 165–172 (1997)
19. J.B. Hiriat-Urruty, B. Lemarechal, *Convex Analysis and Minimization Algorithms* (Springer, New York, 1993)
20. W. Hock, K. Schittkowski, *Test Examples for Nonlinear Programming Codes*, Lecture notes in economics and mathematical systems (Springer, New York, 1987)
21. H.A. Jensen, M.A. Catalan, On the effects of non-linear elements in the reliability-based optimal design of stochastic dynamical systems. Int. J. Non-Linear Mech. **42**(5), 802–816 (2007)
22. H.A. Jensen, M.A. Valdebenito, G.I. Schuëller, D.S. Kusanovic, Reliability-based optimization of stochastic systems using line search. Comput. Methods Appl. Mech. Eng. **198**(49–52), 3915–3924 (2009)
23. H.A. Jensen, D.S. Kusanovic, M.A. Valdebenito, G.I. Schuëller, Reliability-based design optimization of uncertain stochastic systems: a gradient-based scheme. J. Eng. Mech. **138**(1), 60–70 (2012)
24. H.A. Jensen, L. Becerra, M.A. Valdebenito, On the use of a class of interior point algorithms in stochastic structural optimization. Comput. Struct. **126**, 69–85 (2013)
25. J. Kanda, B. Ellingwood, Formulation of load factors based on optimum reliability. Struct. Saf. **9**(3), 197–210 (1991)
26. P.S. Koutsourelakis, Design of complex systems in the presence of large uncertainties: a statistical approach. Comput. Methods Appl. Mech. Eng. **197**(49–50), 4092–4103 (2008)
27. P.S. Koutsourelakis, H.J. Pradlwarter, G.I. Schuëller, Reliability of structures in high dimensions, part I: algorithms and applications. Probabilistic Eng. Mech. **19**(4), 409–417 (2004)
28. N. Kuschel, R. Rackwitz, Two basic problems in reliability-based structural optimization. Math. Methods Oper. Res. **46**(3), 309–333 (1997)
29. H.J. Kushner, G.G. Yin, *Stochastic Approximation and Recursive Algorithms and Applications* (Springer, New York, 2003)
30. D.G. Luenberger, *Linear and Nonlinear Programming*, 2nd edn. (Addison-Wesley, London, 1984)
31. H.G. Matthies, C.E. Brenner, C.G. Bucher, C.G. Soares, Uncertainties in probabilistic numerical analysis of structures and solids - stochastic finite elements. Struct. Saf. **19**(3), 283–336 (1997)
32. O. Möller, R.O. Foschi, L.M. Quiroz, M. Rubinstein, Structural optimization for performance-based design in earthquake engineering: applications of neural networks. Struct. Saf. **31**(6), 490–499 (2009)
33. R.M. Neal, Annealed importance sampling. Stat. Comput. **11**(2), 125–139 (2001)
34. J. Nocedal, S.J. Wright, *Numerical Optimization* (Springer, New York, 1999)
35. M. Papadrakakis, N.D. Lagaros, Reliability-based structural optimization using neural networks and Monte Carlo simulation. Comput. Methods Appl. Mech. Eng. **191**, 3491–3507 (2002)
36. E. Polak, *Optimization Algorithms and Consistent Approximations* (Springer, Berlin, 1997)
37. J.D. Powell, Variable metric methods for constrained optimization, in *Mathematical Programming - The State of the Art*, ed. by A. Bachem, M. Grotschet, B. Korte (Springer, Berlin, Germany, 1983)
38. H.J. Pradlwarter, Deterministic integration algorithms for stochastic response computations of FE-systems. Comput. Struct. **80**(18–19), 1489–1505 (2002)
39. H.J. Pradlwarter, G.I. Schuëller, P.S. Koutsourelakis, D.C. Champris, Application of line sampling simulation method to reliability benchmark problems. Struct. Saf. **29**(3), 208–221 (1998)
40. M.R. Rajashekhar, B. Ellingwood, A new look at the response surface approach for reliability analysis. Struct. Saf. **12**(3), 205–220 (1993)
41. J.O. Royset, E. Polak, Reliability-based optimal design using sample average approximations. Probabilistic Eng. Mech. **19**(4), 331–343 (2004)
42. A. Ruszczynski, A. Shapiro, *Stochastic Programming* (Elsevier, New York, 2003)

43. G.I. Schuëller, Computational methods in optimization considering uncertainties - Special Issue. Comput. Methods Appl. Mech. Eng. **198**(1), 1–164 (2008)
44. G.I. Schuëller, H.A. Jensen, Computational methods in optimization considering uncertainties—an overview. Comput. Methods Appl. Mech. Eng. **198**(1), 2–13 (2008)
45. J.C. Spall, *Introduction to Stochastic Search and Optimization: Estimation, Simulation and Control* (Wiley, Hoboken, 2003)
46. A.A. Taflanidis, Robust stochastic design of viscous dampers for base isolation applications, in *ECCOMAS Thematic Conference on Computational Methods in Structural Dynamics and Earthquake Engineering* (Rhodes, Greece, 2009)
47. A.A. Taflanidis, J.L. Beck, Stochastic subset optimization for optimal reliability problems. Probabilistic Eng. Mech. **23**(2–3), 324–338 (2008)
48. M.A. Valdebenito, G.I. Schuëller, Efficient strategies for reliability-based optimization involving non linear, dynamical structures. IfM-Int. Work. Report No. 3–61 (Institute of Engineering Mechanics, University of Innsbruck, Austria, EU, 2009)
49. M.A. Valdebenito, G.I. Schuëller, Efficient strategies for reliability-based optimization involving non-linear, dynamical structures. Comput. Struct. **89**(19–20), 1797–1811 (2011)
50. F. Van Keulen, K. Vervenne, Gradient-enhanced response surface building. Struct. Muldisciplinary Optim. **27**(5), 337–351 (2004)
51. J. Wang, L.S. Katafygiotis, Reliability-based optimal design of linear structures subjected to stochastic excitations. Struct. Saf. **47**, 29–38 (2014)
52. T. Weise, Global optimization algorithms. Theory and application (2009) e-book. http://www.it-weise.de/
53. A. Zerva, *Spatial Variation of Seismic Ground Motions: Modeling and Engineering Applications* (CRC Press, Boca Raton, FL, USA, 2009)
54. K. Zuev, J.L. Beck, Global optimization using the asymptotically independent Markov sampling method. Comput. Struct. **126**, 107–119 (2013)

Part III
Application to Identification Problems

Chapter 7
Bayesian Finite Element Model Updating

Abstract In this chapter, the implementation of the reduced-order models within Bayesian finite element model updating is explored. The Bayesian framework for model parameter estimation, model selection, and robust predictions of output quantities of interest is first presented. Bayesian asymptotic approximations and sampling algorithms are then outlined. The framework is implemented for updating linear and nonlinear finite element models in structural dynamics using vibration measurements consisting of either identified modal frequencies or measured response time histories. For asymptotic approximations based on modal properties, the formulation for the posterior distribution is presented with respect to the modal properties of the reduced-order model. In addition, analytical expressions for the required gradients with respect to the model parameters are provided using adjoint methods. Two applications demonstrate that drastic reductions in computational demands can be achieved without compromising the accuracy of the model updating results. In the first application, a high-fidelity linear finite element model of a full-scale bridge with hundreds of thousands of degrees-of-freedom (DOFs) is updated using experimentally identified modal properties. In the second application, a nonlinear model of a base-isolated building is updated using acceleration response time histories.

7.1 Motivation

Probability distributions are often used to quantify uncertainties, and probability calculus is employed to propagate these uncertainties through the computational model of a structure in order to make prior robust predictions of output quantities of interest (QoI). To improve the mathematical models and the probability models of uncertainties of both the system and loads, one can exploit the valuable information contained in measured data collected from system component tests or system operation through monitoring. The resulting data-driven updated model, when used for simulations, yields updated or posterior robust predictions, constituting improved and more reliable estimates of the system performance. However, the computational science tools for handling uncertainties in simulations based on test/monitoring data are conceptually and computationally much more challenging than conventional computing tools

H. Jensen and C. Papadimitriou, *Sub-structure Coupling for Dynamic Analysis*,
Lecture Notes in Applied and Computational Mechanics 89,
https://doi.org/10.1007/978-3-030-12819-7_7

[48]. This chapter presents a comprehensive Bayesian probabilistic framework for uncertainty quantification and propagation (UQ+P) in complex structural dynamics simulations based on test data.

Bayesian analysis [13, 15, 64] is used as the logical and computational framework for combining knowledge from test/monitoring data and models in a consistent way. The Bayesian framework exploits the available measured data and any prior information based on engineering experience to (a) select the most probable mathematical models among a competitive family of mathematical models (linear vs nonlinear models; elastic vs hysteretic models; friction/impact models) introduced to represent the behavior of mechanical components; (b) calibrate the parametric uncertainties involved in structural and mechanical models; (c) propagate uncertainties in simulations for updating robust predictions, taking into account the validated models and calibrated uncertainties.

Bayesian finite element model updating is a demanding computational procedure due to the large number of forward dynamic analyses required. The solution for these types of problems requires the evaluation of the system responses at a large number of points in the uncertain parameter space [1, 27, 33, 35, 49]. Consequently, the computational cost may become excessive when the computational time for performing a dynamic analysis is significant. The model reduction techniques discussed in this book can be used to alleviate the computational effort by reducing the number of physical coordinates to a much smaller number of generalized coordinates. However, the construction of reduced-order models at each point in the parameter space implies a recalculation of the fixed-interface normal modes and the interface constraint modes for each substructure. This procedure is computationally expensive due to the substantial computational overhead that arises at the substructure level [49]. An efficient finite element model parametrization scheme was considered in Chaps. 2 and 3 to cope with this difficulty. When the division of the structure into substructures is guided by such a parametrization scheme, dramatic computational savings are achieved.

This chapter is organized as follows. A Bayesian probabilistic framework is developed in Sect. 7.2 for parameter estimation and model selection in structural dynamics using vibration measurements collected during system operation. The Bayesian tools used to carry out the computations are presented in Sect. 7.3. Such tools include asymptotic approximations and sampling algorithms. In Sect. 7.4, the implementation of the Bayesian framework in structural dynamics is presented. The formulation for linear models based on modal frequencies and mode shapes is given in Sect. 7.4.1. For nonlinear models, the Bayesian formulation is presented in Sect. 7.4.2, based on full response time histories (accelerations, displacements, and strains). The applicability, effectiveness, and accuracy of the proposed techniques are demonstrated using high-fidelity linear FE models and field measurements for a motor-way bridge as well as nonlinear models and simulated measurements for a base-isolated structure.

7.2 Bayesian Inference Framework

7.2.1 Finite Element Model and Uncertainty

Consider a finite element model class of a structure, denoted by M, a set of model parameters $\boldsymbol{\theta} \in R^{n_\theta}$, and let $\boldsymbol{r}(\boldsymbol{\theta}|M)$ be the model predictions of output QoI, given a value of the parameter set $\boldsymbol{\theta}$. Probability distribution functions (PDF) are used to quantify the uncertainty in the parameters $\boldsymbol{\theta}$. A prior probability distribution $\pi(\boldsymbol{\theta}|M)$ is assigned to the model parameters to incorporate prior information on the plausibility of each possible value of the model parameters $\boldsymbol{\theta}$. Uncertainty propagation algorithms (e.g., efficient Monte Carlo variants) can be used to propagate the uncertainties through the structural model to predict output QoI. However, the prior probability distribution $\pi(\boldsymbol{\theta}|M)$ is subjective, based on previous knowledge and user experience. In Bayesian finite element model updating, experimental data are used to update the uncertainty in the model parameters $\boldsymbol{\theta}$, and then uncertainty propagation techniques are used to update the uncertainty in the predictions of important QoI. These predictions are referred to as data-informed robust predictions.

A family of competitive finite element model classes M_i, $i = 1, \ldots, m_i$, is often introduced to represent the structure. A prior probability $Pr(M_i)$ is assigned to the model class M_i to reflect our preference to this model class in relation to the rest of the model classes in the family. Experimental data can also be used to update our preference, select the most appropriate model class that best explains the data, and rank the rest of the model classes based on the data. The ranking can then be used to update the uncertainties in the predictions, taking into account all competitive model classes. Such predictions are referred to as data-informed hyper robust predictions, a terminology introduced in [14].

7.2.2 Bayesian Model Parameter Estimation

Let $D \equiv \{\hat{y} \in R^{N_0}\}$ be a set of observations available from experiments, where N_0 is the number of observations. In light of the information contained on these data, the interest lies in updating the probability distribution of the model parameters $\boldsymbol{\theta}$, selecting the best model class, and then propagating uncertainties through the structural dynamics model to quantify the uncertainty in the output QoI. The updated PDF $p(\boldsymbol{\theta}|D, M)$ of the parameters $\boldsymbol{\theta}$, given the data D and the model class M, results from the application of the Bayes' theorem

$$p(\boldsymbol{\theta}|D, M) = \frac{p(D|\boldsymbol{\theta}, M)\,\pi(\boldsymbol{\theta}|M)}{p(D|M)} \tag{7.1}$$

where $p(D|\boldsymbol{\theta}, M)$ is the likelihood of observing the data from the model class M and $p(D|M)$ is the evidence of the model class, given by the multi-dimensional integral

$$p(D|M) = \int_{\Theta} p(D|\boldsymbol{\theta}, M)\, \pi(\boldsymbol{\theta}|M)\, d\boldsymbol{\theta} \tag{7.2}$$

over the space of the uncertain model parameters.

The structure of the likelihood is derived by building a probabilistic model for the discrepancy between the model predictions $\boldsymbol{r}(\boldsymbol{\theta}|M)$ obtained from a particular value of the model parameters $\boldsymbol{\theta}$ and the corresponding data $\hat{\boldsymbol{y}}$. This discrepancy is the result of measurement, model, and computational errors. Introducing an error term $\boldsymbol{e}$ to quantify this discrepancy, the observation data and the model predictions satisfy the prediction error equation

$$\hat{\boldsymbol{y}} = \boldsymbol{r}(\boldsymbol{\theta}|M) + \boldsymbol{e} \tag{7.3}$$

A normal distribution $\boldsymbol{e} \sim N(\boldsymbol{\mu}, \boldsymbol{\Sigma})$, where $\boldsymbol{\mu}$ is the mean and $\boldsymbol{\Sigma}$ is the covariance matrix, is chosen for the prediction error term. Using the maximum entropy principle, this choice is justifiable, as it is the least informative distribution among all distributions with the lowest two moments specified. A zero mean model error $\boldsymbol{\mu} = \boldsymbol{0}$ is assumed in this work. A non-zero mean $\boldsymbol{\mu} \neq \boldsymbol{0}$ could also be used to shift the model predictions $\boldsymbol{r}(\boldsymbol{\theta}|M)$ to account for the bias in these predictions and attempt to reconcile conflicting predictions. The structure of the covariance matrix $\boldsymbol{\Sigma}$ may depend on prediction error parameters with unknown values to be inferred using the Bayesian identification framework. For this, it is common to augment the structural model parameter set $\boldsymbol{\theta}$, so that it also includes these prediction error parameters.

Using the prediction error equation (7.3), the measured quantities conditioned on the value of the parameter set $\boldsymbol{\theta}$ follow a normal distribution with mean $\boldsymbol{r}(\boldsymbol{\theta}|M)$ and covariance matrix $\boldsymbol{\Sigma}(\boldsymbol{\theta})$. Consequently, the likelihood $p(D|\boldsymbol{\theta}, M)$ of observing the data follows the multi-variable normal distribution given by

$$p(D|\boldsymbol{\theta}, M) = \frac{|\boldsymbol{\Sigma}(\boldsymbol{\theta})|^{-1/2}}{(2\pi)^{N_0/2}} \exp\left[-\frac{1}{2} J(\boldsymbol{\theta}, M)\right] \tag{7.4}$$

where

$$J(\boldsymbol{\theta}, M) = [\hat{\boldsymbol{y}} - \boldsymbol{r}(\boldsymbol{\theta}|M)]^T \boldsymbol{\Sigma}^{-1}(\boldsymbol{\theta})[\hat{\boldsymbol{y}} - \boldsymbol{r}(\boldsymbol{\theta}|M)] \tag{7.5}$$

is a weighted measure of fit between the measured data and the prediction of the model, and $|\cdot|$ denotes the determinant.

The selection of the prior distribution $\pi(\boldsymbol{\theta}|M)$ affects the posterior distribution of the model parameters for the case of a relatively small number of data. Usually, a non-informative prior can be used. For example a uniform distribution of the model parameters does not give any preference to the values of the model parameters prior to the data. For cases of large numbers of model parameters where problems of ill-conditioning may arise, a non-uniform distribution, such as a Gaussian prior, can avoid unidentifiability issues and enable the estimation of the most probable value of the model parameters, avoiding convergence problems that may arise from gradient and stochastic optimization techniques used in Bayesian asymptotic approximations (see Sect. 7.3.1).

7.2.3 Bayesian Model Selection

Bayesian model class selection is used to compare alternative model classes within a family $M_{Fam} = \{M_i, i = 1, \ldots, m_c\}$ and select the best model class in light of the experimental data [16, 45]. Let $\boldsymbol{\theta}^{(i)} \in R^{n_{\theta_i}}$ be the parameters of the model class M_i, including prediction error model parameters. The posterior probability $Pr(M_i|D)$ of the ith model class given the data D is [16, 64]

$$Pr(M_i|D) = \frac{p(D|M_i)\, Pr(M_i)}{\sum_{i=1}^{m_c} p(D|M_i)\, Pr(M_i)} \tag{7.6}$$

where $Pr(M_i)$ is the prior probability of the model class M_i and $p(D|M_i)$ is the evidence of the model class M_i given by (7.2) with $\boldsymbol{\theta}$ and M replaced by $\boldsymbol{\theta}^{(i)}$ and M_i, respectively. The optimal model class M_{best} is selected as the one that maximizes $Pr(M_i|D)$, based on (7.6). The probability $Pr(M_i|D)$ ranks the rest of the model classes in terms of their suitability in modeling the structure. This rank can be used for posterior hyper robust predictions [14] of output QoI that take into account all model classes with a significant probability of explaining the available data.

7.2.4 Data-Driven Robust Posterior Predictions

7.2.4.1 PDF of Output QoI

Let q be a scalar output QoI of the system. Before the measured data become available, prior robust predictions are carried out by propagating the prior uncertainties in the model parameters, quantified by the prior PDF $\pi(\boldsymbol{\theta}|M)$. Posterior robust predictions are obtained by taking into account the updated uncertainties in the model parameters, given the measurements D. Let $p(q|\boldsymbol{\theta}, M)$ be the conditional probability distribution of q, given the values of the parameters. Using the total probability theorem, the prior and posterior robust probability distribution $p(q|M)$ of q, taking into account the model M, is given by [14, 53]

$$p(q|M) = \int p(q|\boldsymbol{\theta}, M)\, p(\boldsymbol{\theta}|M)\, d\boldsymbol{\theta} \tag{7.7}$$

as an average of the conditional probability distribution $p(q|\boldsymbol{\theta}, M)$ weighed by the PDF $p(\boldsymbol{\theta}|M)$ of the model parameters, where $p(\boldsymbol{\theta}|M) \equiv \pi(\boldsymbol{\theta}|M)$ for the prior estimate in the absence of data or $p(\boldsymbol{\theta}|M) \equiv p(\boldsymbol{\theta}|D, M)$ for the posterior estimate given the data D.

7.2.4.2 Simplified Measures of Uncertainty in Output QoI

The mean $m_1 = E_{\boldsymbol{\theta}}[q(\boldsymbol{\theta})]$ and the variance $\sigma_q^2 = E_{\boldsymbol{\theta}}[q^2(\boldsymbol{\theta})] - m_1^2 = m_2 - m_1^2$ with respect to $\boldsymbol{\theta}$ are often used as simplified measures of uncertainty in the output QoI q. These are derived from the first two moments m_k of $q(\boldsymbol{\theta})$, $k = 1,2$, given by the multi-dimensional integrals

$$m_k(D, M) = \int [q(\boldsymbol{\theta})]^k \, p(\boldsymbol{\theta}|M) \, d\boldsymbol{\theta} \tag{7.8}$$

over the uncertain parameter space. Computational tools for estimating the multi-dimensional integrals are presented in Sect. 7.3.

7.2.4.3 Prior and Posterior Robust Reliability

A more challenging problem in uncertainty propagation is the estimation of rare events. This is important in analyzing probability of failure or unacceptable performance of a system. The probability of failure is the probability that at least one output QoI exceeds certain threshold levels or, more generally, as the probability that the system performance falls within a failure domain F, usually defined by one or more inequality equations.

Let $Pr(F|\boldsymbol{\theta}, M)$ be the probability of failure of the system conditioned on the value of the parameter set $\boldsymbol{\theta}$. The robust prior or robust posterior failure probability [14, 53] is given by the multi-dimensional probability integral

$$P_F(M) = \int Pr(F|\boldsymbol{\theta}, M) \, p(\boldsymbol{\theta}|M) \, d\boldsymbol{\theta} \tag{7.9}$$

where $p(\boldsymbol{\theta}|M) \equiv \pi(\boldsymbol{\theta}|M)$ for prior probability of failure estimate or $p(\boldsymbol{\theta}|M) \equiv p(\boldsymbol{\theta}|D, M)$ for posterior estimate given the data D. Assuming that a set of independent random variables z is used to quantify input and system uncertainties that are not associated with the ones involved in $\boldsymbol{\theta}$, the failure probability can also be written in the form

$$P_F(M) = Pr(\boldsymbol{\zeta} \in F|M) = \int I_F(\boldsymbol{\zeta}) \, p(\boldsymbol{\zeta}|M) \, d\boldsymbol{\zeta} \tag{7.10}$$

where $\boldsymbol{\zeta} = (z, \boldsymbol{\theta})$ is the augmented set of uncertain parameters, F is a failure region in the augmented parameter space, and I_F is an indicator function that is 1 if $\boldsymbol{\zeta} \in F$

and 0 elsewhere over the space of feasible system parameters $\boldsymbol{\zeta}$. The estimation of the failure probability integrals (7.9) and (7.10) can be done using sampling techniques. For further details, the reader is referred to Sect. 7.3.4 and Ref. [34].

7.3 Bayesian Computational Tools

7.3.1 Asymptotic Approximations

For a large enough number of measured data, the posterior distribution of the model parameters in (7.1) can be asymptotically approximated by a Gaussian distribution [13]

$$p(\boldsymbol{\theta}|D, M) \approx \frac{|\boldsymbol{C}(\hat{\boldsymbol{\theta}})|^{-1/2}}{(2\pi)^{n_\theta/2}} \exp\left[-\frac{1}{2}(\boldsymbol{\theta} - \hat{\boldsymbol{\theta}})^T \boldsymbol{C}^{-1}(\hat{\boldsymbol{\theta}})(\boldsymbol{\theta} - \hat{\boldsymbol{\theta}})\right] \quad (7.11)$$

centered at the most probable value $\hat{\boldsymbol{\theta}}$ of the model parameters obtained by maximizing the posterior PDF $p(\boldsymbol{\theta}|D, M)$ or, equivalently, minimizing the negative of the log posterior PDF function

$$\begin{aligned} g(\boldsymbol{\theta}, M) &= -\ln p(\boldsymbol{\theta}|D, M) \\ &= \frac{1}{2}J(\boldsymbol{\theta}, M) + \frac{1}{2}\ln|\boldsymbol{\Sigma}(\boldsymbol{\theta})| - \ln\pi(\boldsymbol{\theta}|M) + \frac{N_0}{2}ln(2\pi) \end{aligned} \quad (7.12)$$

with a covariance matrix $\boldsymbol{C}(\hat{\boldsymbol{\theta}}) = \boldsymbol{H}^{-1}(\hat{\boldsymbol{\theta}})$ equal to the inverse of the Hessian $\boldsymbol{H}(\boldsymbol{\theta}) = \nabla\nabla^T g(\boldsymbol{\theta}, M)$ of the function $g(\boldsymbol{\theta}, M)$ in (7.12) evaluated at the most probable value $\hat{\boldsymbol{\theta}}$. This approximation, known as the Bayesian central limit theorem, is asymptotically correct for a large number of data. The asymptotic result (7.11), although approximate, provides a good representation of the posterior PDF for a number of applications involving even a relatively small number of data.

The asymptotic approximation (7.11) provides an adequate representation of the posterior probability distribution in the case of unimodal distributions. For multi-modal distributions, the asymptotic approximation can be improved by considering a weighted contribution of each mode with weights based on the probability volume of the PDF in the neighborhood of each mode [13]. The weighted estimate is reasonable, provided that the modes are separable. For interacting modes or closely spaced modes, this estimate is inaccurate due to the overlapping of the regions of high probability volume involved in the interaction. However, implementation problems exist in multi-modal cases, due to the inconvenience in estimating all modes of the multimodal distribution [40]. Asymptotic approximations have also been introduced to handle the unidentifiable cases [38, 39] manifested for a relatively large number of model parameters in relation to the information contained in the data.

For model selection, an asymptotic approximation [16, 52, 64] based on the Laplace method provides an estimate of the evidence integral in (7.2) that appears in the model selection equation (7.6). The asymptotic estimate for $Pr(M_i|D)$ is given in the form [16, 50]

$$Pr(M_i|D) = \left(\sqrt{2\pi}\right)^{n_{\theta_i}} \frac{p(D|\hat{\boldsymbol{\theta}}^{(i)}, M_i)\, \pi(\hat{\boldsymbol{\theta}}^{(i)}|M_i)}{\sqrt{|\boldsymbol{H}_i(\hat{\boldsymbol{\theta}}^{(i)})|}} \frac{Pr(M_i)}{\sum_{i=1}^{m_c} p(D|M_i)\, Pr(M_i)} \tag{7.13}$$

where $\hat{\boldsymbol{\theta}}^{(i)}$ is the most probable value of the parameters of the model class M_i and $\boldsymbol{H}_i(\boldsymbol{\theta}) = \nabla\, \nabla^T g_i(\boldsymbol{\theta}, M)$ is the Hessian of the function $g_i(\boldsymbol{\theta}^{(i)}, M_i)$ given in (7.12) for the model class $M \equiv M_i$. It should be noted that the asymptotic estimate for the probability of a model class M_i can readily be obtained with the most probable value and the Hessian of the particular mode. For the multi-modal case, the expression (7.13) can be generalized by adding the contributions from all modes.

For the posterior robust prediction integrals (7.7) or (7.8), similar asymptotic approximations are available to simplify the integrals. Details are presented in [18] using the asymptotic approximations developed in [59]. However, for each quantity of interest, an extra optimization problem needs to be solved along with evaluating the Hessian of an objective function at the new optimum. The approximations require the solution of as many optimization problems and Hessian evaluations as the number of output quantities of interest. For several output QoI, these approximations can be computationally costly. In light of the high computational cost and the lack of accuracy of the asymptotic approximations, alternative sampling techniques, discussed in Sect. 7.3.4, are used to estimate the multi-dimensional integrals.

7.3.2 *Gradient-Based Optimization Algorithms*

The optimization problems that arise in the asymptotic approximations are solved using available single objective optimization algorithms. The optimization of $g(\boldsymbol{\theta}, M)$ given in (7.12) with respect to $\boldsymbol{\theta}$ can readily be carried out numerically using any available algorithm for optimizing a nonlinear function of several variables. Gradient-based optimization algorithms can be conveniently used to achieve fast convergence to the optimum. However, to guarantee the convergence of the gradient-based algorithms for models involving a relatively large number of DOFs, analytical equations for the gradients of the response QoI involved in the objective function $g(\boldsymbol{\theta}, M)$ are required. The computational effort scales with the number of parameters in $\boldsymbol{\theta}$.

Adjoint methods provide a computationally effective way to estimate the gradients of the objective function with respect to all parameters by solving a single adjoint problem, making the computational effort independent of the number of variables in the set $\boldsymbol{\theta}$. A review of a model non-intrusive adjoint method for the case of Bayesian parameter estimation based on modal frequencies and mode shapes is given in [47].

For nonlinear models of structures, the techniques for computing gradients of the objectives with respect to the parameters are model intrusive, requiring tedious algorithmic and software developments that, in most cases, are not easily integrated within the commercial software packages. Selected examples of model intrusiveness include the sensitivity formulation for hysteretic-type nonlinearities in structural dynamics and earthquake engineering [11, 12]. Gradient and adjoint formulations require considerable algorithmic development overhead associated with developing analytical expressions and implementing them in software. Finally, it should be noted that there are systems and types of nonlinearities (e.g., contact, sliding, and impact) where the development of an adjoint formulation or analytical equations for the sensitivity of objective functions to parameters is not possible. Derivative-free local optimization techniques are more appropriate to use in such cases.

7.3.3 Stochastic Optimization Algorithms

Stochastic optimization algorithms, such as evolutionary algorithms, are random search algorithms that better explore the parameter space for detecting the neighborhood of the global optimum, avoiding premature convergence to a local optimum. In addition, stochastic optimization algorithms do not require the evaluation of the gradient of the objective function with respect to the parameters. Thus, they are model non-intrusive, since there is no need to formulate the equations for the derivatives either by direct or adjoint techniques. Despite their slow convergence, evolutionary algorithms are highly parallelizable, so the time-to-solution in an HPC environment is often comparable to conventional gradient-based optimization methods [30].

Stochastic optimization algorithms can be used with parallel computing environments to find the optimum for non-smooth functions or for models for which an adjoint formulation cannot be developed. In the absence of an HPC environment, the disadvantage of the stochastic optimization algorithms arises from the high number of system re-analyses, which may make the computational effort excessive for real-world problems for which a simulation may take minutes, hours, or even days to complete.

A parallelized version of the covariance matrix adaptation evolutionary strategy (CMA-ES) [31] can be used to solve the single-objective optimization problems arising in the Bayesian asymptotic approximations. The CMA-ES algorithm exhibits fast convergence properties among several classes of evolutionary algorithms, especially when searching for a single global optimum. The Hessian estimation required in Bayesian asymptotic approximations can be computed using the Romberg method [41]. This procedure is based on a number of system re-analyses at the neighborhood of the optimum, which can all be independently performed for problems involving either calibration or propagation and are, thus, highly parallelizable. Details can be found in [30].

7.3.4 Sampling Algorithms

Sampling algorithms are non-local methods capable of providing accurate representations for the posterior PDF and accurate robust predictions of output QoI. Sampling algorithms refer mostly to Markov Chain Monte Carlo (MCMC) [20, 32, 44] variants. They are used to generate samples $\boldsymbol{\theta}_i, i = 1, \ldots, N$, for populating the posterior PDF in (7.1), estimating the model evidence and computing the uncertainties in output QoI. Among the sampling algorithms available, the transitional MCMC algorithm (TMCMC) [21] is one of the most promising algorithms for finding and populating the important region of interest with samples, even in challenging unidentifiable cases and multi-modal posterior distributions. The TMCMC algorithm is well suited for parallel implementation in a computer cluster. HPC techniques are used to reduce the time-to-solution of the TMCMC algorithm. Details of the parallel implementation are given in [1, 30]. The parallelized version of the TMCMC algorithm for Bayesian UQ has been implemented in software and is available at https://github.com/cselab/pi4u, while an improved version of the TMCMC algorithm, termed BASIS [60], is available in Matlab at https://gitlab.ethz.ch/mavt-cse/BASIS_1.1. Approximate methods based on kernels are used to estimate marginal distributions of the parameters from the samples $\boldsymbol{\theta}_i, i = 1, \ldots, N$. An advantage of the TMCMC algorithm is that it yields an estimate of the evidence in (7.2) of the model class M_i based on the samples already generated by the algorithm.

Sampling methods can be conveniently used to estimate the multi-dimensional integrals (7.7) and (7.8) from the samples $\boldsymbol{\theta}_i, i = 1, \ldots, N$, generated from the posterior probability distribution $p(\boldsymbol{\theta}|D, M)$. Specifically, the integrals (7.7) and (7.8), respectively, can be approximated by the sample estimates

$$p(q|M) \approx \frac{1}{N} \sum_{i=1}^{N} p(q|\boldsymbol{\theta}_i, M) \tag{7.14}$$

$$m_k(D, M) \approx \frac{1}{N} \sum_{i=1}^{N} [q(\boldsymbol{\theta}_i)]^k \tag{7.15}$$

The sample estimates (7.14) and (7.15) require independent forward system simulations that can be executed in a perfectly parallel fashion.

For rare events, the subset simulation [4] is computationally the most efficient sampling algorithm to provide an accurate estimate of the multi-dimensional failure probability integral (7.10) with the lowest number of samples. The subset simulation was first introduced to handle the conditional probability of failure integrals $P_F(M)$ formulated by (7.10) with $\boldsymbol{\zeta} = z$ and then the robust prior reliability integral (7.10) with $\boldsymbol{\zeta} = (z, \boldsymbol{\theta})$ and $p(\boldsymbol{\zeta}|M) = p(z|M)\pi(\boldsymbol{\theta}|M)$. Certain improvements on the MCMC sampling within subset simulation have recently been proposed by Papaioannou et al. [55]. In Jensen et al. [34], subset simulation was extended to treat the robust posterior reliability integrals of the form (7.10) with $\boldsymbol{\zeta} = (\boldsymbol{z}, \boldsymbol{\theta})$ and $p(\boldsymbol{\zeta}|M) = p(\boldsymbol{\zeta}|D, M)$, the posterior PDF. It should be noted that, usually, due to independence between $\boldsymbol{z}$ and $\boldsymbol{\theta}$, the PDF of $\boldsymbol{\zeta}$ is $p(\boldsymbol{\zeta}|M) = p(\boldsymbol{z}|M)p(\boldsymbol{\theta}|D, M)$, which simplifies the evaluation of the integral with subset simulation [34]. Subset simulation is highly parallelizable, as indicated in previous chapters, and its parallel implementation for heterogeneous computer architectures is discussed in [30].

7.4 Implementation in Structural Dynamics

The implementation of the Bayesian inference framework in structural dynamics has been outlined separately for linear and nonlinear finite element models. For linear models, the inference can be based on experimentally identified modal frequencies and mode shape components at sensor locations. Usually, it is convenient to measure the vibration of the structure under operational conditions by placing sensors at various locations to measure only output response time histories. There are a number of techniques for estimating the modal frequencies and mode shapes. Notably, the Bayesian modal parameter estimation methods proposed in [6, 7] are intended for output-only vibration measurements. In addition to the most probable values of the modal characteristics, the uncertainty in these characteristics is also estimated and asymptotically approximated by Gaussian distributions. For nonlinear models, the inference cannot be based on modal properties.The formulation can be based directly on measured response time histories. Details in the implementation of the Bayesian framework for each case will be next presented. In all cases, one needs to develop the equation for the prediction error, which is used to formulate the likelihood function involved in the Bayes' theorem (7.3).

7.4.1 Likelihood Formulation for Linear Models Based on Modal Properties

The data D consist of the square of the modal frequencies, $\hat{\lambda}_r = \hat{\omega}_r^2$, and the mode shapes, $\hat{\boldsymbol{\phi}}_r \in R^{N_{0,r}}$, $r = 1, \ldots, m$, experimentally estimated using vibration measurements, where m is the number of identified modes and $N_{0,r}$ is the number of measured components for mode r. Let $\omega_r(\boldsymbol{\theta})$ and $\boldsymbol{\phi}_r(\boldsymbol{\theta}) \in R^{N_{0,r}}$ be the rth modal frequency and mode shape at $N_{0,r}$ DOFs, respectively, predicted by the linear finite element model for a given value of the model parameter set $\boldsymbol{\theta}$.

The formulation of the likelihood $p(D|\boldsymbol{\theta}, M)$ in (7.1) can be found in a number of published papers (e.g., [19, 26, 56, 57, 61, 65]). Without loss of generality and for demonstration purposes, two formulations are discussed here. Alternative formulations can easily be incorporated by noting that the whole framework depends on the model parameters through the modal properties involved in the likelihood function and not on the detailed form of the likelihood function that arises from alternative formulations.

7.4.1.1 Formulation 1

The prediction error equation for the rth modal frequency is introduced as

$$\hat{\omega}_r^2 = \tilde{\omega}_r^2(\boldsymbol{\theta}) + \varepsilon_{\lambda_r} \tag{7.16}$$

with

$$\tilde{\omega}_r^2(\boldsymbol{\theta}) = \omega_r^2(\boldsymbol{\theta}) \tag{7.17}$$

where ε_{λ_r} is the modal frequency error taken to be Gaussian with zero mean and standard deviation $\sigma_{\omega_r}\hat{\omega}_r$, and σ_{ω_r} is a prediction error parameter. The prediction error equation for the rth mode shape is

$$\hat{\boldsymbol{\phi}}_r = \tilde{\boldsymbol{\phi}}_r(\boldsymbol{\theta}) + \boldsymbol{\varepsilon}_{\phi_r} \tag{7.18}$$

with

$$\tilde{\boldsymbol{\phi}}_r(\boldsymbol{\theta}) = \beta_r(\boldsymbol{\theta})\boldsymbol{\phi}_r(\boldsymbol{\theta}) \tag{7.19}$$

where $\boldsymbol{\varepsilon}_{\phi_r}$ is the mode shape error taken to be Gaussian with zero mean and covariance matrix $diag(\sigma_{\phi_r}^2 ||\hat{\boldsymbol{\phi}}_r||^2)$, $\sigma_{\phi_r}^2$ is a prediction error parameter,

$$\beta_r(\boldsymbol{\theta}) = \hat{\boldsymbol{\phi}}_r^T \boldsymbol{\phi}_r(\boldsymbol{\theta}) / \|\boldsymbol{\phi}_r(\boldsymbol{\theta})\|^2 \tag{7.20}$$

is a normalization constant that guarantees that the measured mode shape $\hat{\boldsymbol{\phi}}_r$ at the measured DOFs is closest to the model mode shape $\tilde{\boldsymbol{\phi}}_r(\boldsymbol{\theta})$ predicted by the particular value of $\boldsymbol{\theta}$, and $|| \cdot ||^2$ represents the usual Euclidian norm. The structural model parameter set $\boldsymbol{\theta}$ is augmented to include the unknown prediction error parameters σ_{ω_r} and σ_{ϕ_r}.

The squares of the modal frequencies $\omega_r^2(\boldsymbol{\theta}) = \lambda_r(\boldsymbol{\theta})$ and the mode shape components $\boldsymbol{\phi}_r(\boldsymbol{\theta}) = \mathbf{L}_r\boldsymbol{\psi}_r(\boldsymbol{\theta}) \in R^{N_{0,r}}$ at the $N_{0,r}$ measured DOFs are computed from the full mode shapes $\boldsymbol{\psi}_r(\boldsymbol{\theta}) \in R^n$ that satisfy the eigenvalue problem

$$[\boldsymbol{K}(\boldsymbol{\theta}) - \lambda_r(\boldsymbol{\theta})\boldsymbol{M}(\boldsymbol{\theta})]\boldsymbol{\psi}_r(\boldsymbol{\theta}) = \mathbf{0} \tag{7.21}$$

where $\boldsymbol{K}(\boldsymbol{\theta}) \in R^{n\times n}$ and $\boldsymbol{M}(\boldsymbol{\theta}) \in R^{n\times n}$ are the stiffness and mass matrices, respectively, of the FE model of the structure, n is the number of model DOFs, and $\mathbf{L}_r \in R^{N_{0,r}\times n}$ is an observation matrix, usually comprised of zeros and ones, that maps the n model DOFs to the $N_{0,r}$ observed DOFs for mode r. For a model with a large number of DOFs, $N_{0,r} \ll n$.

Assuming equal variances $\sigma^2_{\omega_r^2} = \sigma^2$ for the model prediction error ε_{λ_r} of all modal frequencies and equal variances $\sigma^2_{\boldsymbol{\phi}_r} = \sigma^2/w$ for the model prediction error $\varepsilon_{\boldsymbol{\phi}_r}$ of all mode shapes, the likelihood function can then be readily obtained in the form

$$p(D|\boldsymbol{\theta}, M) = \frac{1}{\left(\sqrt{2\pi}\sigma\right)^{N_0}} \exp\left[-\frac{1}{2\sigma^2} J(\boldsymbol{\theta}, w)\right] \tag{7.22}$$

where $N_0 = m + \sum_{r=1}^{m} N_{0,r}$, w is a weighting factor and

$$J(\boldsymbol{\theta}, w) = J_1(\boldsymbol{\theta}) + w J_2(\boldsymbol{\theta}) \tag{7.23}$$

In (7.23) the following modal frequency residuals

$$J_1(\boldsymbol{\theta}) = \sum_{r=1}^{m} \varepsilon^2_{\lambda_r}(\boldsymbol{\theta}) = \sum_{r=1}^{m} \frac{\left[\lambda_r(\boldsymbol{\theta}) - \hat{\lambda}_r\right]^2}{\hat{\lambda}_r^2} \tag{7.24}$$

and mode shape residuals

$$J_2(\boldsymbol{\theta}) = \sum_{r=1}^{m} \varepsilon^2_{\boldsymbol{\phi}_r}(\boldsymbol{\theta}) = \sum_{r=1}^{m} \frac{\left\|\tilde{\boldsymbol{\phi}}_r(\boldsymbol{\theta}) - \hat{\boldsymbol{\phi}}_r\right\|^2}{\left\|\hat{\boldsymbol{\phi}}_r\right\|^2} \tag{7.25}$$

measure the discrepancies for the modal frequencies and mode shape components, respectively, between the identified modal data and the model-predicted modal data. Using (7.19) and the fact that $\boldsymbol{\phi}_r(\boldsymbol{\theta}) = \mathbf{L}_r \boldsymbol{\psi}_r(\boldsymbol{\theta})$, the mode shape vector $\tilde{\boldsymbol{\phi}}_r(\boldsymbol{\theta})$ in (7.25) takes the form

$$\tilde{\boldsymbol{\phi}}_r(\boldsymbol{\theta}) = \mathbf{L}_r \boldsymbol{\psi}_r(\boldsymbol{\theta}) \beta_r(\boldsymbol{\theta}) \tag{7.26}$$

where, using (7.20), $\beta_r(\boldsymbol{\theta})$ is given by

$$\beta_r(\boldsymbol{\theta}) = \frac{\left[\mathbf{L}_r \boldsymbol{\psi}_r(\boldsymbol{\theta})\right]^T \hat{\boldsymbol{\phi}}_r}{\left\|\mathbf{L}_r \boldsymbol{\psi}_r(\boldsymbol{\theta})\right\|^2} \tag{7.27}$$

It can be shown that the square of the mode shape residuals in (7.25) is related to the modal assurance criterion (MAC) value of the mode r by [54]

$$\varepsilon^2_{\phi_r}(\boldsymbol{\theta}) = 1 - MAC^2_r(\boldsymbol{\theta}) = 1 - \left[\frac{[\mathbf{L}_r\boldsymbol{\psi}_r]^T\hat{\boldsymbol{\phi}}_r}{\|\mathbf{L}_r\boldsymbol{\psi}_r\|\,\|\hat{\boldsymbol{\phi}}_r\|}\right]^2 \geq 0 \tag{7.28}$$

since $0 \leq MAC_r \leq 1$. Thus, $\varepsilon^2_{\phi_r}(\boldsymbol{\theta})$ in (7.25) is also a measure of the distance of the square MAC-value from one, or equivalently, a measure of the correlation of the model-predicted mode shape and the measured mode shape for mode r.

One issue to keep in mind is that, to apply the formulation, one needs to estimate the correspondence between the experimental and model-predicted modes. This correspondence accounts for the fact that some modes of the system are not identified experimentally and also for the fact that the model-predicted modes may switch order for different values of the model parameters. Although this correspondence can be predicted in a number of cases using the MAC-values between experimentally identified and model-predicted mode shapes, there are a number of cases where the procedure may completely fail. For example, for modal frequencies with an algebraic multiplicity of greater than one, the mode shapes span a higher than one-dimensional subspace. In this case, an experimentally identified mode shape may be a linear combination of model-predicted mode shapes. This situation arises in practice when the modes are very closely spaced. In this case, the experimental mode shape is not close to one of the closely spaced mode shapes. To properly account for the prediction error in this case, an experimental mode shape has to be compared to the best linear combination of the closely spaced mode shapes of the FE model. The prediction error equation (7.18) can be modified to account for this case as described in the next section.

7.4.1.2 Formulation 2

The formulation presented in the current subsection generalizes the prediction error equation (7.18) to avoid the need for mode correspondence, as well accounts for the case of closely spaced modes. This is achieved by letting $\tilde{\boldsymbol{\phi}}_r(\boldsymbol{\theta})$ in Eq. (7.26) be a linear combination of the FE model mode shapes

$$\tilde{\boldsymbol{\phi}}_r(\boldsymbol{\theta}) = \sum_{k=1}^{m} \beta_{kr}(\boldsymbol{\theta})\boldsymbol{\phi}_k(\boldsymbol{\theta}) = \mathbf{L}_r\boldsymbol{\Psi}_r(\boldsymbol{\theta})\boldsymbol{\beta}_r(\boldsymbol{\theta}) \tag{7.29}$$

where $\boldsymbol{\Psi}_r(\boldsymbol{\theta}) = \{\boldsymbol{\psi}_1(\boldsymbol{\theta}), \ldots, \boldsymbol{\psi}_m(\boldsymbol{\theta})\}$, and selecting the vector of coefficients $\boldsymbol{\beta}_r(\boldsymbol{\theta}) = [\beta_{1r}(\boldsymbol{\theta}), \ldots, \beta_{mr}(\boldsymbol{\theta})]^T$, so that the distance $\left\|\hat{\boldsymbol{\phi}}_r - \tilde{\boldsymbol{\phi}}_r(\boldsymbol{\theta})\right\|$ is minimal, guaranteeing that the measured mode shape $\hat{\boldsymbol{\phi}}_r$ is closest to the model-predicted mode shape $\tilde{\boldsymbol{\phi}}_r(\boldsymbol{\theta})$. This minimum is achieved by selecting

$$\boldsymbol{\beta}_r(\boldsymbol{\theta}) = \left[\boldsymbol{\Psi}^T_r(\boldsymbol{\theta})\mathbf{L}^T_r\mathbf{L}_r\boldsymbol{\Psi}_r(\boldsymbol{\theta})\right]^{-1}[\mathbf{L}_r\boldsymbol{\Psi}_r(\boldsymbol{\theta})]^T\hat{\boldsymbol{\phi}}_r \tag{7.30}$$

Also, the square of the modal frequencies, $\tilde{\lambda}_r$, is evaluated from the Rayleigh quotient

$$\tilde{\lambda}_r(\boldsymbol{\theta}) = \tilde{\omega}_r^2(\boldsymbol{\theta}) = \frac{\tilde{\boldsymbol{\psi}}_r^T(\boldsymbol{\theta})\boldsymbol{K}(\boldsymbol{\theta})\tilde{\boldsymbol{\psi}}_r(\boldsymbol{\theta})}{\tilde{\boldsymbol{\psi}}_r^T(\boldsymbol{\theta})\boldsymbol{M}(\boldsymbol{\theta})\tilde{\boldsymbol{\psi}}_r(\boldsymbol{\theta})} = \frac{\boldsymbol{\beta}_r^T(\boldsymbol{\theta})\boldsymbol{\Lambda}(\boldsymbol{\theta})\boldsymbol{\beta}_r(\boldsymbol{\theta})}{\boldsymbol{\beta}_r^T(\boldsymbol{\theta})\boldsymbol{\beta}_r(\boldsymbol{\theta})} \tag{7.31}$$

where $\tilde{\boldsymbol{\psi}}_r$ is defined similarly to (7.29) as $\tilde{\boldsymbol{\psi}}_r = \left[\boldsymbol{\psi}_1(\boldsymbol{\theta}), \ldots, \boldsymbol{\psi}_m(\boldsymbol{\theta})\right]\boldsymbol{\beta}_r(\boldsymbol{\theta})$.

A similar version of this formulation was originally presented in [61]. Formulation 2 coincides with Formulation 1 for the case $\beta_{kr} = \beta_r\delta_{kr}$, where δ_{kr} is the Kronecker delta. This occurs for a sufficiently large number of sensors where the model mode shapes are obtained to be approximately orthogonal, resulting in a MAC-value between non-corresponding mode shapes closer to zero and, thus, coefficients that follow the aforementioned condition. For a small number of sensors, there might be more than one mode shape that significantly contributes to the model mode shape $\tilde{\boldsymbol{\phi}}_r(\boldsymbol{\theta})$, resulting in discrepancies between Formulations 1 and 2. As a special case, one can use (7.29) to associate an experimentally identified mode shape of closely spaced modes as a linear combination of the corresponding closely spaced modes predicted by the FE model. This, however, requires the identification of the corresponding closely spaced modes using the MAC-value. Further discussion on this issue falls outside the scope of this monograph.

Extensions of Formulation 2 can be found in [63]. Although these extensions seem to suggest that there is no need to compute the modal properties, there are a number of issues that may deteriorate the performance of the extended formulations. One such issue is the multi-modality of the posterior distribution when the number of sensors is not sufficiently large. This multi-modality significantly deteriorates the effectiveness of the FE model updating method.

7.4.1.3 Formulations Using Model Reduction

Bayesian tools for estimating the parameters of FE models based on modal properties require repeated eigenvalue analyses to be performed for a moderate to very large number of $\boldsymbol{\theta}$ values. Thus, the computational demands highly depend on the time required to build the FE model and perform the eigenvalue analysis. For FE models with a large number of DOFs, this can substantially increase the computational effort to excessive levels. The fast, accurate, and parameterization-consistent model reduction techniques that have been developed in this book can be integrated with the Bayesian techniques to substantially reduce the model. In such schemes, the computationally expensive repeated solutions of the component eigenvalue problems are completely avoided. By solving substantially reduced eigenvalue problems and avoiding the construction of the reduced matrices, drastic reductions in computational demands are achieved without compromising the solution accuracy [33, 49].

The model reduction methods are applicable to both asymptotic and stochastic simulation tools used in the Bayesian framework. The formulation for integrating model reduction techniques into the Bayesian algorithm will be next described.

Using the transformation $\boldsymbol{u}(t) = \mathbf{T}_G\bar{\mathbf{T}}(\boldsymbol{\theta})\bar{\mathbf{q}}(t)$ in (1.96), the physical mode shapes $\boldsymbol{\psi}_r(\boldsymbol{\theta})$ of the unreduced model can be written in terms of the generalized mode shapes $\boldsymbol{\varphi}_r(\boldsymbol{\theta})$ of the reduced model through the equation (1.99) as follows

$$\boldsymbol{\psi}_r(\boldsymbol{\theta}) = \mathbf{T}_G\bar{\mathbf{T}}(\boldsymbol{\theta})\boldsymbol{\varphi}_r(\boldsymbol{\theta}) \tag{7.32}$$

where the matrices $\mathbf{T}_G$ and $\bar{\mathbf{T}}(\boldsymbol{\theta})$ are defined in Chap. 1. The eigenvectors $\boldsymbol{\varphi}_r(\boldsymbol{\theta})$ and the eigenvalues $\bar{\omega}_r^2(\boldsymbol{\theta})$ of the reduced finite element model are given by solving the reduced eigenvalue problem

$$[\bar{\mathbf{K}}(\boldsymbol{\theta}) - \bar{\omega}_r^2(\boldsymbol{\theta})\bar{\mathbf{M}}(\boldsymbol{\theta})]\boldsymbol{\varphi}_r(\boldsymbol{\theta}) = \mathbf{0} \tag{7.33}$$

where $\bar{\mathbf{M}}(\boldsymbol{\theta})$ and $\bar{\mathbf{K}}(\boldsymbol{\theta})$ are the reduced mass and stiffness matrices. Equation (7.33) is exactly the same as Eq. (1.100). The dependence of the matrices $\bar{\mathbf{M}}(\boldsymbol{\theta})$ and $\bar{\mathbf{K}}(\boldsymbol{\theta})$ on $\boldsymbol{\theta}$ is fully specified in the formulation presented in Chaps. 2 and 3 for the different model reduction cases. The advantage of the model reduction is that the evaluation of the reduced matrices $\bar{\mathbf{M}}(\boldsymbol{\theta})$, $\bar{\mathbf{K}}(\boldsymbol{\theta})$ and $\bar{\mathbf{T}}(\boldsymbol{\theta})$ does not require repeating online the reduction procedure that was outlined in Chaps. 1 and 2 for different values of the parameter set $\boldsymbol{\theta}$. These matrices are given by efficient expansions at the space of reduced generalized coordinates, which are exact for the parameterized schemes examined in Chap. 2.

Substituting the form of $\boldsymbol{\psi}_r(\boldsymbol{\theta})$ given in Eq. (7.32) into $\boldsymbol{\phi}_r(\boldsymbol{\theta}) = \mathbf{L}_r\boldsymbol{\psi}_r(\boldsymbol{\theta})$, the mode shape components of mode r at the measured DOFs are finally obtained in the form

$$\boldsymbol{\phi}_r(\boldsymbol{\theta}) = \bar{\mathbf{L}}_r(\boldsymbol{\theta})\boldsymbol{\varphi}_r(\boldsymbol{\theta}) \tag{7.34}$$

where

$$\bar{\mathbf{L}}_r(\boldsymbol{\theta}) = \mathbf{L}_r\mathbf{T}_G\bar{\mathbf{T}}(\boldsymbol{\theta}) \tag{7.35}$$

Thus, in the case where model reduction is used in Formulation 1, the square of the modal frequencies and the mode shapes involved in the objective functions (7.24) and (7.25) have exactly the same form as in (7.17), (7.26), and (7.27) defined for the unreduced models, with $\omega_r^2(\boldsymbol{\theta})$ in (7.17), $\boldsymbol{\psi}_r(\boldsymbol{\theta})$ in (7.26), and the constant matrix $\mathbf{L}_r$ in (7.26) being replaced by $\bar{\omega}_r^2(\boldsymbol{\theta})$, $\boldsymbol{\varphi}_r(\boldsymbol{\theta})$ and the parameter-dependent matrix $\bar{\mathbf{L}}_r(\boldsymbol{\theta}) = \mathbf{L}_r\mathbf{T}_G\bar{\mathbf{T}}(\boldsymbol{\theta})$, respectively. Available model updating formulations and software can, thus, be used to handle the Bayesian parameter estimation by merely replacing the eigenvalue problem (7.21) of the original unreduced mass and stiffness matrices with the eigenvalue problem (7.33) of the reduced mass and stiffness matrices, as well as replacing the constant matrices $\mathbf{L}_r$ of zeros and ones by the parameter-dependent matrices $\bar{\mathbf{L}}_r(\boldsymbol{\theta}) = \mathbf{L}_r\mathbf{T}_G\bar{\mathbf{T}}(\boldsymbol{\theta})$.

Using model reduction, a similar simplification holds for Formulation 2. Specifically, using $\boldsymbol{\psi}_r(\boldsymbol{\theta})$ from (7.32), the matrix of eigenvectors takes the form

$$\boldsymbol{\Psi}_r(\boldsymbol{\theta}) = \{\boldsymbol{\psi}_1(\boldsymbol{\theta}), \ldots, \boldsymbol{\psi}_m(\boldsymbol{\theta})\} = \mathbf{T}_G \bar{\mathbf{T}}(\boldsymbol{\theta}) \bar{\boldsymbol{\Psi}}_r(\boldsymbol{\theta}) \tag{7.36}$$

where

$$\bar{\boldsymbol{\Psi}}_r(\boldsymbol{\theta}) = \{\varphi_1(\boldsymbol{\theta}), \ldots, \varphi_m(\boldsymbol{\theta})\} \tag{7.37}$$

is the matrix of the mode shapes corresponding to the reduced model. Substituting (7.36) into (7.29) and (7.30), one readily derives that

$$\tilde{\boldsymbol{\phi}}_r(\boldsymbol{\theta}) = \bar{\mathbf{L}}_r(\boldsymbol{\theta}) \bar{\boldsymbol{\Psi}}_r(\boldsymbol{\theta}) \boldsymbol{\beta}_r(\boldsymbol{\theta}) \tag{7.38}$$

where

$$\boldsymbol{\beta}_r(\boldsymbol{\theta}) = \left[\bar{\boldsymbol{\Psi}}_r^T(\boldsymbol{\theta}) \bar{\mathbf{L}}_r^T(\boldsymbol{\theta}) \bar{\mathbf{L}}_r(\boldsymbol{\theta}) \bar{\boldsymbol{\Psi}}_r(\boldsymbol{\theta})\right]^{-1} [\bar{\mathbf{L}}_r(\boldsymbol{\theta}) \bar{\boldsymbol{\Psi}}_r(\boldsymbol{\theta})]^T \hat{\boldsymbol{\phi}}_r \tag{7.39}$$

Similarly, using $\boldsymbol{\psi}_r(\boldsymbol{\theta})$ from (7.32), the vector $\tilde{\boldsymbol{\psi}}_r(\boldsymbol{\theta})$ defined in (7.31) has the form

$$\tilde{\boldsymbol{\psi}}_r(\boldsymbol{\theta}) = \mathbf{T}_G \bar{\mathbf{T}}(\boldsymbol{\theta}) \bar{\boldsymbol{\Psi}}_r(\boldsymbol{\theta}) \boldsymbol{\beta}_r(\boldsymbol{\theta}) \tag{7.40}$$

Taking into account the reduced eigenvalue problem (7.33) and the fact that

$$\begin{aligned}
\tilde{\boldsymbol{\psi}}_r^T(\boldsymbol{\theta}) \mathbf{K}(\boldsymbol{\theta}) \tilde{\boldsymbol{\psi}}_r(\boldsymbol{\theta}) &= \boldsymbol{\beta}_r^T(\boldsymbol{\theta}) \bar{\boldsymbol{\Psi}}_r^T(\boldsymbol{\theta}) [\mathbf{T}_G \bar{\mathbf{T}}(\boldsymbol{\theta})]^T \boldsymbol{K}(\boldsymbol{\theta}) [\mathbf{T}_G \bar{\mathbf{T}}(\boldsymbol{\theta})] \bar{\boldsymbol{\Psi}}_r(\boldsymbol{\theta}) \boldsymbol{\beta}_r(\boldsymbol{\theta}) \\
&= \boldsymbol{\beta}_r^T(\boldsymbol{\theta}) \bar{\boldsymbol{\Psi}}_r^T(\boldsymbol{\theta}) \bar{\boldsymbol{K}}(\boldsymbol{\theta}) \bar{\boldsymbol{\Psi}}_r(\boldsymbol{\theta}) \boldsymbol{\beta}_r(\boldsymbol{\theta}) \\
&= \boldsymbol{\beta}_r^T(\boldsymbol{\theta}) \bar{\boldsymbol{\Lambda}}_r(\boldsymbol{\theta}) \boldsymbol{\beta}_r(\boldsymbol{\theta})
\end{aligned} \tag{7.41}$$

and similarly

$$\begin{aligned}
\tilde{\boldsymbol{\psi}}_r^T(\boldsymbol{\theta}) \mathbf{M}(\boldsymbol{\theta}) \tilde{\boldsymbol{\psi}}_r(\boldsymbol{\theta}) &= \boldsymbol{\beta}_r^T(\boldsymbol{\theta}) \bar{\boldsymbol{\Psi}}_r^T(\boldsymbol{\theta}) [\mathbf{T}_G \bar{\mathbf{T}}(\boldsymbol{\theta})]^T \boldsymbol{M}(\boldsymbol{\theta}) [\mathbf{T}_G \bar{\mathbf{T}}(\boldsymbol{\theta})] \bar{\boldsymbol{\Psi}}_r(\boldsymbol{\theta}) \boldsymbol{\beta}_r(\boldsymbol{\theta}) \\
&= \boldsymbol{\beta}_r^T(\boldsymbol{\theta}) \bar{\boldsymbol{\Psi}}_r^T(\boldsymbol{\theta}) \bar{\boldsymbol{M}}(\boldsymbol{\theta}) \bar{\boldsymbol{\Psi}}_r(\boldsymbol{\theta}) \boldsymbol{\beta}_r(\boldsymbol{\theta}) \\
&= \boldsymbol{\beta}_r^T(\boldsymbol{\theta}) \boldsymbol{\beta}_r(\boldsymbol{\theta})
\end{aligned} \tag{7.42}$$

the eigenvalues $\tilde{\lambda}_r(\boldsymbol{\theta}) = \tilde{\omega}_r^2(\boldsymbol{\theta})$ are provided by the last expression in Eq. (7.31), with the matrix of eigenvalues $\boldsymbol{\Lambda}_r(\boldsymbol{\theta})$ replaced by the matrix of eigenvalues $\bar{\boldsymbol{\Lambda}}_r(\boldsymbol{\theta})$, computed by solving the eigenvalue problem (7.33) for the reduced finite element model.

It should be noted that, in the special case of reducing only the internal DOFs for each component and assuming that the mass matrix of each substructure is independent of $\boldsymbol{\theta}$, the transformation matrix $\bar{\mathbf{T}}(\boldsymbol{\theta})$ takes the form $\bar{\mathbf{T}}(\boldsymbol{\theta}) = \mathbf{T}_D$, which is independent of $\boldsymbol{\theta}$ and, thus, the matrix $\bar{\mathbf{L}}_r(\boldsymbol{\theta})$ in (7.35) simplifies to a constant matrix

$$\bar{\mathbf{L}}_r = \mathbf{L}_r \mathbf{T}_G \bar{\mathbf{T}}_D \tag{7.43}$$

independent of the parameter set $\boldsymbol{\theta}$ (see Chap. 2).

7.4.1.4 Gradient Estimation Using Adjoint Techniques

For Bayesian asymptotic approximations, first-order and second-order adjoint techniques are available [47] using the Nelson's method [46] to efficiently compute the required first- and second-order sensitivities in the optimization problems and the Hessian. In contrast to the Fox and Kapoor [23] method for estimating sensitivities, in Nelson's method, the gradients of the modal frequencies and the mode shape vector of a specific mode are computed only from the value of the modal frequency and the mode shape vector of the same mode, independently of the values of the modal frequencies and mode shape vectors of the rest of the modes. For structural model classes with a large number of degrees of freedom and very few contributing modes, this representation of the gradients clearly presents significant computational advantages over methods that represent mode shape gradients as a weighted, usually arbitrarily truncated, sum of all system mode shape vectors [23].

Following [47] and using the formulation for the unreduced system matrices, the gradient of the square error $\varepsilon^2_{\lambda_r}(\boldsymbol{\theta})$ in (7.24) (Formulation 1) is given by

$$\begin{aligned}\frac{\partial \varepsilon^2_{\lambda_r}(\boldsymbol{\theta})}{\partial \theta_j} &= \frac{\partial \varepsilon^2_{\lambda_r}(\boldsymbol{\theta})}{\partial \lambda_r}\frac{\partial \lambda_r}{\partial \theta_j} \\ &= \left[\frac{\partial \varepsilon^2_{\lambda_r}(\boldsymbol{\theta})}{\partial \lambda_r}\boldsymbol{\psi}_r^T(\boldsymbol{\theta})\right]\left[\mathbf{K}_j(\boldsymbol{\theta}) - \omega_r^2 \mathbf{M}_j(\boldsymbol{\theta})\right]\boldsymbol{\psi}_r(\boldsymbol{\theta}) \end{aligned}\tag{7.44}$$

and the gradient of the square error $\varepsilon^2_{\boldsymbol{\phi}_r}(\boldsymbol{\theta})$ in (7.25) (Formulation 1) is given by

$$\begin{aligned}\frac{\partial \varepsilon^2_{\boldsymbol{\phi}_r}(\boldsymbol{\theta})}{\partial \theta_j} &= \nabla^T_{\boldsymbol{\phi}_r}\varepsilon^2_{\boldsymbol{\phi}_r}(\boldsymbol{\theta})\frac{\partial \boldsymbol{\phi}_r}{\partial \theta_j} = \nabla^T_{\boldsymbol{\phi}_r}\varepsilon^2_{\boldsymbol{\phi}_r}(\boldsymbol{\theta})\frac{\partial\left(\mathbf{L}_r\boldsymbol{\psi}_r(\boldsymbol{\theta})\right)}{\partial \theta_j} \\ &= \nabla^T_{\boldsymbol{\phi}_r}\varepsilon^2_{\boldsymbol{\phi}_r}(\boldsymbol{\theta})\mathbf{L}_r\frac{\partial \boldsymbol{\psi}_r}{\partial \theta_j} \\ &= -\left[\mathbf{x}_r^T(\boldsymbol{\theta})(\mathbf{I} - \mathbf{M}\boldsymbol{\psi}_r(\boldsymbol{\theta})\boldsymbol{\psi}_r^T(\boldsymbol{\theta}))\right]\left[\mathbf{K}_j(\boldsymbol{\theta}) - \omega_r^2\mathbf{M}_j(\boldsymbol{\theta})\right]\boldsymbol{\psi}_r(\boldsymbol{\theta}) \end{aligned}\tag{7.45}$$

where $\mathbf{x}_r(\boldsymbol{\theta})$ is given by the solution of the linear system of equations

$$\mathbf{A}_r^*(\boldsymbol{\theta})\mathbf{x}_r(\boldsymbol{\theta}) = \mathbf{L}_r^T \nabla_{\boldsymbol{\phi}_r}\varepsilon^2_{\boldsymbol{\phi}_r}(\boldsymbol{\theta}) \tag{7.46}$$

with

$$\frac{\partial \varepsilon_{\omega_r}^2(\boldsymbol{\theta})}{\partial \omega_r^2} = \frac{2\varepsilon_{\omega_r}(\boldsymbol{\theta})}{\hat{\omega}_r^2} \tag{7.47}$$

$$\nabla_{\boldsymbol{\phi}_r}^T \varepsilon_{\boldsymbol{\phi}_r}^2(\boldsymbol{\theta}) = \frac{2\beta_r(\boldsymbol{\theta})}{\left\|\hat{\boldsymbol{\phi}}_r\right\|^2}\left[\beta_r(\boldsymbol{\theta})\boldsymbol{\phi}_r(\boldsymbol{\theta}) - \hat{\boldsymbol{\phi}}_r\right] \tag{7.48}$$

while $\mathbf{K}_j(\boldsymbol{\theta}) \equiv \partial \mathbf{K}(\boldsymbol{\theta})/\partial \theta_j$ and $\mathbf{M}_j \equiv \partial \mathbf{M}/\partial \theta_j$. In (7.46), the matrix $\mathbf{A}_r^*(\boldsymbol{\theta})$ is used to denote the modified matrix derived from $\mathbf{A}_r(\boldsymbol{\theta}) = \mathbf{K}(\boldsymbol{\theta}) - \omega_r^2\mathbf{M}(\boldsymbol{\theta})$ by replacing the elements of the kth column and the kth row by zeroes and the (k, k) element of $\mathbf{A}_r$ by one, where k denotes the element of the mode shape vector $\boldsymbol{\psi}_r(\boldsymbol{\theta})$ with the highest absolute value [46].

Specifically, applying the formulation for the reduced model, the gradient of the square error $\varepsilon_{\lambda_r}^2(\boldsymbol{\theta})$ is simplified to

$$\begin{aligned}\frac{\partial \varepsilon_{\lambda_r}^2(\boldsymbol{\theta})}{\partial \theta_j} &= \frac{\partial \varepsilon_{\lambda_r}^2(\boldsymbol{\theta})}{\partial \lambda_r}\frac{\partial \lambda_r}{\partial \theta_j} \\ &= \left[\frac{\partial \varepsilon_{\lambda_r}^2(\boldsymbol{\theta})}{\partial \lambda_r}\boldsymbol{\varphi}_r^T(\boldsymbol{\theta})\right]\left[\bar{\mathbf{K}}_j(\boldsymbol{\theta}) - \omega_r^2\bar{\mathbf{M}}_j(\boldsymbol{\theta})\right]\boldsymbol{\varphi}_r(\boldsymbol{\theta})\end{aligned} \tag{7.49}$$

while the gradient of the square error $\varepsilon_{\boldsymbol{\phi}_r}^2(\boldsymbol{\theta})$ in (7.25) is simplified to

$$\begin{aligned}\frac{\partial \varepsilon_{\boldsymbol{\phi}_r}^2(\boldsymbol{\theta})}{\partial \theta_j} &= \nabla_{\boldsymbol{\phi}_r}^T \varepsilon_{\boldsymbol{\phi}_r}^2(\boldsymbol{\theta})\frac{\partial \boldsymbol{\phi}_r}{\partial \theta_j} = \nabla_{\boldsymbol{\phi}_r}^T \varepsilon_{\boldsymbol{\phi}_r}^2(\boldsymbol{\theta})\frac{\partial\left(\bar{\mathbf{L}}_r(\boldsymbol{\theta})\boldsymbol{\varphi}_r(\boldsymbol{\theta})\right)}{\partial \theta_j} \\ &= \nabla_{\boldsymbol{\phi}_r}^T \varepsilon_{\boldsymbol{\phi}_r}^2(\boldsymbol{\theta})\frac{\partial \bar{\mathbf{L}}_r(\boldsymbol{\theta})}{\partial \theta_j}\boldsymbol{\varphi}_r(\boldsymbol{\theta}) + \nabla_{\boldsymbol{\phi}_r}^T \varepsilon_{\boldsymbol{\phi}_r}^2(\boldsymbol{\theta})\bar{\mathbf{L}}_r(\boldsymbol{\theta})\frac{\partial \boldsymbol{\varphi}_r(\boldsymbol{\theta})}{\partial \theta_j} \\ &= \left\{\nabla_{\boldsymbol{\phi}_r}^T \varepsilon_{\boldsymbol{\phi}_r}^2(\boldsymbol{\theta})\mathbf{L}_r\mathbf{T}_G\right\}\frac{\partial \bar{\mathbf{T}}(\boldsymbol{\theta})}{\partial \theta_j}\boldsymbol{\varphi}_r(\boldsymbol{\theta}) - \\ &\quad \left[\bar{\mathbf{x}}_r^T(\boldsymbol{\theta})(\bar{\mathbf{I}} - \bar{\mathbf{M}}\boldsymbol{\varphi}_r(\boldsymbol{\theta})\boldsymbol{\varphi}_r^T(\boldsymbol{\theta}))\right]\left[\bar{\mathbf{K}}_j(\boldsymbol{\theta}) - \bar{\omega}_r^2\bar{\mathbf{M}}_j(\bar{\boldsymbol{\theta}})\right]\boldsymbol{\varphi}_r(\boldsymbol{\theta})\end{aligned} \tag{7.50}$$

where $\bar{\mathbf{x}}_r(\boldsymbol{\theta})$ is given by the solution of the reduced linear system of equations

$$\bar{\mathbf{A}}_r^*(\boldsymbol{\theta})\bar{\mathbf{x}}_r(\boldsymbol{\theta}) = \bar{\mathbf{L}}_r^T(\boldsymbol{\theta})\nabla_{\boldsymbol{\phi}_r}\varepsilon_{\boldsymbol{\phi}_r}^2(\boldsymbol{\theta}) \tag{7.51}$$

while $\bar{\mathbf{K}}_j(\boldsymbol{\theta}) \equiv \partial \bar{\mathbf{K}}(\boldsymbol{\theta})/\partial \theta_j$ and $\bar{\mathbf{M}}_j(\boldsymbol{\theta}) \equiv \partial \bar{\mathbf{M}}(\boldsymbol{\theta})/\partial \theta_j$. In (7.51), the matrix $\bar{\mathbf{A}}_r^*(\boldsymbol{\theta})$ is used to denote the modified matrix derived from $\bar{\mathbf{A}}_r(\boldsymbol{\theta}) = \bar{\mathbf{K}}(\boldsymbol{\theta}) - \bar{\omega}_r^2\bar{\mathbf{M}}(\boldsymbol{\theta})$ by replacing the elements of the kth column and the kth row by zeroes and the (k, k) element of $\bar{\mathbf{A}}_r(\boldsymbol{\theta})$ by one, where k denotes the element of the mode shape vector $\boldsymbol{\varphi}_r$ with the highest absolute value [46]. Significant computational savings are obtained from the fact that the linear system (7.46) is reduced to the linear system (7.51).

The computation of the derivatives of the square errors for the modal properties of the rth mode with respect to the parameters in $\boldsymbol{\theta}$ requires only one solution of the reduced linear system (7.51), independent of the number of parameters in $\boldsymbol{\theta}$. For a large number of parameters in the set $\boldsymbol{\theta}$, the above formulation for the gradient of the mean error in modal frequencies given in (7.44) and the gradient of the mean error of the mode shape components in (7.45) are computationally efficient and informative. The dependence on θ_j comes through the term $\mathbf{K}_j$ and the term $\mathbf{M}_j$. For the case where the mass matrix is independent of $\boldsymbol{\theta}$, $\mathbf{M}_j = 0$, and the formulation is further simplified. The end result of the proposed adjoint method is the solution of as many linear systems of equations as the number of model-predicted modes. The size of the linear systems equals the number of DOFs of the structural model, which adds to the computational burden. However, the linear systems are independent of each other and can be carried out in parallel, significantly accelerating the time-to-solution. The integration of model reduction techniques with the adjoint methods can significantly reduce the size of the linear system (7.46) to the reduced size defined in (7.51) and, thus, substantially reduce the computational effort.

It should be noted that a similar analysis is available to obtain the Hessian of the objective functions $\varepsilon^2_{\omega_r}(\boldsymbol{\theta})$ and $\varepsilon^2_{\varphi_r}(\boldsymbol{\theta})$ from the second derivatives of the eigenvalues and the eigenvectors, respectively. Details can be found in [47] and fall outside the scope of this monograph.

7.4.2 Likelihood Formulation Based on Response Time Histories

For nonlinear FE models, one can formulate the likelihood directly using response time histories. Modeling nonlinearities arises from various sources, such as material constitutive laws, contact, sliding, and impact between structural components, nonlinear isolation devices (e.g., nonlinear dampers) in civil infrastructure and nonlinear suspension models in vehicles. Often, the nonlinearities are localized in isolated parts of a structure, while the rest of the structure manifests a linear behavior. For example, in vehicles, localized nonlinearities are activated at the suspension mainly due to the nonlinear dampers, while the frame usually behaves linearly. In civil engineering applications, the nonlinearities are often localized at the joints and connections with isolation damper devices (e.g., bridges) designed to resist the dynamic loads through nonlinear dissipation, leaving the rest of the structure at its linear elastic range during system operation.

Consider the measured response time histories $D = \{\hat{x}_j(k) \in R,\ j = 1, \ldots, N_0,\ k = 1, \ldots, N_D\}$ at time instances $t = k\Delta t$, of N_0 response quantities (displacements, velocities, accelerations, and strains) at different points in the structure, where N_D is the number of the samples data using a sampling period Δt. The response time history predictions from the nonlinear model corresponding to a particular value of the parameter set $\boldsymbol{\theta}$ of the same quantities and points in the structure are denoted by $\{x_j(k, \boldsymbol{\theta}) \in R, j = 1, \ldots, N_0,\ k = 1, \ldots, N_D\}$. The likelihood is formulated by introducing the error $e_j(k)$ between the measured and the model-predicted response time histories, associated with the jth DOF and the time instant k, to satisfy the prediction error equation

$$\hat{x}_j(k) = x_j(k, \boldsymbol{\theta}) + e_j(k) \tag{7.52}$$

$j = 1, \ldots, N_0$ and $k = 1, \ldots, N_D$. The difference between the measured and model-predicted response is attributed to both experimental and modeling errors.

The prediction errors $e_j(k)$, measuring the fit between the measured and the model-predicted response time histories are modeled by zero-mean Gaussian distributions. In this work, it is assumed that the model prediction errors are uncorrelated in time and that the variances at different time instants are equal for all sampling data of the jth response time history, i.e., $e_j(k) \sim N(0, \sigma_j^2)$, $k = 1, \ldots, N_D$. Each measured time history is generally obtained from a different sensor (displacement, velocity, acceleration, or strain sensor) with a different accuracy and noise level, giving rise to as many prediction error variances σ_j^2 as the number of measured time histories. The prediction error parameters $\sigma_j, j = 1, \ldots, N_0$, are considered unknown and are included in the parameter set $\boldsymbol{\theta}$ to be estimated along with the structural model parameters.

The likelihood function $p(D|\boldsymbol{\theta}, M)$, which quantifies the probability of obtaining the data given a specific set of structural parameters and prediction error parameters, is derived by noting that the measured time histories $\hat{x}_j(k)$ are independent Gaussian variables with mean $x_j(k, \boldsymbol{\theta})$ and variance σ_j^2. Taking advantage of the independence of the measured quantities both at different time instants of the same time history as well as between different time histories, the likelihood takes the form

$$p(D|\boldsymbol{\theta}) = \prod_{j=1}^{N_0} \prod_{k=1}^{N_D} p\left(\hat{x}_j(k)\,|\boldsymbol{\theta}\right) \tag{7.53}$$

Substituting $p\left(\hat{x}_j(k)\,|\boldsymbol{\theta}\right)$ by a Gaussian PDF and rearranging terms, one obtains that

$$p(D|\boldsymbol{\theta}) = \frac{1}{\left(\sqrt{2\pi}\right)^{N_D N_0} \prod_{j=1}^{N_0} \sigma_j^{N_D}} \exp\left\{-\frac{N_0 N_D}{2} J(\boldsymbol{\theta})\right\} \tag{7.54}$$

where

$$J(\boldsymbol{\theta}) = \frac{1}{N_0} \sum_{j=1}^{N_0} \frac{1}{\sigma_j^2} J_j(\boldsymbol{\theta}) \tag{7.55}$$

The quantity

$$J_j(\boldsymbol{\theta}) = \frac{1}{N_D} \sum_{k=1}^{N_D} \left[\hat{x}_j(k) - x_j(k, \boldsymbol{\theta})\right]^2 \quad (7.56)$$

represents the measure of fit between the measured and the model-predicted response time history at the jth measured DOF.

Formulations of the likelihood for the case where full measured response time histories are available can be found in [13] for linear models and in [24, 25, 29, 33, 42, 43] for nonlinear models. The likelihood and the posterior of the parameters of a FE model are functions of the response time histories predicted by the FE model. Thus, each posterior evaluation requires a model simulation run involving the numerical integration of nonlinear equations of motion for the structure.

7.5 Numerical Examples

Two example problems are considered in this chapter. The objective of the first example is to evaluate the effectiveness of the proposed model reduction technique in a linear model identification problem using modal characteristics. The goal of the second application is to evaluate the performance of the proposed technique in a nonlinear model updating problem by employing dynamic response data.

7.5.1 Example 1: Updating of Linear Model

A high fidelity linear finite element model of the Metsovo bridge is used to demonstrate the accuracy and computational effectiveness of the proposed model reduction techniques when applied within the Bayesian parameter inference framework. The parameter estimation is based on modal characteristics identified from ambient vibration measurements. The derived updated model is considered as representative of the initial structural condition of the bridge, which can be further used for structural health monitoring purposes and for updating structural reliability and safety.

The description of the bridge, its instrumentation, and the procedure for experimentally estimating the modal frequencies and mode shapes are outlined in Sect. 7.5.1.1. Details for the high fidelity linear FE model of the bridge are presented in Sect. 7.5.1.2. The model components, the model parameterization, and the model reduction results are presented in Sect. 7.5.1.3, while the model parameter estimation results are outlined and discussed in Sects. 7.5.1.4, 7.5.1.5, and 7.5.1.6. Finally, computational issues are discussed in Sect. 7.5.1.7.

Fig. 7.1 Metsovo bridge

7.5.1.1 Bridge Description, Instrumentation, and Modal Identification

The Metsovo bridge, shown in Fig. 7.1, is made of reinforced concrete. The total length of the bridge deck is 537 m. The bridge has 4 spans of lengths 44.78 m, 117.87 m, 235 m, and 140 m, and 3 piers through which the left pier (45 m) supports the box beam superstructure through pot bearings (movable in both horizontal directions), while the central pier (110 m) and the right pier (35 m) monolithically connect to the structure. The total width of the deck is 13.95 m, for each carriageway. The superstructure is composed of a single box beam section of height varying from 4 m to 13.5 m. The central and the right piers are founded on huge circular 12 m-rock sockets at a depth of 25 and 15 m, respectively.

Acceleration measurements collected under normal operating conditions of the bridge were used to identify its modal properties (natural frequencies, mode shapes, damping ratios). The excitation of the bridge during the measurements was primarily due to road traffic, which ranged from motorcycles to heavy trucks, and environmental excitation, such as wind loading and ground micro-tremor. The measured data were collected using 5 triaxial and 3 uniaxial accelerometers paired with a 24-bit data logging system and an internal SD flash drive for data storage. The synchronization of the sensors was achieved by using a GPS module in each of the sensors. Thirteen different sensor configurations are used to cover the entire length of the deck, with the available limited number of sensor units and assemble mode shapes in

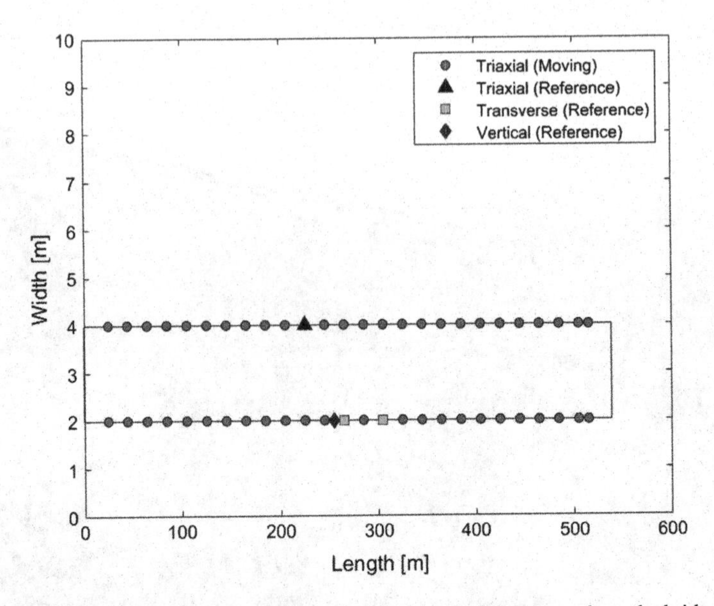

Fig. 7.2 Locations of moving and reference triaxial and uniaxial sensors along the bridge deck

as many as 159 vertical, transverse, and longitudinal directions using reference and moving sensors. The sensors are located approximately 20 m apart, so that the types of as many of the lowest modes as possible are correctly identified. All 159 sensor locations and measurement directions are shown in Fig. 7.2.

Details of the instrumentation and the identification of modal properties are given in [3]. For each sensor configuration, the measurements lasted 20 min with a sampling rate of 200 Hz. The modal properties are identified using a Bayesian modal identification methodology [6, 7]. The type and the mean values of the modal frequencies are shown in Table 7.1 for the lowest 20 modes. The full mode shapes of the bridge at all 105 sensor locations along the transverse and vertical directions of the bridge deck were assembled by combining the partial mode shapes identified for all 13 sensor configurations. Measurements along the longitudinal direction of the bridge deck are ignored since they will not be included in the model updating process. A mode shape assembling algorithm similar to [5, 62] was used. The mode shapes of the lowest 12 modes were successfully identified, except the 10th mode shape which was very poorly identified and excluded from the modal data. Representative assembled mode shapes are shown in Figs. 7.3 and 7.4, and compared with the mode shapes predicted by the nominal FE model of the bridge.

Table 7.1 Experimentally identified and nominal model predicted modal frequencies, and MAC-values (T: deck bending in transverse direction, V: deck bending in vertical direction)

Mode	Type	Experimental mean (Hz)	Nominal model (Hz)	MAC-values
1	T	0.306	0.293	0.998
2	T	0.603	0.574	0.922
3	V	0.623	0.619	0.953
4	T	0.965	0.849	0.986
5	V	1.047	1.050	0.919
6	T	1.139	1.070	0.997
7	V	1.428	1.388	0.983
8	T	1.697	1.578	0.985
9	V	2.005	1.690	0.061
10	V	2.303	1.966	–
11	T	2.367	2.156	0.891
12	T	2.590	2.316	0.858
13	–	2.723	2.500	–
14	–	3.086	2.745	–
15	–	3.127	2.815	–
16	–	3.480	2.876	–
17	–	3.861	2.950	–
18	–	4.059	3.320	–
19	–	4.210	3.381	–
20	–	4.410	3.520	–

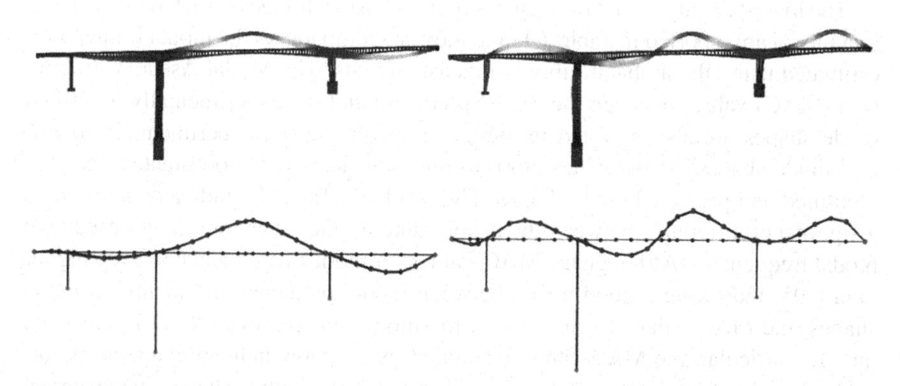

Fig. 7.3 Model-predicted and experimentally identified mode shapes for first bending - vertical (left) and third bending - vertical (right)

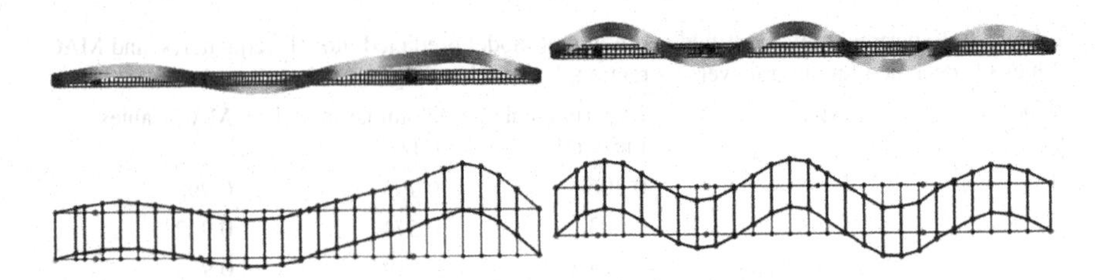

Fig. 7.4 Model-predicted and experimentally identified mode shapes for third transverse (left) and fifth transverse (right)

7.5.1.2 Linear Finite Element Model

A detailed FE model of the bridge is created using three-dimensional tetrahedron quadratic Lagrange finite elements. The nominal values of the modulus of elasticity of the deck and the three piers are selected to be the design values. To take the soil-structure interaction into account, large blocks of material are used to model the soil and embed the piers and abutments into these blocks. The nominal values of the soil stiffness are selected based on soil tests where a large uncertainty in soil stiffness is reported. The largest size of the elements in the mesh is of the order of the thickness of the deck cross-section. Several mesh sizes in the deck, piers, and soil blocks were explored, and an accuracy analysis was performed to find a reasonable trade-off between the number of degrees of freedom of the model and the accuracy in predicting the lowest 20 modal properties. A mesh of 830,115 DOFs was kept for the bridge-foundation-soil model of the structure. This mesh provides sufficiently accurate results for the lowest 20 modal frequencies, with errors of the order of 0.1–0.5% compared to the smallest possible mesh sizes involving approximately 3 million DOFs.

The lowest twenty modal frequencies of the Metsovo bridge predicted by the nominal model are reported in Table 7.1, and they are compared to the modal frequencies estimated using the ambient vibration measurements. The Modal Assurance Criterion (MAC) values between the model-predicted and the experimentally identified mode shapes are also reported in Table 7.1 for the eleven experimentally identified mode shapes. MAC-values close to one indicate a very good match between identified and predicted mode shapes. The results in Table 7.1 indicate that there is a significant difference between the experimentally identified and model-predicted modal frequencies. Although the MAC-values for modes 1, 3, 4 and 6 to 8 are higher than 0.95, indicating a good match between model-predicted and identified mode shapes, the MAC-values for the rest of the modes are significantly different from one. In particular, the MAC-value for mode 9 is very low indicating a lack of correspondence between the model-predicted and identified mode shape. The observed discrepancies between the experimentally identified modal properties and the ones predicted by the nominal FE model necessitate the use of FE model updating to calibrate the FE model parameters. In this manner, the predictive capabilities of the finite element model can be improved.

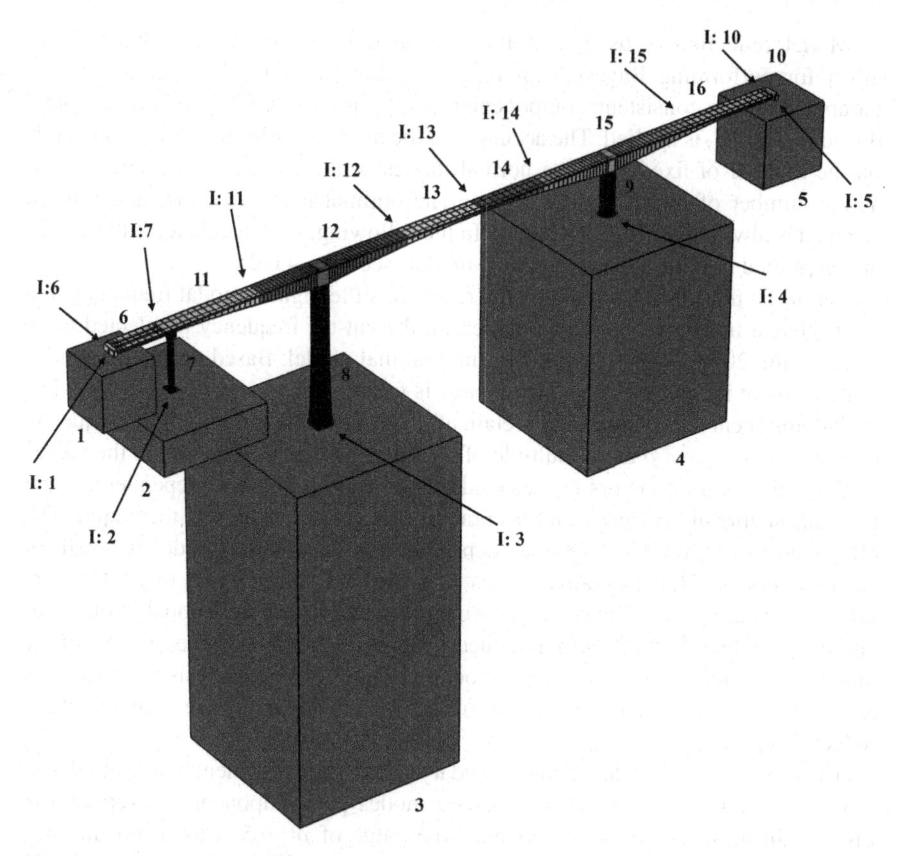

Fig. 7.5 Substructures and interfaces of FE model of the bridge

7.5.1.3 Reduction of the Finite Element Model

The nominal finite element model is used to check in detail the model reduction technique and its effectiveness in terms of size reduction and accuracy. To demonstrate the model reduction technique and its effectiveness, the bridge is divided into 16 physical substructures or components, as shown in Fig. 7.5. This partition results in 15 interfaces between the physical components. All components and interfaces are numbered, so that for each component and interface, the corresponding neighborhood components/interfaces are clearly defined. A linear finite element model of the bridge is considered. For a component-consistent parameterization, one parameter per component is assumed. This parameter is selected to be the modulus of elasticity for each component, so that the stiffness matrix of each component linearly depends on the component parameter, while the mass matrix is constant, independent of the parameters.

Model reduction is used to reduce the model and, thus, the computational effort for performing Bayesian inference to manageable levels. Specifically, the parameterization-consistent component mode synthesis (CMS) technique, introduced in Chap. 2, is applied. The accuracy of the model reduction technique depends on the number of fixed-interface normal modes kept for each component, as well as the number of interface modes kept. The optimal number of the modes to be retained is always problem-dependent. In the following, some guidance will be used and exploited to reduce the original FE model (see Sect. 4.6.2).

Let ω_c be the cut-off frequency that represents the highest modal frequency that is of interest in FE model updating. Herein, the cut-off frequency is selected to be equal to the 20th modal frequency of the nominal model. Based on the unreduced finite element model, the cut-off frequency is found to be $\omega_c = \omega_{20} = 3.51$ Hz. For each component s, it is selected to retain all fixed-interface modes with a frequency of less than $\omega_{s,\max} = \rho^s \omega_c$, a multiple of the cut-off frequency ω_c, where the values of the multiplication factors ρ^s, selected herein to be component dependent, affect the computational efficiency and accuracy of the model reduction technique. The effectiveness of the CMS technique depends on the number of modes retained for each component. Representative values of ρ^s usually range from 1 to 10. One can select the values of ρ^s offline by comparing the modal properties (modal frequencies and mode shapes) estimated by a reduced model for different values of ρ^s and the modal properties of the original FE model. Such an offline analysis is carried out as an effort to minimize the number of modes kept, maintaining the accuracy of the reduced model.

The total number of internal and boundary DOFs per component of the unreduced model, as well as the number of retained modes per component, are reported in Fig. 7.6. In most structural components, the value of $\rho^s = 5$ is used. For the five soil components, the value of $\rho^s = 1$ is found to be sufficient to maintain a high accuracy in the values of the modal frequencies predicted by the reduced-order model. As a result, fewer modes were retained for the soil components, without compromising the accuracy of the predicted modal frequencies. It is clear from these results that more than three orders of a magnitude reduction in the number of DOFs per component is achieved using model reduction. A total of 170 fixed-interface modes (dominant fixed-interface normal modes) out of the 830,115 DOFs are retained for all 16 components. The total number of DOFs of the reduced model is 16,205, which includes 16,035 interface DOFs for all components (see Table 7.2).

Figure 7.7 shows the fractional error between the modal frequencies computed using the complete FE model and the ones computed using the reduced-order model based on dominant fixed-interface normal modes and the entire set of interface DOFs. It is seen that the error for the modal frequencies falls below 0.18%, which ensures high levels of accuracy for the ρ^s values selected for each component.

From the information provided in Table 7.2, it is evident that a large number of generalized coordinates for the reduced system arises from the interface DOFs. The large number of interface DOFs can be reduced by retaining only a fraction of the interface modes [49]. The approach based on constant local interface modes is considered in the present application (see Sects. 1.8 and 3.5). Following a

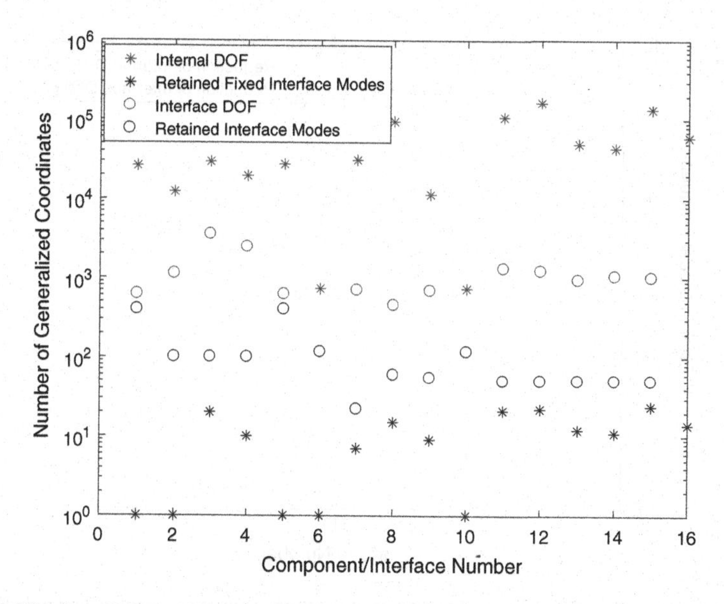

Fig. 7.6 Number of internal and boundary DOFs, retained fixed-interface modes and retained interface modes for each substructure and interface

Table 7.2 Number of generalized DOFs and percentage modal frequency error for the full (unreduced) and reduced models

DOF	Full model	$\rho^s = 5$	$\rho^s = 5$, $\nu^s = 200$
Internal	814,080	170	170
Interface	16,035	16,035	1,721
Total	830,115	16,205	1,891
Highest percentage error (%)	0.00	0.18	0.23

procedure similar to the fixed-interface modes, for each interface, the interface modes that have a frequency of less than $\omega^s_{\max} = \nu^s \omega_c$, a multiple of the cut-off frequency, are retained, where the multiplication factor ν^s is problem-dependent. For most interfaces, the value of $\nu^s = 200$ is used. Much higher values of $\nu^s = 5{,}000$ are used for the four interfaces that connect the physical components 6 and 10 of the right and left abutments (see Fig. 7.5) to the bridge deck (Interfaces 6 and 10) and the soil blocks (Interfaces 1 and 5). Results for the values of ν^s used and the modes kept per interface are given in Fig. 7.6. With the exception of interfaces 6, 10, 1, and 5, the number of retained interface modes is a fraction of the interface DOFs for each interface.

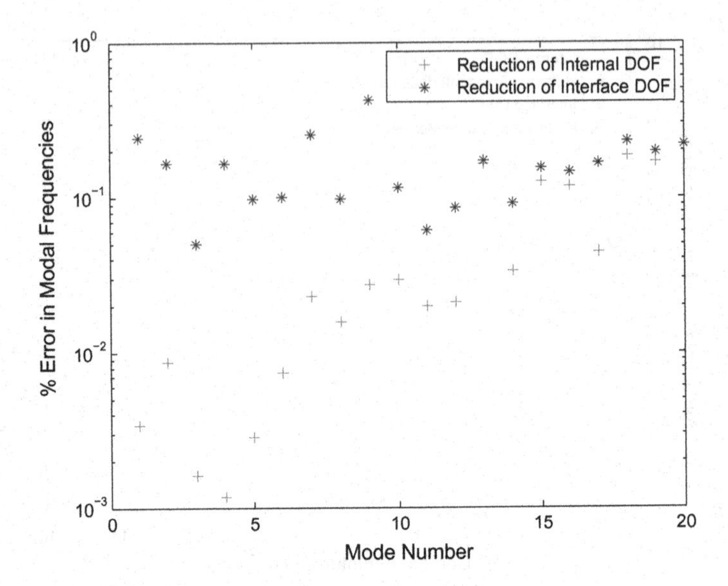

Fig. 7.7 Fractional error in the modal frequencies computed using the complete and reduced finite element model

Figure 7.7 presents the results for the fractional error in the modal frequencies as a function of the mode number, computed using the finite element model with fixed-interface normal modes and interface modes. It can be seen that the fractional error for the lowest 20 modes of the structure falls below 0.23%. Thus, sufficiently accurate results are obtained for the case of reducing the fixed-interface normal modes and interface modes. The reduced system has 1,891 DOFs, from which 170 generalized coordinates are fixed-interface modes for all components and the other 1,721 generalized coordinates are interface modes (see Table 7.2). The number of generalized coordinates is drastically reduced by almost three orders of magnitude compared to the number of DOFs of the original unreduced FE model. The significant reduction in the number of generalized coordinates of the reduced system and the increased accuracy of the results are promising to use the proposed model reduction method in Bayesian FE model updating, which requires a large number of finite element model runs.

In conclusion, using model reduction techniques, a drastic reduction in the number of generalized coordinates is obtained, which can exceed three orders of magnitude, without sacrificing the accuracy with which the lowest model frequencies are computed. The time-to-solution for one run of the reduced model is of the order of a few seconds, which should be compared to approximately two minutes required for solving the unreduced FE model. Further reductions are expected to result by retaining a lower number of modes per component and employing the formulation based on static correction [36].

7.5.1.4 Finite Element Model Updating

The FE model of the bridge-foundation-soil system is updated next, based on the experimentally identified modal frequencies and mode shapes. The updating is based on the formulation presented in Sect. 7.4.1.1 for $w = 1$. The FE model is parameterized using three parameters associated with the modulus of elasticity of one or more structural components. Specifically, the first parameter, θ_1, accounts for the modulus of elasticity of the deck (Components 11–16) of the bridge, as shown in Fig. 7.5. A perfect correlation of the modulus of elasticity of the deck components is assumed. The second parameter, θ_2, accounts for the modulus of elasticity of the three piers (Components 7–9), assumed to be perfectly correlated, while the third parameter, θ_3, accounts for the modulus of elasticity of the soil (Components 1–5), assumed to be the same at all bridge supports. The model parameters in the set $\boldsymbol{\theta}$ scale the nominal values of the properties that they model. The corresponding parametrization functions are similar to the ones used in Chap. 4.

The prior distribution is assumed to be uniform with bounds in the domain $[0.1,10] \times [0.1,10] \times [0.1,200]$ for the deck, pier, and soil model parameters, and in the domain $[0.001,1]$ for the prediction error parameter σ. The larger size of the soil parameter was chosen to account for the higher uncertainty reported in the values of the soil stiffness from soil tests. In addition, the effect of soil stiffness on the model behavior for the low amplitude vibrations recorded during the ambient vibration tests is investigated.

The model updating is performed using a subset of the experimentally identified modes. The rest of the identified modes are used to validate the performance of the updated model. Specifically, the lowest 15 modal frequencies are used for parameter inference, while the remaining 5 modal frequencies of modes 16–20 are used to validate the updated model. Two model updating cases are considered. Model updating Case 1 is based on using the lowest 15 experimentally identified modal frequencies, while model updating Case 2 is based on using the 11 available mode shapes, in addition to the lowest 15 modal frequencies. Approximately 105 observed degrees of freedom are used, corresponding to measurements along the transverse and vertical directions of the bridge deck. The measurements along the longitudinal directions of the bridge deck are ignored.

Results are obtained using the parallelized TMCMC algorithm [21] to generate samples from the posterior PDF of the structural model and prediction error model parameters. One thousand samples per TMCMC stage are used, resulting in a total runtime of approximately 30 min using the reduced 1,891 DOFs model in a 4-core computer. In the final TMCMC stage, the 1,000 samples populate the support of the posterior PDF and quantify the uncertainty in the parameters given the experimental data.

The projections of the TMCMC samples in the two-dimensional parameter space, along with the marginal posterior distribution of the model parameters, are shown in Figs. 7.8 and 7.9 for the model updating Cases 1 and 2, respectively. Table 7.3 reports the statistics of the model parameters summarized in the mean, and the 5% and 95% quantile values of the model parameters. For the model updating Case 1,

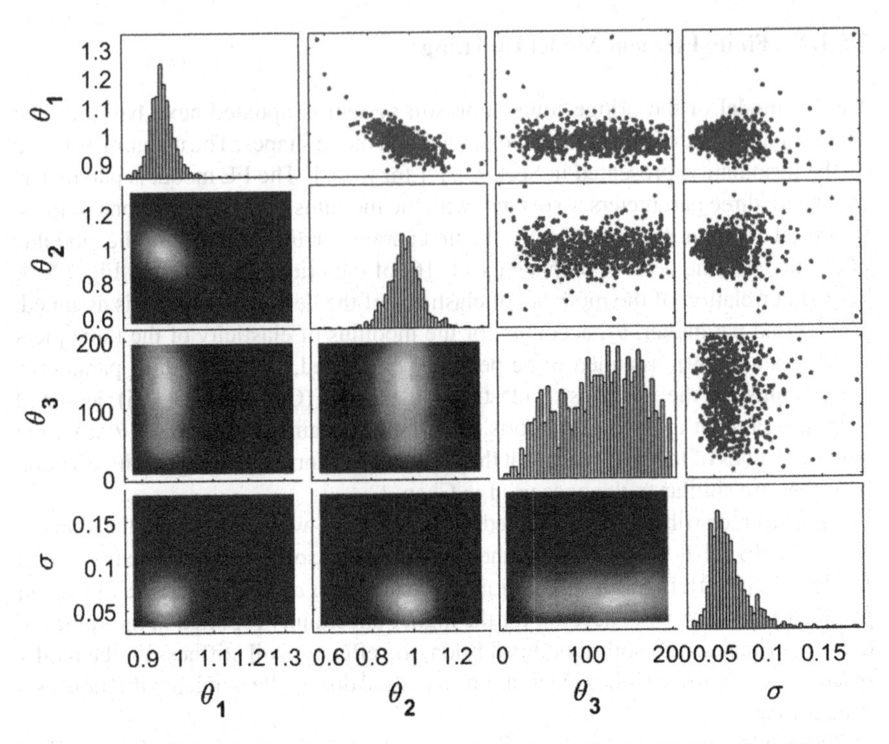

Fig. 7.8 Two-dimensional sample projections and marginal posterior distributions of model parameters (model updating Case 1). θ_1: Deck, θ_2: Piers, θ_3: Soil, σ: Prediction error

the updated mean values for the deck and pier stiffness parameters are approximately 0.96 and 0.97 times their nominal values, with spread of uncertainty about the mean values of the order of 5% and 12%, respectively. The marginal posterior distribution of the soil stiffness parameter in Fig. 7.8 and Table 7.3 indicates that any value of the soil higher than approximately 35 times the nominal soil stiffness value is equally likely. Specifically, values of the soil parameter θ_3 from 35 to 200 (the higher bound of the prior PDF for the soil parameter θ_3) provide an equally good fit between the experimentally identified and the model-predicted modal frequencies. Results suggest that the soil is very stiff, since its stiffness values higher than 35 times the nominal stiffness value do not substantially affect the modal properties. The physical implication of this is that the bridge can be considered as fixed at the base for the low vibration levels considered in the finite element model updating. From the (θ_1, θ_2) projection of the samples shown in Fig. 7.8, a negative correlation is observed between the deck and pier stiffnesses, which can be justified since an increase in the deck stiffness is counterbalanced by a decrease in the stiffness of the piers to maintain the fit between the model-predicted and experimentally identified modal frequencies.

For the model updating Case 2, the uncertainties in the model parameters are significantly reduced compared to the uncertainty estimated for the model updating

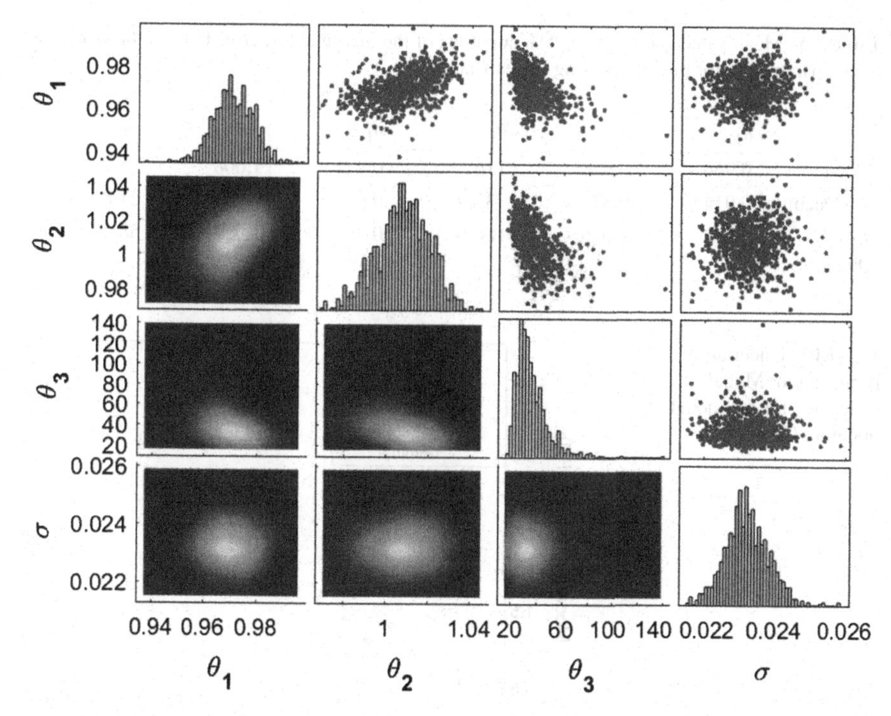

Fig. 7.9 Two-dimensional sample projections and marginal posterior distributions of model parameters (model updating Case 2). θ_1: Deck, θ_2: Piers, θ_3: Soil, σ: Prediction error

Case 1. This is due to the large number of data associated with the mode shape components. The updated mean values for the deck and pier stiffness parameters are approximately 0.97 and 1.00 times their nominal values with uncertainties of the order of 1.5% and 2%, respectively. Results suggest that the pier stiffness values are exactly the same as the design values, while the deck appears to be slightly less stiff that the one used for design. From the results in Fig. 7.9 and Table 7.3 it can be concluded that the uncertainty in the soil stiffness parameter, ranging from approximately 20 to 60 times its nominal value with spread of uncertainty of approximately 40% the mean value, is relatively much larger than the uncertainty in the deck and pier stiffness parameters. This signifies that, for the low amplitude vibration levels considered, the soil appears to be significantly stiffer than the soil tests suggest. From the (θ_1, θ_2), (θ_1, θ_3), and (θ_2, θ_3) projections of the samples shown in Fig. 7.9, a negative correlation is observed between the soil and pier stiffnesses, which can be justified since an increase in the soil stiffness is counterbalanced by a decrease in the pier stiffnesses to maintain the fit between the model-predicted and experimentally identified modal properties. For comparison purposes, note that the scales used in Figs. 7.8 and 7.9 are different.

Table 7.3 Mean value and 5% and 95% quantiles of the marginal posterior PDF of the structural model parameters for model updating Cases 1 and 2

	Case 1			Case 2		
	θ_1	θ_2	θ_3	θ_1	θ_2	θ_3
Mean	0.966	0.968	116.5	0.971	1.008	35.7
5% Quantile	0.914	0.842	37.9	0.958	0.987	21.8
95% Quantile	1.023	1.096	186.6	0.983	1.027	58.4

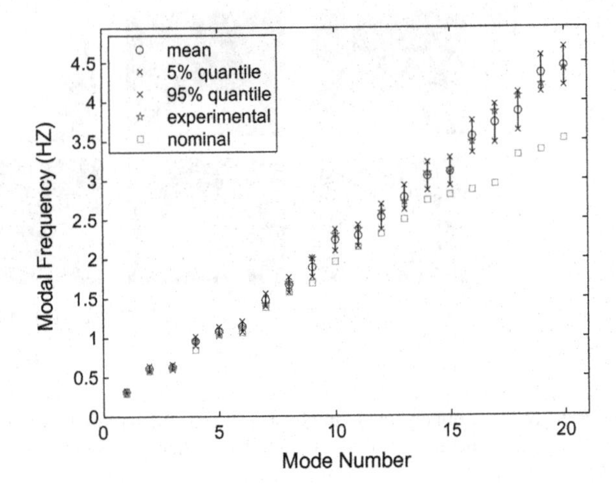

Fig. 7.10 Uncertainty propagation: Modal frequency fits for model updating Case 1

7.5.1.5 Uncertainty Propagation

The posterior parameter uncertainty is propagated next through the model to estimate the uncertainties predicted by the updated-model for the lowest 20 modal frequencies. The mean values of the modal frequencies and the uncertainty in the values of the modal frequencies in terms of the 5% and 95% quantiles are shown in Figs. 7.10 and 7.11 for the model updating Cases 1 and 2, respectively. For comparison purposes, the predictions from the nominal FE model and the experimentally identified values of the modal frequencies are also shown in these figures. The improvement of the updated model compared to the nominal model is evident. The updated model significantly improves, compared to the nominal model, the fit between the model-predicted and the experimentally identified modal frequencies. In particular, the higher the mode, the larger the discrepancy between the nominal model and experimentally identified modal frequencies. In addition, the uncertainty in the predictions of the updated model increases as the mode number increases, indicating that higher modes are more sensitive to variations in the values of the model parameters compared to lower ones.

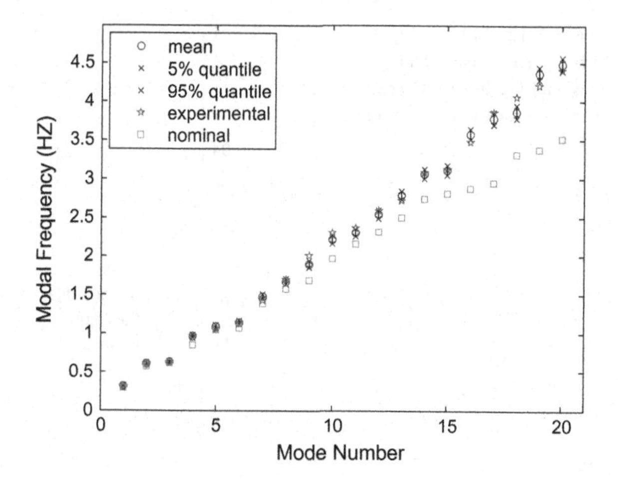

Fig. 7.11 Uncertainty propagation: Modal frequency fits for model updating Case 2

For the model updating Case 1, it is observed that most of the experimental modal frequency values for the lowest 20 modes fall well within the 90% uncertainty interval of the model predictions. This is a strong indication of the accuracy of the updated model and its predictive capability, as far as the modal frequencies are concerned. For the model updating Case 2, the size of the uncertainty interval for the modal frequencies decreases substantially due to the significantly smaller uncertainty identified for the model parameters. The relatively large uncertainty in the soil stiffness observed in Fig. 7.9, and the small uncertainty in the modal frequency predictions indicate that the modal frequencies are insensitive to changes in the soil stiffness values over the support of the posterior distribution. It is also worth noting that the updated model inferred from the lowest 15 modal frequencies and 11 mode shapes provides significantly better predictions of the last 5 modal frequencies (modes 16 to 20), which were not included in the model updating process.

Similar results for the MAC-values, between the experimentally identified mode shapes and the mode shapes predicted by the updated finite element models, along with the 5% and 95% percentiles, are presented in Figs. 7.12 and 7.13 for the model updating Cases 1 and 2, respectively. For the model updating Case 2, it can be seen that most of the MAC-values are greater than 0.95 for almost all modes, confirming the adequacy of the selected model class. Although for mode 9 a lower MAC-value in the range of 0.82 to 0.87 is obtained, the updating models resulted in a substantial improvement in the MAC-values for mode 9 as compared to the low MAC-value of 0.061 (see Table 7.1) predicted by the nominal finite element model. Also, the uncertainty in the MAC-values is very small, signifying that the MAC-values are insensitive to changes in the values of the model parameters over the support of the posterior distribution. In contrast, the model updating Case 1 resulted in MAC-values lower than 0.95 and large uncertainties in the MAC-values for modes 11 and 12. Comparing Cases 1 and 2, a clear improvement in the MAC-values is observed when the mode shapes of the 11 modes are used for updating the model classes. This

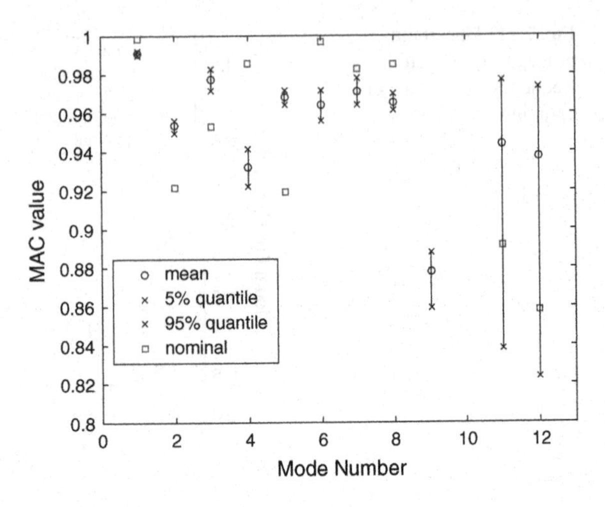

Fig. 7.12 MAC-values between measured and model-predicted mode shapes for model updating Case 1

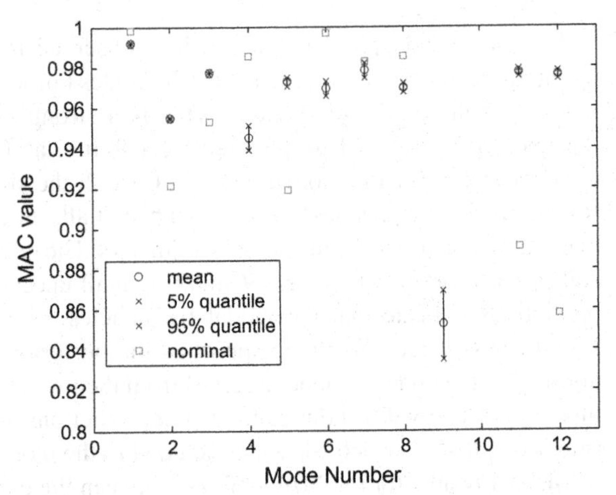

Fig. 7.13 MAC-values between measured and model-predicted mode shapes for model updating Case 2

indicates that predictions of output QoI from the updated FE model based on modal frequencies may be less reliable.

Furthermore, the two Cases 1 and 2, which differ by the measurement data used in model updating, provide conflicting values of the soil stiffness uncertainties, although both cases give qualitatively similar results suggesting a much higher soil stiffness value than the nominal one estimated by soil tests. It is believed that Case 2 reflects better the uncertainties in the values of the soil stiffness in light of the data available for both modal frequencies and mode shapes.

7.5.1.6 Alternative Parameterized Model Classes

In the following, alternative model classes are introduced in order to perform a model class selection. They correspond to the same finite element model but with different parameterization schemes involving higher number of parameters. In particular, three parameterization schemes are investigated: a five-parameter model class obtained from the existing three-parameter model class by introducing one independent stiffness parameter per pier; a seven-parameter model class obtained from the existing three-parameter model class by introducing one independent parameter per soil block; and a nine-parameter model class obtained from the existing three-parameter model class by introducing one independent parameter per pier and soil block. Results show that the alternative model classes do not improve the fit between measured and model-predicted modal properties. Bayesian model selection is used to rank the 3 to 9 parameter model classes based on the measured data. The highest evidence is obtained for the three-parameter model class. In particular, based on the TMCMC estimates, a positive difference $\Delta = \ln Pr(M_3|D) - \ln Pr(M_5|D)$ between the log-evidence values of the three- and the five-parameter model classes M_3 and M_5 of magnitude 1.41 and 4.00 is estimated for the Cases 1 and 2, respectively. Thus, between the three-parameter model class M_3 and the five-parameter model class M_5, the Bayesian model selection method rewards the three-parameter model class with final ranking probabilities $Pr(M_3|D)/Pr(M_5|D) = \exp(\Delta) = 4.1$ and 54.4 for the Cases 1 and 2, respectively, provided that the priors for the two model classes are equal. These aforementioned results are consistent with our expectations, since the fit that the alternative model classes provide to the data is almost the same as the three-parameter model class, and so the more complex model classes, in terms of their number of parameters, are penalized [16].

7.5.1.7 Computational Issues

The TMCMC algorithm used for parameter estimation based on the reduced 1,891 DOFs model involves 1,000 samples per TMCMC stage and results in 14 TMCMC stages for the model updating Case 2. As a result, the number of finite element analyses and eigenvalue analyses for computing the lowest 20 modal frequencies and mode shapes is 14,000. It should be noted that the computing time scales linearly with the number of available cores when the parallelized TMCMC algorithm is activated. Herein, the analysis was performed in a 4-core double-threaded computer. The total time-to-solution is approximately 30 min using the reduced 1,891 DOFs model. For the large-order FE model developed for the Metsovo bridge with more than 800,000 DOFs, the computational demands involved are excessive due to the approximately two minutes required to complete one model simulation run to estimate the lowest 20 modal frequencies and mode shapes. Using the fact that the number of eigenvalue analyses are approximately 14,000, the total time-to-solution in the same 4-core double-threaded computer is expected to be of the order of three days. Overall, more than two orders of a magnitude reduction in computational time is achieved to perform the model updating using the reduced 1,892 DOFs model. Thus, it is clear that a drastic reduction in computational effort is obtained.

7.5.2 Example 2: Updating of Nonlinear Model

7.5.2.1 Structural Model

The base-isolated structural system shown in Fig. 7.14 is considered in the second example. The superstructure consists of a 10-floor, three-dimensional reinforced concrete building model. Material properties of the reinforced concrete structure have been assumed as follows: Young's modulus $E = 2.34 \times 10^{10}$ N/m^2; Poisson's ratio $\nu = 0.3$; and mass density $\rho = 2{,}500$ kg/m^3. The height of each floor is 3.5 m, leading to a total height of 35.0 m for the structure. The floors are modeled with shell elements with a thickness of 0.3 m and beam elements of a rectangular cross section of dimension 0.3 m $\times$ 0.6 m from floors 1 to 5 and 0.25 m $\times$ 0.5 m from floors 6 to 10. Each floor is supported by 48 columns of rectangular cross section of

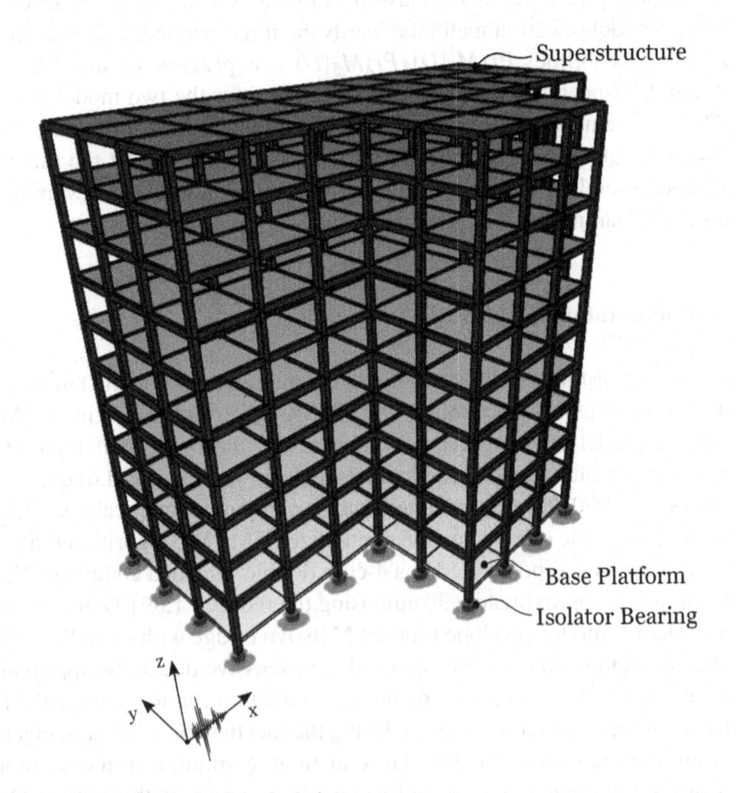

Fig. 7.14 Building model with base isolation system

dimension 0.8 m × 0.9 m. The corresponding finite element model has approximately 40,000 degrees of freedom. A 2% of critical damping for the modal damping ratios is introduced in the model. The isolation system is composed of a platform and 48 rubber bearings. The total mass of the platform is equal to 6.0×10^5 kg. As stated in Chap. 5, the isolation devices (rubber bearings) consist of layers of rubber and steel, with the rubber being vulcanized to the steel plates. A detailed description of the analytical model that characterizes the structural behavior of the rubber bearings is given in Sect. 5.8.2.

7.5.2.2 Response of the Combined System

In general, base-isolated buildings are designed such that the superstructure remains elastic. Hence, the superstructure is modeled as a three-dimensional linear elastic system, while the base is assumed to be rigid in plane, and it is modeled using three degrees of freedom. Let $\mathbf{u}_e(t)$ be the nth dimensional vector of relative displacements of the superstructure with respect to the base and $\mathbf{M}_e$, $\mathbf{C}_e$, $\mathbf{K}_e$ be the corresponding mass, damping, and stiffness matrices, respectively. Also, let $\mathbf{u}_b(t)$ be the vector of base displacements with three components and $\mathbf{G}_e$ be the matrix of earthquake influence coefficients of dimension $n \times 3$, that is, the matrix that couples the excitation components of the vector $\ddot{\mathbf{u}}_g(t)$ to the degrees of freedom of the superstructure. The equation of motion of the superstructure is expressed in the form

$$\mathbf{M}_e\ddot{\mathbf{u}}_e(t) + \mathbf{C}_e\dot{\mathbf{u}}_e(t) + \mathbf{K}_e\mathbf{u}_e(t) = -\mathbf{M}_e\mathbf{G}_e[\ddot{\mathbf{u}}_b(t) + \ddot{\mathbf{u}}_g(t)] \tag{7.57}$$

where $\ddot{\mathbf{u}}_b(t)$ is the vector of base accelerations relative to the ground. The equation of motion for the base platform is written as

$$\begin{aligned}(\mathbf{G}_e^T\mathbf{M}_e\mathbf{G}_e + \mathbf{M}_b)(\ddot{\mathbf{u}}_b(t) + \ddot{\mathbf{u}}_g(t)) + \mathbf{G}_e^T\mathbf{M}_e\ddot{\mathbf{u}}_e(t)\\ +\mathbf{C}_b\dot{\mathbf{u}}_b(t) + \mathbf{K}_b\mathbf{u}_b(t) + \mathbf{f}_{rb}(t) = \mathbf{0}\end{aligned} \tag{7.58}$$

where $\mathbf{M}_b$ is the mass matrix of the rigid base, $\mathbf{C}_b$ is the resultant damping matrix of viscous isolation components, $\mathbf{K}_b$ is the resultant stiffness matrix of linear elastic isolation components, and $\mathbf{f}_{rb}(t)$ is the vector containing the nonlinear forces activated on the isolators. It is noted that the set of equations (7.57) and (7.58) can be written in a compact form as in Eq. (1.1) (see Sect. 1 of Chap. 1).

If the dynamic response of the superstructure is represented by a linear combination of its mode shapes, that is, $\mathbf{u}_e(t) = \boldsymbol{\varphi}_e\boldsymbol{v}_e(t)$, where $\boldsymbol{\varphi}_e$ is the matrix of mode shapes of the superstructure and $\boldsymbol{v}_e(t)$ is the corresponding vector of modal response functions, the combined equation of motion of the base-isolated structural system can be written as

$$\begin{bmatrix}\mathbf{I} & \boldsymbol{\varphi}_e^T\mathbf{M}_e\mathbf{G}_e\\ \mathbf{G}_e{}^T\mathbf{M}_e\boldsymbol{\varphi}_e & \mathbf{M}_b + \mathbf{G}_e{}^T\mathbf{M}_e\mathbf{G}_e\end{bmatrix}\begin{Bmatrix}\ddot{\boldsymbol{v}}_e(t)\\ \ddot{\mathbf{u}}_b(t)\end{Bmatrix} + \begin{bmatrix}\mathbf{C}_{ev} & \mathbf{0}\\ \mathbf{0} & \mathbf{C}_b\end{bmatrix}\begin{Bmatrix}\dot{\boldsymbol{v}}_e(t)\\ \dot{\mathbf{u}}_b(t)\end{Bmatrix} +$$

$$\begin{bmatrix} \mathbf{K}_{ev} & \mathbf{0} \\ \mathbf{0} & \mathbf{K}_b \end{bmatrix} \begin{Bmatrix} \boldsymbol{v}_e(t) \\ \mathbf{u}_b(t) \end{Bmatrix} = - \begin{bmatrix} \boldsymbol{\varphi}_e{}^T \mathbf{M}_e \mathbf{G}_e \\ \mathbf{M}_b + \mathbf{G}_e{}^T \mathbf{M}_e \mathbf{G}_e \end{bmatrix} \ddot{\mathbf{u}}_g(t) - \begin{Bmatrix} \mathbf{0} \\ \mathbf{f}_{rb}(t) \end{Bmatrix} \tag{7.59}$$

where the matrices $\mathbf{C}_{ev}$ and $\mathbf{K}_{ev}$ are given by

$$\mathbf{C}_{ev} = \begin{bmatrix} 2\xi_{e1}\omega_{e1} & & & & \\ & \cdot & & & \\ & & 2\xi_{er}\omega_{er} & & \\ & & & \cdot & \\ & & & & 2\xi_{em}\omega_{em} \end{bmatrix}, \quad \mathbf{K}_{ev} = \begin{bmatrix} \omega_{e1}^2 & & & & \\ & \cdot & & & \\ & & \omega_{er}^2 & & \\ & & & \cdot & \\ & & & & \omega_{em}^2 \end{bmatrix}$$

in which ω_{er}, $r = 1, ..., m$, are the natural frequencies of the original system, $\xi_{er}, r = 1, ..., m$, are the corresponding damping ratios and $m \ll n$ is the number of modes considered. Note that the natural frequencies and mode shapes of the original model of the superstructure are obtained from the reduced-order system model (see Sect. 1.10 of Chap. 1). The combined equation of motion (superstructure and base-isolation system) constitutes a nonlinear system of equations due to the nonlinearity of the isolation forces. The solution of the equation of motion (7.59) is obtained in an iterative manner by using any suitable step-by-step nonlinear integration scheme [10].

7.5.2.3 Model Updating Problem

The structural system previously described is used for model updating using simulated response data. In particular, the external diameter of the isolators D_e and the total height of rubber in the devices H_r are estimated. These parameters are parameterized as $D_e = \theta_1 \bar{D}_e$ and $H_r = \theta_2 \bar{H}_r$, where $\bar{D}_e = 0.85$m and $\bar{H}_r = 0.14$m are the corresponding nominal values. These parameters control the value of the parameters α and A of the hysteretic model affecting the nonlinear behavior of the isolators (see Sect. 5.8.2). In addition, the stiffness of the columns of the first floor of the superstructure in the x direction is also estimated. This property is parameterized by the dimensionless parameter θ_3.

The superstructure is divided into two linear substructures as shown in Fig. 7.15. Substructure 1 is composed of the column elements of the first floor, while Substructure 2 contains the rest of the superstructure components. In this context the base isolation system, that is, the isolation devices and the base platform can be considered as a nonlinear substructure. It is assumed that the base-isolated structural system is built and the response data are available to update the isolator parameters and the stiffness of the columns of the first floor. The model updating is based on the measurements of the ground acceleration at the support of the isolation system, the acceleration response in the x direction at the base platform, and the acceleration response in the x direction on the first and second floors. To this end, the original unreduced finite element model of the combined system (superstructure and base-isolation system) is excited horizontally (in the x direction) with the Santa Lucia ground-motion record

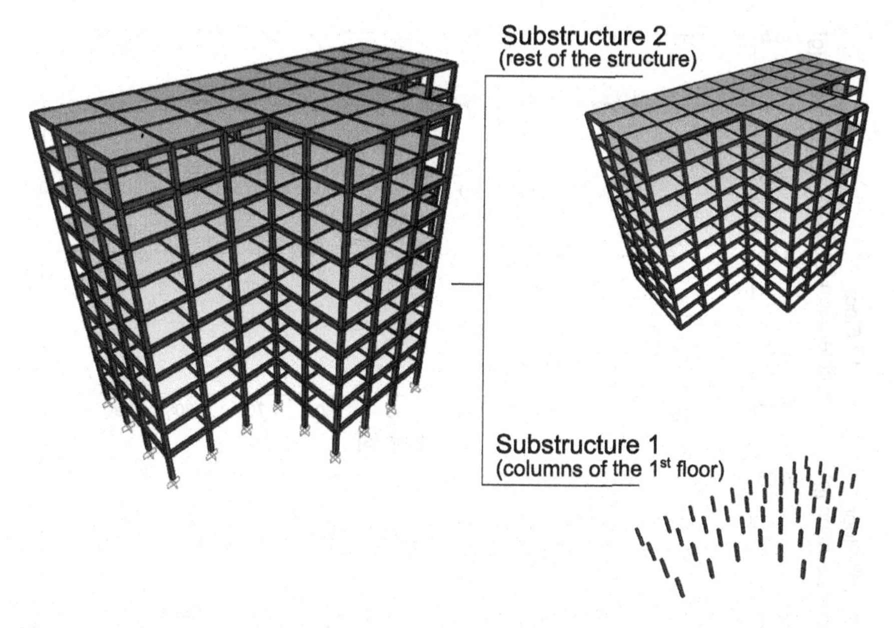

Fig. 7.15 Linear substructures of the superstructure finite element model used for model updating

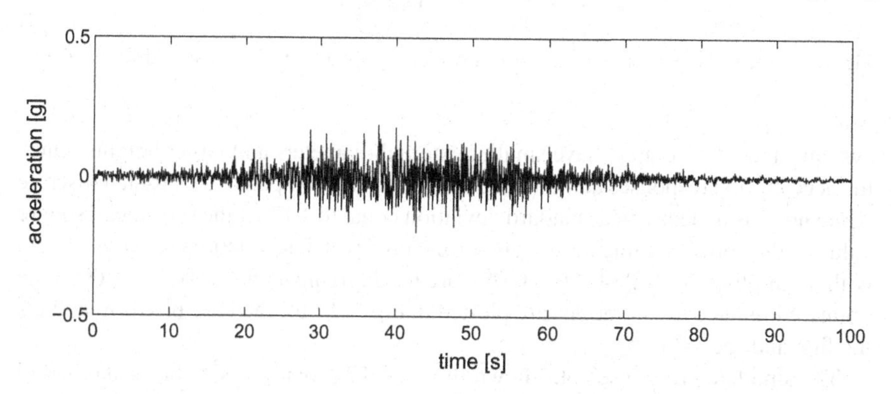

Fig. 7.16 Santa Lucia ground-motion record (2010 Chilean earthquake)

recorded during the 2010 Chilean earthquake. The input ground acceleration time history is shown in Fig. 7.16. It corresponds to a ground motion of moderate intensity.

The actual base-isolated system used to generate the simulated data is characterized as follows. The isolator parameters are set equal to $D_e = 0.78$m, $H_r = 0.17$m, and $D_i = 0.10$m, and the stiffness of the superstructure columns of the first floor in the x direction is either kept fixed to its nominal value or is reduced by 20% with respect to its nominal stiffness value, depending on the model updating case considered (see Sect. 7.5.2.4). It is noted that this model is more flexible than the nominal

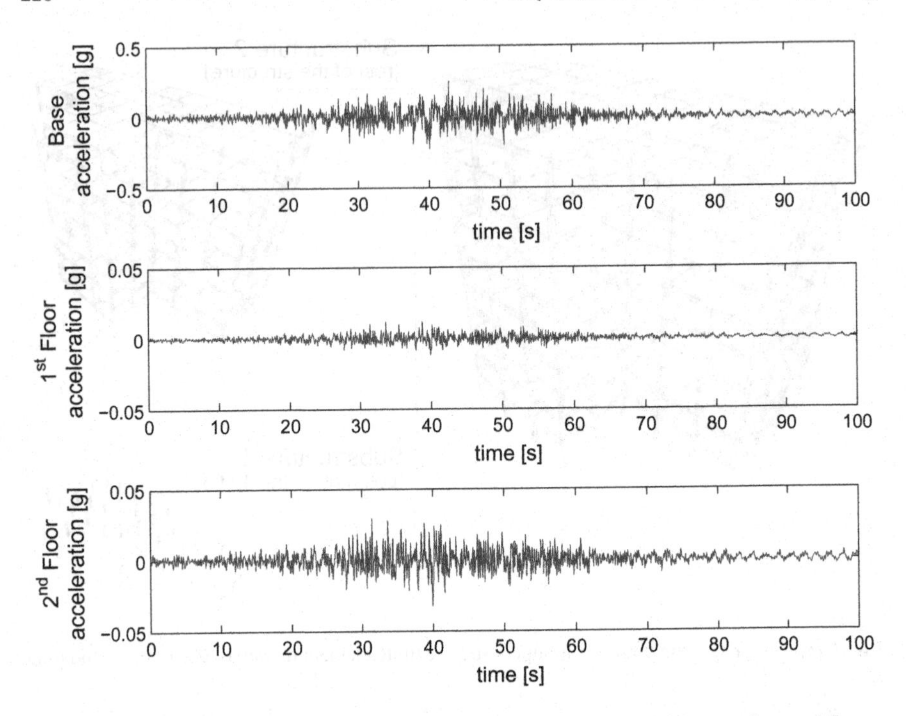

Fig. 7.17 Simulated acceleration time histories at the base platform, first floor, and second floor

system, since the isolators have smaller external diameters and larger heights. Once the acceleration responses at the various locations are computed, a Gaussian discrete white noise is added with a standard deviation equal to 10% of the root-mean square value of the corresponding acceleration time histories. One hundred seconds of data with a sampling interval of $\Delta t = 0.05$ s are used, giving a total of $N_D = 2{,}000$ data points. Simulated accelerations are plotted in Fig. 7.17 for the base platform and for the first and second floors.

The simulated response data shown in Fig. 7.17 provides the data for the model updating process. The actual implementation of this process is carried out by using a reduced-order system model based on dominant fixed-interface normal modes. The reduced model is defined as follows. For each linear substructure of the superstructure, all fixed-interface normal modes that have a frequency less than a given cut-off frequency are selected to be retained. The cut-off frequency is set to be proportional to the 12th modal frequency of the original unreduced superstructure finite element model. Validation calculations show that retaining 48 generalized coordinates for Substructure 1 and 174 generalized coordinates for Substructure 2 is adequate in the context of this application. In fact, with this number of generalized coordinates, the fractional error (in percentage) between the modal frequencies using the complete finite element model and the modal frequencies computed using the reduced-order model falls below 0.1% for the lowest 12 modes. Then, a total of 222 generalized coordinates, corresponding to the fixed-interface normal modes, out of 39,712 DOFs of the original model, are retained for the two linear substructures in this application. The number of interface degrees of freedom is equal to 288 in this case. Note that no interface reduction is considered. The total number of degrees of freedom of the reduced model represents more than 98% reduction with respect to the unreduced model in this case.

7.5.2.4 Model Updating Results

Two cases are considered for the model updating problem. In the first case, only the isolator parameters are identified with the provided data. To this end, the dimensionless parameter θ_3 is set equal to one, that is, there is no reduction in the stiffness of the first floor superstructure columns. Simulated data are generated for $\theta_3 = 1$. To test the capabilities of the updating process in the framework of the combined system, the identification of the isolator parameters as well as the stiffness properties of the superstructure are simultaneously considered in the second case. In the first case, independent uniform prior distributions are assumed for parameters θ_1 and θ_2 that are over the range [0.5, 1.4]. As previously pointed out, these parameters affect the nonlinear behavior of the isolators. The number of samples at the different iteration steps of the transitional Markov chain Monte Carlo method is taken as 1,000. The updating process converges in six stages (steps) in this case. The samples from the posterior probability density function are displayed in terms of parameters θ_1 and θ_2 in Fig. 7.18.

The value of the nominal model parameters is also indicated in the figure. The large prior uncertainty about parameters θ_1 and θ_2 (uniformly distributed over the range [0.5, 1.4]) is significantly reduced, which is visible from the decreased range of the posterior samples. In fact, the samples of the external diameter are distributed

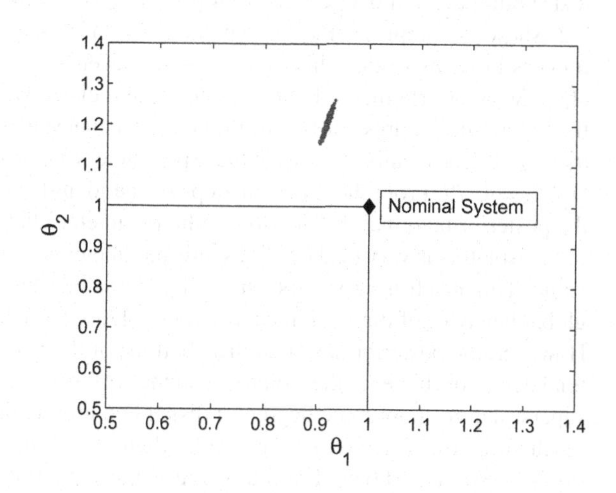

Fig. 7.18 Samples of the posterior probability density function in the $(\theta_1 - \theta_2)$ space generated at the last step of the transitional Markov chain Monte Carlo method

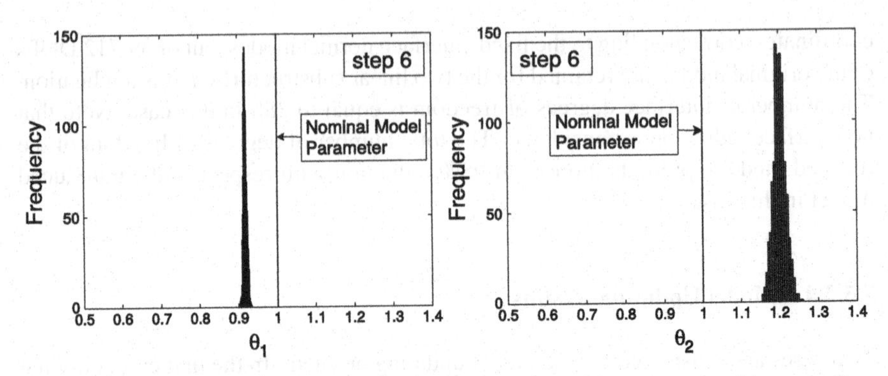

Fig. 7.19 Posterior histograms of model parameters θ_1 and θ_2. Mean estimates: $\bar{\theta}_1 = 0.92$, $\bar{\theta}_2 = 1.20$

around the actual value of 0.78 m ($\theta_1 = 0.92$), while the samples of the total height of rubber are spread around the actual value of 0.17 m ($\theta_2 = 1.20$) as shown in the corresponding histograms of these parameters (Figs. 7.19). The results of Fig. 7.18 also suggest that the data for the updated isolation model result in uncertainties that are correlated along a certain direction. The correlation structure is consistent with the fact that the base isolation system becomes stiffer as the rubber diameter is increased. Contrarily, the isolation system becomes more flexible as the height of the rubber is increased [37]. Therefore, an increase in the rubber diameter is compensated by an increase in the height of the rubber during the updating process, and all points along that direction correspond to isolation system models that have similar base drift responses.

In the second case, the stiffness of the superstructure columns of the first floor in the x direction is identified together with the isolator parameters. Simulated data for this case are produced for $\theta_3 = 0.8$ (actual value). An independent uniform prior distribution defined over the interval [0.5, 1.5] is assumed for θ_3. Figures 7.20 and 7.21 show the samples of the parameters θ_1, θ_2, and θ_3 converging during the updating process in terms of their histograms. After seven steps of the transitional Markov chain Monte Carlo method, the parameters associated with the isolation system and the superstructure are distributed around their actual values $\theta_1 = 0.92, \theta_2 = 1.20$, and $\theta_3 = 0.80$. The mean estimates of these parameters are equal to $\bar{\theta}_1 = 0.92, \bar{\theta}_2 = 1.22$, and $\bar{\theta}_3 = 0.79$. From the different steps of the identification process, it is clear that the posterior marginal distribution of the parameter related to the external diameter of the isolators is very peaked. Thus, this parameter is identifiable to almost a unique value. This result is reasonable since this parameter has a significant effect on the global behavior of the combined system (isolation system and superstructure) [37]. However, the posterior marginal distributions of the parameters associated with the total height of rubber in the isolators and the stiffness of the first floor columns of the superstructure show some degree of dispersion around their mean estimates. These results are also observed in Fig. 7.22, where the projection of the samples in the $(\theta_2 - \theta_3)$ space is shown. From this figure, the support of the unidentifiable domain

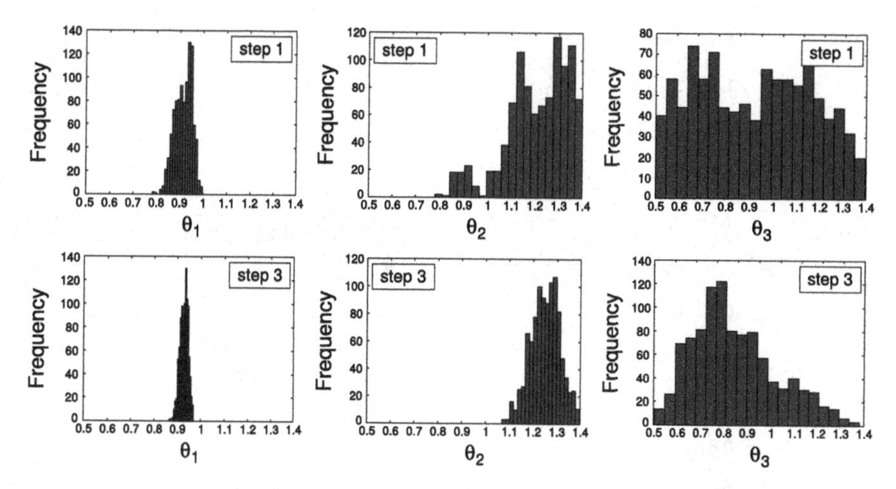

Fig. 7.20 Histograms of model parameters θ_1, θ_2 and θ_3 during the first stages of the identification process (steps 1 and 3)

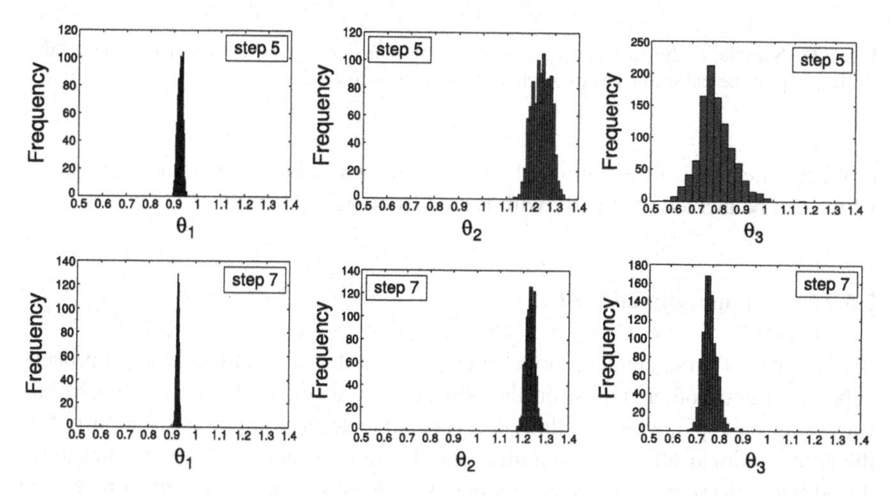

Fig. 7.21 Histograms of model parameters θ_1, θ_2 and θ_3 during the last stages of the identification process (steps 5 and 7)

is quite clear. Furthermore, there is no clear interaction between these two parameters. This result is reasonable from a structural point of view, since numerical validations have shown that the response of the base platform is not very sensitive to the total height of rubber in the isolators over the range considered here. Similarly, numerical simulations have demonstrated that the overall response of the combined system is not very sensitive to the 20% stiffness reduction of the first floor superstructure columns. The previous results illustrate some important advantages of Bayesian updating procedures over traditional techniques that attempt to identify one best model when there is a limited number of data available. In this case, the identification

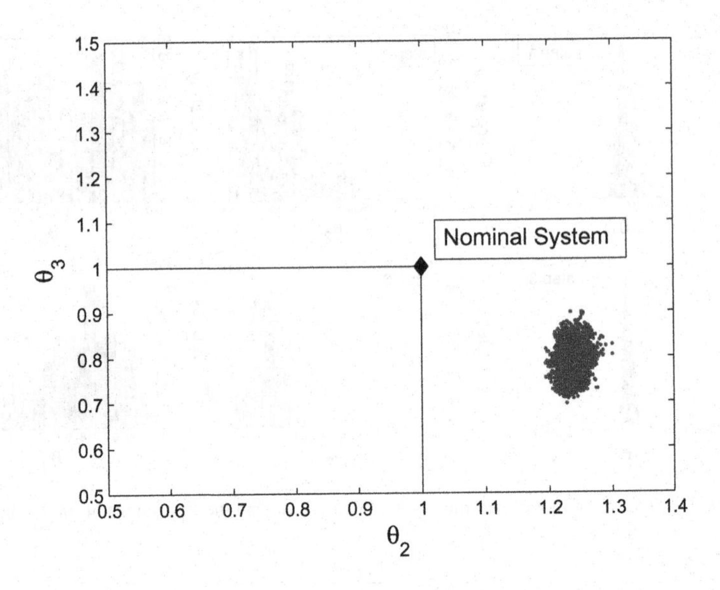

Fig. 7.22 Samples of the posterior probability density function in the $(\theta_2 - \theta_3)$ space generated at the last step of the transitional Markov chain Monte Carlo method

problem emerges as ill-conditioned without a unique solution for some parameters, which can be tackled efficiently with Bayesian model updating.

7.5.2.5 Computational Cost

The number of finite element model runs required during the identification process depends, among other things, on the number of transitional Markov chain Monte Carlo stages (steps), which, in this case, is equal to seven. For comparison purposes, the computational effort for obtaining one dynamic response of the original unreduced finite element model is approximately 1.20 min. Multiplying this time by the total number of dynamic analyses required by the transitional Markov chain Monte Carlo method, the total time is expected to be of the order of 143 h (almost six days). In contrast, the updating process of the reduced-order combined system (isolation system and superstructure) takes approximately 20 h. Thus, a drastic reduction in computational efforts is achieved without compromising the predictive capability of the proposed identification methodology. The previous computational efforts are based on the consideration of some parallelization features of the identification process. In particular eight transitional Markov Chain Monte Carlo samples are simultaneously run in the present implementation.

References

1. P. Angelikopoulos, C. Papadimitriou, P. Koumoutsakos, Bayesian uncertainty quantification and propagation in molecular dynamics simulations: a high performance computing framework. J. Chem. Phys. **137**(14) (2012)
2. Y. Aoyama, G. Yagawa, Component mode synthesis for large scale structural eigenanalysis. Comput. Struct. **79**(6), 605–615 (2001)
3. C. Argyris, Bayesian Uncertainty Quantification and Optimal Experimental Design in Data-Driven Simulations of Engineering Systems, Ph.D. thesis, Department of Mechanical Engineering, University of Thessaly, 2017
4. S.K. Au, J.L. Beck, Estimation of small failure probabilities in high dimensions by subset simulation. Probab. Eng. Mech. **16**, 263–277 (2011)
5. S.K. Au, Assembling mode shapes by least squares. Mech. Syst. Signal Process. **25**, 163–179 (2010)
6. S.K. Au, Fast Bayesian ambient modal identification in the frequency domain, part II: posterior uncertainty. Mech. Syst. Signal Process. **26**, 76–90 (2012)
7. S.K. Au, *Operational Modal Analysis: Modeling, Bayesian Inference, Uncertainty Laws* (Springer, Berlin, 2017)
8. R.R. Craig Jr, M.C.C. Bampton, Coupling of substructures for dynamic analysis. AIAA J **6**(5), 678–685 (1965)
9. R.M. Bampton, A modal combination program for dynamic analysis of structures. Technical memorandum 33–920 (Jet Propulsion Laboratory, Pasadena, CA, 1 July 1967)
10. K.J. Bathe, *Finite Element Procedures* (Prentice Hall, New Jersey, 2006)
11. M. Barbato, A. Zona, J.P. Conte, Finite element response sensitivity analysis using three-field mixed formulation: general theory and application to frame structures. Int. J. Numer. Methods Eng. **69**(1), 114–161 (2007)
12. M. Barbato, J.P. Conte, Finite element response sensitivity analysis: a comparison between force-based and displacement-based frame element models. Comput. Methods Appl. Mech. Eng. **194**(12–16), 1479–1512 (2005)
13. J.L. Beck, L.S. Katafygiotis, Updating models and their uncertainties. I: Bayesian statistical framework. ASCE J. Eng. Mech. **124**(4), 455–461 (1998)
14. J.L. Beck, A. Taflanidis, Prior and posterior robust stochastic predictions for dynamical systems using probability logic. Int. J. Uncertain. Quantif. **3**(4), 271–288 (2013)
15. J.L. Beck, Bayesian system identification based on probability logic. Struct. Control. Health Monit. **17**(7), 825–847 (2010)
16. J.L. Beck, K.V. Yuen, Model selection using response measurements: Bayesian probabilistic approach. ASCE J. Eng. Mech. **130**(2), 192–203 (2004)
17. J.L. Beck, S.K. Au, Bayesian updating of structural models and reliability using Markov chain Monte Carlo simulation. ASCE J. Eng. Mech. **128**(4), 380–391 (2002)
18. E.N. Chatzi, C. Papadimitriou (eds.), *Identification Methods for Structural Health Monitoring*, Series: CISM-International Centre for Mechanical Sciences (Springer, Berlin, 2016)
19. K. Christodoulou, C. Papadimitriou, Structural identification based on optimally weighted modal residuals. Mech. Syst. Signal Process. **21**(1), 4–23 (2007)
20. S.H. Cheung, J.L. Beck, Bayesian model updating using hybrid Monte Carlo simulation with application to structural dynamic models with many uncertain parameters. ASCE J. Eng. Mech. **135**(4), 243–255 (2009)
21. J.Y. Ching, Y.C. Chen, Transitional Markov chain Monte Carlo method for Bayesian model updating, model class selection, and model averaging. ASCE J. Eng. Mech. **133**(7), 816–832 (2007)
22. O. Ditlevsen, H.O. Madsen, *Structural Reliability Methods* (Wiley, Chichester, 1996)
23. R.L. Fox, M.P. Kapoor, Rate of change of eigenvalues and eigenvectors. AIAA J. **6**(12), 2426–2429 (1968)
24. D. Giagopoulos, C. Salpistis, S. Natsiavas, Effect of nonlinearities in the identification and fault detection of Gear-pair systems. Int. J. Non-Linear Mech. **41**, 213–230 (2006)

25. D. Giagopoulos, D.-C. Papadioti, C. Papadimitriou, S. Natsiavas, Bayesian uncertainty quantification and propagation in nonlinear structural dynamics. in *International Modal Analysis Conference (IMAC), Topics in Model Validation and Uncertainty Quantification* (2013), pp. 33–41
26. B. Goller, J.L. Beck, G.I. Schueller, Evidence-based identification of weighting factors in Bayesian model updating using modal data. ASCE J. Eng. Mech. **138**(5), 430–440 (2012)
27. B. Goller, H.J. Pradlwarter, G.I. Schueller, An interpolation scheme for the approximation of dynamical systems. Comput. Methods Appl. Mech. Eng. **200**(1–4), 414–423 (2011)
28. P.L. Green, E.J. Cross, K. Worden, Bayesian system identification of dynamical systems using highly informative training data. Mech. Syst. Signal Process. **56**, 109–122 (2015)
29. P.L. Green, Bayesian system identification of a nonlinear dynamical system using a novel variant of simulated annealing. Mech. Syst. Signal Process. **52**, 133–146 (2015)
30. P.E. Hadjidoukas, P. Angelikopoulos, C. Papadimitriou, P. Koumoutsakos, Π4U: A high performance computing framework for Bayesian uncertainty quantification of complex models. J. Comput. Phys. **284**(1), 1–21 (2015)
31. N. Hansen, S.D. Muller, P. Koumoutsakos, Reducing the time complexity of the derandomized evolution strategy with covariance matrix adaptation (CMA-ES). Evol. Comput. **11**(1), 1–18 (2003)
32. W.K. Hastings, Monte-Carlo sampling methods using Markov chains and their applications. Biometrika **57**(1), 97–109 (1970)
33. H.A. Jensen, E. Millas, D. Kusanovic, C. Papadimitriou, Model reduction techniques for Bayesian finite element model updating using dynamic response data. Comput. Methods Appl. Mech. Eng. **279**, 301–324 (2014)
34. H.A. Jensen, C. Vergara, C. Papadimitriou, E. Millas, The use of updated robust reliability measures in stochastic dynamical systems. Comput. Methods Appl. Mech. Eng. **267**, 293–317 (2013)
35. H.A. Jensen, F. Mayorga, C. Papadimitriou, Reliability sensitivity analysis of stochastic finite element models. Comput. Methods Appl. Mech. Eng. **296**, 327–351 (2015)
36. H.A. Jensen, A. Munoz, C. Papadimitriou, C. Vergara, An enhanced substructure coupling technique for dynamic re-analyses: application to simulation-based problems. Comput. Methods Appl. Mech. Eng. **307**, 215–234 (2016)
37. H.A. Jensen, D.S. Kusanovic, M. Papadrakakis, Reliability-based characterization of base-isolated structural systems, in *European Congress on Computational Methods in Applied Sciences and Engineering* (Vienna, Austria, 2012)
38. L.S. Katafygiotis, H.F. Lam, Tangential-projection algorithm for manifold representation in unidentifiable model updating problems. Earthq. Eng. Struct. Dyn. **31**(4), 791–812 (2002)
39. L.S. Katafygiotis, H.F. Lam, C. Papadimitriou, Treatment of unidentifiability in structural model updating. Adv. Struct. Eng.-Int. J. **3**(1), 19–39 (2000)
40. L.S. Katafygiotis, J.L. Beck, Updating models and their uncertainties. II: model identifiability. ASCE J. Eng. Mech. **124**(4), 463–467 (1998)
41. J.N. Lyness, C.B. Moler, Generalized Romberg methods for integrals of derivatives. Numer. Math. **14**(1), 1–13 (1969)
42. P. Metallidis, G. Verros, S. Natsiavas, C. Papadimitriou, Fault detection and optimal sensor location in vehicle suspensions. J. Vib. Control. **9**(3–4), 337–359 (2003)
43. P. Metallidis, S. Natsiavas, Parametric identification and health monitoring of complex ground vehicle models. J. Vib. Control. **14**(7), 1021–1036 (2008)
44. N. Metropolis, A.W. Rosenbluth, M.N. Rosenbluth, A.H. Teller, E. Teller, Equation of state calculations by fast computing machines. J. Chem. Phys. **21**(6), 1087–1092 (1953)
45. M. Muto, J.L. Beck, Bayesian updating and model class selection for hysteretic structural models using stochastic simulation. J. Vib. Control. **14**(1–2), 7–34 (2008)
46. R.B. Nelson, Simplified calculation of eigenvector derivatives. AIAA J. **14**(9), 1201–1205 (1976)
47. E. Ntotsios, C. Papadimitriou, Multi-objective optimization algorithms for finite element model updating, in *Proceedings of International Conference on Noise and Vibration Engineering (ISMA)* (2008), 1895–1909

48. J.T. Oden, T. Belytschko, J. Fish, T.J.R. Hughes, C. Johnson, D. Keyes, A. Laud, L. Petzold, D. Srolovitz, S. Yip, Simulation-Based Engineering Science (SBES) Revolutionizing Engineering Science through Simulation, Report of the NSF, Blue Ribbon Panel on SBES (2006)
49. C. Papadimitriou, D.C. Papadioti, Component mode synthesis techniques for finite element model updating. Comput. Struct. **126**, 15–28 (2013)
50. C. Papadimitriou, L.S. Katafygiotis, Bayesian modeling and updating, in *Engineering Mechanics Reliability Handbook*, ed. by E. Nikolaidis, D.M. Ghiocel, S. Singhal (CRC Press, Boca Raton, 2004)
51. C. Papadimitriou, J.L. Beck, L.S. Katafygiotis, Asymptotic expansions for reliability and moments of uncertain dynamic systems. ASCE J. Eng. Mech. **123**(12), 1219–1229 (1997)
52. C. Papadimitriou, L.S. Katafygiotis, A Bayesian methodology for structural integrity and reliability assessment. Int. J. Adv. Manuf. Syst. **4**(1), 93–100 (2001)
53. C. Papadimitriou, J.L. Beck, L.S. Katafygiotis, Updating robust reliability using structural test data. Probab. Eng. Mech. **16**(2), 103–113 (2001)
54. C. Papadimitriou, E. Ntotsios, D. Giagopoulos, S. Natsiavas, Variability of updated finite element models and their predictions consistent with vibration measurements. Struct. Control. Health Monit. **19**(5), 630–654 (2011)
55. I. Papaioannou, W. Betz, K. Zwirglmaier, D. Straub, MCMC algorithms for subset simulation. Probab. Eng. Mech. **41**, 89–103 (2015)
56. E. Simoen, B. Moaveni, J.L. Conte, G. Lombaert, Uncertainty quantification in the assessment of progressive damage in a 7-story full-scale building slice. ASCE J. Eng. Mech. **139**(12), 1818–1830 (2013)
57. E. Simoen, C. Papadimitriou, G. Lombaert, On prediction error correlation in Bayesian model updating. J. Sound Vib. **332**(18), 4136–4152 (2013)
58. Y.C Tan, M.P. Castanier, C. Pierre. Characteristic mode based component mode synthesis for power flow analysis in complex structures, in *Proceedings of the 41st AIAA/ASME/ASCE/AHS/ASC structures, Structural Dynamics and Materials Conference and Exhibit, Reston VA*, vol. 1908–917 (2000)
59. L. Tierney, J.B. Kadane, Accurate approximations for posterior moments and marginal densities. J. Am. Stat. Assoc. **81**(393), 82–86 (1986)
60. S. Wu, P. Angelikopoulos, C. Papadimitriou, P. Koumoutsakos, Bayesian Annealed Sequential Importance Sampling (BASIS): an unbiased version of Transitional Markov Chain Monte Carlo. ASCE-ASME J. Risk Uncert. Eng. Sys. Part B Mech. Eng. **4**, 011008–1 (2018)
61. M.W. Vanik, J.L. Beck, S.K. Au, Bayesian probabilistic approach to structural health monitoring. ASCE J. Eng. Mech. **126**(7), 738–745 (2000)
62. W.-J. Yan, L.S. Katafygiotis, A novel Bayesian approach for structural model updating utilizing statistical modal information from multiple setups. Struct. Saf. **52**(Part B), 260–271 (2015)
63. K.V. Yuen, J.L. Beck, L.S. Katafygiotis, Efficient model updating and health monitoring methodology using incomplete modal data without mode matching. Struct. Control. Health Monit. **13**, 91–107 (2006)
64. K.-V. Yuen, *Bayesian Methods for Structural Dynamics and Civil Engineering* (Wiley, New York, 2010)
65. K.V. Yuen, J.L. Beck, L.S. Katafygiotis, Efficient model updating and health monitoring methodology using incomplete modal data without mode matching. Struct. Control. Health Monit. **13**(1), 91–107 (2006)